ERGEBNISSE DER BIOLOGIE

HERAUSGEGEBEN VON

H. AUTRUM · E. BÜNNING · K. v. FRISCH
E. HADORN · A. KÜHN · E. MAYR · A. PIRSON
J. STRAUB · H. STUBBE · W. WEIDEL

REDIGIERT VON

HANSJOCHEM AUTRUM

ZWEIUNDZWANZIGSTER BAND

MIT 55 ABBILDUNGEN

SPRINGER-VERLAG
BERLIN · GÖTTINGEN · HEIDELBERG
1960

ISBN-13: 978-3-540-02510-8 e-ISBN-13: 978-3-642-94769-8
DOI: 10.1007/978-3-642-94769-8

Inhaltsverzeichnis

Berichtigung zum Band XXI

Im Inhaltsverzeichnis des Bandes XXI muß es heißen:

[1] Der 2. Teil dieses Artikels, in dem Reaktionssysteme besprochen werden, bei denen ausschließlich kurzwelliges Licht und nahes Ultraviolett wirksam sind, folgt im nächsten Band.

Comparative aspects of hypothalamic-hypophysial relationships

By C. Barker Jørgensen and Lis Olesen Larsen

Zoophysiological Laboratory, University of Copenhagen, Denmark

With 3 Figures

Contents

Introduction

Until recently studies on hypothalamic-hypophysial relationships almost exclusively dealt with the hypothalamic control of the release of the antidiuretic and oxytocic hormones from the pars nervosa of the neurohypophysis.[1] However, since the late 1940's a rapidly growing literature has unequivocally demonstrated a hypothalamic control of at least some of the functions of the pars distalis in higher vertebrates, for instance gonadotropic, adrenocorticotropic and thyrotropic functions. Strong evidence supports the theory that this control is mediated by substances

[1] The division and nomenclature of the hypophysial region is similar to that used by Green (1951) and Wingstrand (1951). The vertebrate hypophysis consists of the adenohypophysis and the neurohypophysis. The adenohypophysis can generally be subdivided into a pars distalis and a pars intermedia. The neurohypophysis is closely connected with the hypothalamus through the hypothalamic-hypophysial tract. It is undivided in cyclostomes and most groups of fishes. In the higher vertebrates, the neurohypophysis is differentiated into a pars nervosa and a median eminence (see p. 13).

The following abbreviations and synonyms have been used in the text: ACTH = adrenocorticotropic hormone = corticotropin; CRF = corticotropin-releasing factor; TSH = thyrotropic hormone.

released in the median eminence and carried with the blood to the pars distalis. Further evidence obtained during the last few years favours the concept that these substances are chemically closely related to the pars nervosa hormones. In the present paper these new discoveries are considered in the light of previous knowledge of the various aspects of hypothalamic-hypophysial relationships in vertebrates, and some consequences to theories on the evolution of neurohypophysial functions are pointed out.

Evidence of hypothalamic control of pars distalis function

Cyclostomes and fishes. Nothing definite is known about hypothalamic control of pars distalis function in these groups (see p. 15 and 17).

Amphibians. Investigations on the effect of transplantation of the pars distalis to other parts of the body have been of great importance in the elucidation of the hypothalamic control of pars distalis function. A few experiments of this kind have been performed on amphibians. In the toad *Bufo bufo* hypophysectomy immediately causes abnormal appearance of the skin. Moulting is inhibited and the epidermis hyperkeratinizes. The condition of the skin can therefore be used to evaluate hypophysial function. When pars distalis was autografted under the recessus opticus of the brain the toads behaved like hypophysectomized animals, whereas when the gland was exstirpated and regrafted onto the original site, the toads showed normal hypophysial functions as judged from the appearance of the skin. The graft at the heterotopic site was richly vascularized and viable, but cytologically dedifferentiated [Jacobsohn and Jørgensen (1956)]. The loss of function of grafts outside the range of hypothalamic influence is not due to irreversible damage of the pars distalis tissue, because regrafting of non-functional autografts from the eye muscle to the original site under the median eminence can result in resumed functioning of the pars distalis (own unpublished results). Frogs, *Rana temporaria*, with autografted pars distalis in the eye muscle did not reproduce during the spring, whereas frogs with autografts under the median eminence were capable of normal reproduction. Thus gonadotropic secretion of the pars distalis depends upon hypothalamic control. Similar results were obtained with the urodele *Triton cristatus* [Vivian and Schott (1958)].

Hypothalamic influence on the pars distalis has also been revealed in experiments using properly placed hypothalamic lesions. In *Triton cristatus* lesions in the preoptic area were found to inhibit gonadotropic secretion as shown by loss of spermatogenesis [Mazzi (1952)]. Toads with lesions in the same region ceased moulting (Scharrer (1934)].

Birds. Hypothalamic control of pars distalis function has not been investigated in reptiles, but in birds gonadotropic activity has been found to be under nervous control. Increasing light intensity and day length activate the resting gonads in ducks, but light stimulation (of testes) was inhibited by lesions in the anterior hypothalamus [ASSENMACHER (1957)] or by transplanting pars distalis to the anterior eye chamber in hypophysectomized ducks [ASSENMACHER and BENOIT (1958)]. Obviously stimulation by light depends upon intact hypothalamic structures and upon close connection between the hypothalamus and the pars distalis. In the domestic fowl the timing of release of the ovulation-inducing hormone is apparently under control of the central nervous system [FRAPS (1954, 1955)].

Mammals. From the extensive literature on hypothalamic dependance of pars distalis function in mammals, only a few examples will be cited.

Pars distalis grafts placed under the median eminence of hypophysectomized rats often showed normal gonadotropic activity as judged by oestrus cycles and pregnancy in females and spermatogenesis in males. Even pars distalis tissue of immature donors or of adult male donors was capable of maintaining normal oestrus cycles and pregnancy when placed under the median eminence of female rats. By way of contrast, grafts placed under the temporal lobe of the brain, even when richly vascularized, showed no gonadotropic activity [HARRIS and JACOBSOHN (1952)]. Gonadotropic activity likewise ceased in hypophyses transplanted to the kidney in female rats. The grafts, however, could resume functioning, as shown by a recurrence of sexual cycles, when retransplanted under the median eminence. In rats with grafts retransplanted from the kidney to the temporal lobe, sexual cycles never recurred. Histologically, the grafts under the temporal lobe, like the kidney grafts, were dedifferentiated, whereas the grafts under the median eminence became redifferentiated with basophils of normal appearance. The loss of cytological differentiation and gonadotropic secretion in pars distalis transplanted to other parts is therefore not due to injury of the tissue but to its loss of contact with the hypothalamus [NIKITOVITCH-WINER and EVERETT (1957)]. The nature of the hypothalamic control of gonadotropic activity of the pars distalis in the spontaneously ovulating rat has been studied by means of agents capable of blocking the nervous activation of the pars distalis [EVERETT and SAWYER (1950, 1953), EVERETT (1956)]. In order to block gonadotropin release and accordingly normal ovulation a blocking agent, e.g. atropin, has to be administered within a period of two hours on the day before the expected ovulation. Administration prior to or after this period does not prevent the hypothalamic activation of the pars distalis.

Hypothalamic influence on the adrenocorticotropic activity of the pars distalis is obvious from the following examples. Normal rabbits respond to such stimuli as restraint or exposure to cold by releasing ACTH from the pars distalis as is evidenced by a reduction of the number of circulating lymphocytes. This lymphopenic response is reduced or abolished by effective separation of the pars distalis from the hypothalamus by section of the hypophysial stalk and insertion of a waxed-paper plate between the cut ends [Fortier et al. (1957)]. In dogs hypothalamic lesions can reduce the normal rate of adrenal corticoid secretion as is shown by measurements on blood drawn from a permanent canula in the adrenal vein. Less ACTH is released following operative trauma [Hume (1958)]. Appropriate hypothalamic lesions in guinea pigs likewise reduce normal levels of ACTH secretion and abolish the increased secretion which normally follows administration of diphtheria-toxin [Schmid et al. (1957), Winkler et al. (1958)]. Liberation of ACTH can also be induced by electrical stimulation of hypothalamus [Endröczi et al. (1957), rats; Mason (1958), monkeys].

In normal mammals exposure to cold stimulates the thyrotropic activity of the pars distalis, causing increased release of thyroid hormone. This response to cold was abolished in hypophysectomized rabbits with grafts of pars distalis in the anterior eye chamber [von Euler and Holmgren (1956a and b)] and in hamsters with grafts in the cheek pouch [Knigge and Bierman (1958)]. In rabbits, stalk sections that effectively destroyed connections between hypothalamus and pars distalis, also abolished the inhibition of TSH secretion as was shown by normal rabbits in response to restraint ("psychic stress") [Brown-Grant et al. (1957)].

It is thus evident that gonadotropic, adrenocorticotropic and thyrotropic function of the pars distalis is under hypothalamic control. It remains to be discussed whether, and to what extent, these pars distalis functions are independent of the central nervous system. The question is still a matter of controversy and will not be treated in detail here. Probably gonadotropic activity of the pars distalis is low or absent when the gland is deprived of its normal hypothalamic connections. However, it is generally agreed that the transplanted pars distalis or the pars distalis isolated from hypothalamic control by stalk section or hypothalamic lesions, possesses considerable ability to autonomously produce ACTH and TSH. Thus certain forms of stress ("systemic stress") can still induce ACTH release. The thyroids function at a level much above that of hypophysectomized controls [Brown-Grant et al. (1957), d'Angelo (1958), Florsheim (1958), Fortier et al. (1957), Greer (1957), Scow and Greer (1955), von Euler and Holmgren (1956 a and b)].

Several attempts have been made by means of lesions to localize the hypothalamic areas or structures that may take part in the neural control of the release of the various tropic hormones from the pars distalis. Lesions mostly inhibit secretion of more than one of the hormones. The structures therefore seem to overlap. However in the rat, the preferred experimental animal, lesions in the anterior hypothalamus mainly inhibit thyroid function [BOGDANOVE (1957), D'ANGELO (1958), FLORSHEIM (1958), GREER (1957), SLUSHER (1958)], whereas ACTH-regulating structures are generally supposed to be placed in the mid hypothalamus [GREER (1957), SLUSHER (1958)]. In the guinea pig SCHMID et al. (1957) found the ACTH-regulating structures localized around the nucleus hypothalamus ventromedialis and dorsomedialis. Gonadotropin secretion has been inhibited by lesions in the median hypothalamus. These lesions did not influence TSH secretion [BOGDANOVE (1957), D'ANGELO (1958), SLUSHER (1958)]. However, the anterior hypothalamus may also contain structures involved in the sexual cycle of the rat [GREER (1957), FLERKÓ and SZENTAGOTHAI (1957)].

Nervous or humoral hypothalamic control?

Some investigators state that the vertebrate pars distalis is richly innervated and that consequently its function is probably under direct nervous control [METUZALS (1956, 1958), VAZQUES-LOPEZ (1948)]. However, only a scanty nerve supply is generally found, and it is argued that when a dense net of fibres has been observed it has not been substantiated that these fibres are nerves and not reticular fibres [GREEN (1951), HARRIS (1955), SMITH (1956), WINGSTRAND (1951)]. The negative results of direct electrical stimulation of the hypophysis certainly does not support the claim that there are secretory nerves whereas stimulation of the hypothalamus or median eminence can cause release of ACTH, TSH or gonadotropins [HARRIS (1948b), HARRIS and WOODS (1958), MARKEE et al. (1946)].

If the pars distalis does not contain secretory nerves its function must be controlled humorally. Peculiar vascular connections between hypothalamus and pars distalis are, indeed, suitable for such humoral control (Fig. 2 and 3, p. 12 and 14). In higher vertebrates blood passing through a dense capillary plexus in the median eminence is drained through portal vessels directly into the blood sinuses of the pars distalis. POPA and FIELDING (1930) discovered this hypophysial portal system in mammals and WISLOCKI (1937) described the true direction of the blood flow. The portal circulation is also characteristic of birds [WINGSTRAND (1951)] and amphibians [GREEN (1947), HOUSSAY et al. (1935), anurans; MAZZI and PEYROT (1957), urodeles]. In lower vertebrates the vascular connections between hypothalamus and pars distalis are simpler, but fundamentally similar. Blood flows through capillaries of the neurohypophysis into the pars distalis (and pars intermedia) before passing into the systemic circulation [GREEN (1951)].

The hypophysial portal circulation has therefore been assumed to mediate the hypothalamic control by carrying hormone-releasing factors to the pars distalis. In higher vertebrates the factors are supposed to be liberated in the median eminence from the nerve endings of the hypothalamic tract and to diffuse into the primary plexus of the portal system [Harris (1948a)]. A corticotropin-releasing factor (CRF) has in fact been demonstrated in brain blood from rats subjected to stress. CRF was not present in brain blood from non-stressed rats nor in blood drawn from the carotid of stressed rats [Schapiro et al. (1958)]. Blood from the portal region, but not from the carotid, of the dog likewise stimulated ACTH secretion when injected into intact rats. In hypophysectomized rats no effect was observed [Porter and Jones (1956)].

Birds are especially suitable for demonstrating the significance of humoral control of the pars distalis, because it is possible to sever the portal vessels without interrupting the nervous connections between the hypothalamus and hypophysis. Section of the portal vessels abolishes the hypothalamic control of the gonadotropic function of the pars distalis in the duck as judged from the failure of the operated birds to respond to light. Section of the nervous connections to the hypophysis does not prevent light-induced sexual maturation [Benoit and Assenmacher (1953)]. Substances that can release hormones from the pars distalis and thus be the mediators of hypothalamic control are therefore most probably transmitted with the portal blood. Their number, chemical nature, and site of origin are not yet definitely known, but are the subject of much current research.

Chemical nature of hypothalamic hormone-releasing factors

Several well known physiologically active substances such as histamine, adrenaline and serotonine are present in high concentrations in the hypothalamus or neurohypophysis and have been considered as possible natural factors causing hormone release from the pars distalis.

Histamine was found to stimulate ACTH release in the rabbit; injection of histamine solutions caused lymphopenia. The response was abolished by hypophysectomy [Fuche and Kahlson (1957)]. Apparently, however, histamine acts indirectly via the hypothalamus because the ACTH-releasing effect could be abolished in rats by hypothalamic lesions [McCann (1957)] or by transplanting the pars distalis to the anterior eye chamber [Martini (1958)]. *Adrenaline,* too, induced ACTH release from the pars distalis in the rat, but, as in the case of histamine, the response disappeared after lesion of the hypothalamus or median eminence [McCann (1957), Smelik and de Wied (1958)]. Injection of adrenaline or nor-adrenaline into the primary plexus of the portal vessels of female rabbits frequently produced ovulation. The effect was apparently not due to adrenaline itself, however, but to the acidity of the solution injected, because after adjustment to neutrality only one of eight rabbits ovulated after receiving the large dose of 150 γ of adrenaline [DONOVAN and HARRIS (1955)]. Adrenaline is, therefore, most probably not a natural ACTH-or gonadotropin-releasing agent. *Acetylcholine* has also been investigated as to its pars distalis-stimulating activity. No ACTH-releasing effect was found on the pars distalis tissue transplanted to the anterior eye chamber in the rat [MARTINI (1958)]. On theoretical grounds, acetylcholine is unlikely to be a humoral transmitter in the portal blood, because it is rapidly broken down in the blood stream [HARRIS (1955)]. *Serotonine* caused ACTH release in normal rats, but not after destruction of the median eminence [SMELIK and DE WIED (1958)] or after transplantation of the pars distalis to the anterior eye chamber of hypophysectomized rats [MARTINI (1958)]. Also *substance P* [see GUILLEMIN (1957)] is stated to release ACTH, but the response could be abolished by hypothalamic lesions [McCANN (1957)]. Serotonine and substance P are therefore presumably ruled out as natural ACTH-releasing factors.

A lipid extracted from the hypothalamus was found to cause eosinopenia and adrenal ascorbic acid depletion — other symptoms of ACTH release — when injected into intact rats, but not when injected into hypophysectomized rats [SLUSHER and ROBERTS (1954)]. The lipid was not found in the cortex of the brain. It is stated to cause release not only of ACTH, but also of TSH and gonadotropins from the pars distalis of several mammals [CURRI (1958)]. It is doubtful, however, whether this interesting lipid fraction represents natural humoral links between the hypothalamus and the pars distalis since its ACTH-releasing activity can be abolished by destruction of the median eminence [DE WIED et al. (1958)].

Special interest attaches to the mammalian antidiuretic hormone which has been found to stimulate the pars distalis directly. The hormone acts not only in mammals but also in amphibians. Injections of synthetic vasopressin induced moulting in toads (*Bufo bufo)* in which moulting had been inhibited by isolation of the pars distalis from the hypothalamus. Presumably, vasopressin caused moulting by stimulating the inactivated pars distalis, because the hormone had no effect in hypophysectomized toads [JØRGENSEN and NIELSEN (1958)].

In mammals the ACTH-releasing activity of vasopressin has been extensively studied. In rats pitressin or synthetic vasopressin causes adrenal ascorbic acid depletion even in animals with hypothalamic lesions [McCANN (1957), McCANN and FRUIT (1957), SMELIK and DE WIED

(1958)]. The response is abolished by hypophysectomy, but reappears in hypophysectomized rats after transplantation of pars distalis tissue, e.g. to the anterior eye chamber [MARTINI and DE POLI (1956), STUTINSKY et al. (1952)]. ACTH-releasing activity was demonstrated in the guinea pig by the increased urinary excretion of corticosteroids following injection of pitressin into normal, but not into hypophysectomized animals [SOBEL et al. (1955)]. In normal man the corticosteroid level in the plasma was raised after injection of vasopressin, but no effect was observed in a patient suffering from hypophysial insufficiency [McDONALD et al. (1956), WELLER (1957)]. Additional support of the idea of a direct action of vasopressin on the pars distalis appears from the finding that the rate at which injected cortisone disappears from the blood of adrenalectomized rats is not influenced by simultaneous injection of vasopressin. Vasopressin, therefore, does not act indirectly by causing a fall in the concentration of the circulating cortisone. (A fall is known to stimulate ACTH release from the pars distalis), [ESER and TÜZÜNKAM (1958)].

The results obtained with the other mammalian pars nervosa hormone, oxytocin, are equivocal. In man, for example, it was without effect on the pars distalis (ACTH release) [McDONALD & WEISE (1956)]. In rats, by way of contrast, oxytocin is about as effective as vasopressin in causing adrenal ascorbic acid depletion or eosinopenia. However, the adrenal ascorbic acid depletion response is stated to disappear after hypothalamic lesions [McCANN (1957)], whereas oxytocin still produced eosinopenia in hypophysectomized rats bearing intra-ocular grafts of pars distalis tissue [MARTINI and DE POLI (1956)].

The observations that injections of antidiuretic hormone cause ACTH release from the pars distalis and that hypothalamic lesions causing diabetes insipidus would also inhibit stress-induced ACTH secretion from the pars distalis led to the belief that the antidiuretic hormone might be the natural hypothalamic ACTH-releasing factor [FULFORD and McCANN (1955), McCANN (1957), McCANN and BROBECK (1954), McCANN and FRUIT (1957), McCANN et al. (1958)]. The theory is, however, contradicted by other observations of which some may be mentioned. In dogs hypothalamic lesions have been produced which abolished ACTH secretion after stress without causing diabetes insipidus, or which caused diabetes insipidus without interfering with enhanced ACTH secretion after stress [HUME (1958)].[1] Endogenous release of

[1] In toads, lesions in the preoptic region mostly caused degeneration of the pars nervosa and abolished function of pars distalis. However, in some toads in which parts of the preoptic nucleus had been spared the pars nervosa was normal even when the pars distalis was not functioning. In other experiments a complete degeneration of the pars nervosa was obtained by sectioning the preoptic-hypophysial tract immediately behind the chiasma without interfering with the function of the

antidiuretic hormone in dogs caused by infusion of hypertonic saline did not result in changes in the level of plasma 17-hydroxy-corticosteroids. The threshold dose of purified vasopressin that would increase plasma corticosteroids was about 1000 times the dose necessary to produce antidiuresis [NICHOLS Jr. et al. (1958)]. In rats, injections of ethyl alcohol stimulated secretion of ACTH and inhibited simultaneously the secretion of antidiuretic hormone [REZÁBEK (1957)]. Injection of acetic acid extract of the median eminence of the calf caused ACTH release in rats. The response was not abolished by hypothalamic lesions. The effect was apparently not solely due to vasopressin contamination, because the ACTH-releasing activity of the extract could be greatly diminished by treatment with pepsin which does not destroy vasopressin [ROYCE and SAYERS (1958)]. Blood sampled from the sella turcica immediately after hypophysectomy in rats also contained an ACTH-releasing factor different from vasopressin [PORTER and RUMSFELD (1956)].

It is therefore highly improbable that the natural ACTH-releasing factor is identical with the antidiuretic hormone. However, the fact that even synthetic vasopressin is capable of directly stimulating the pars distalis suggests that it is chemically related to the natural factor. This suggestion is strongly supported by investigations of GUILLEMIN, SAFFRAN, and their co-workers. They tested the ACTH-releasing activity of purified extracts of the pars nervosa and hypothalamus on tissue cultures of pars distalis according to the method introduced by SAFFRAN and SCHALLY (1955a and b). Rat hypophyses are cut in halves, one half is incubated with the substance to be tested for ACTH-releasing activity, the other half provides a measure of the spontaneous release of ACTH. After incubation the amounts of ACTH given off to the culture medium by the two halves are determined by bioassay. The method has been criticized [FORTIER (1958), FORTIER and WARD (1958)]. However, when highly purified substances are used the method is probably reliable.

By chromatography of extracts of the hypothalamus or pars nervosa a small peptide has been separated which stimulates release of ACTH when added to the incubation medium. The substance is different from the antidiuretic and oxytocic hormones, from ACTH, histamine, acetylcholine, adrenaline, nor-adrenaline, and serotonine [GUILLEMIN (1957), GUILLEMIN et al. (1957), SCHALLY and SAFFRAN (1956)]. The substance also stimulates ACTH release in human subjects as judged from the increase in plasma concentration of corticosteroids [CLAYTON et al. (1957)]. By serial paper chromatography in different solvent systems the factor has been further concentrated and purified. One millimicrogram or

---

pars distalis. The preoptic-pars nervosa system is apparently not, therefore, controlling the functioning of the pars distalis in the toad (JØRGENSEN, LARSEN and WINGSTRAND unpublished).

less of the most purified material is capable of significantly increasing the release of ACTH from the rat pars distalis *in vitro*. By acid hydrolysis and analysis for amino acids the material was found to have seven or eight amino acids in common with vasopressin [Schally et al. (1958)]. The close chemical relationship between the antidiuretic hormone and CRF is further evidenced by the finding that the threshold dose for ACTH-releasing activity of vasopressin is only some 10 times the threshold dose of the most concentrated CRF preparation which also possessed some antidiuretic, but no oxytocic activity [Saffran et al. (1958)].

The mammalian pars nervosa hormones have also been assayed for gonadotropin- and TSH-releasing activity. Injections of oxytocin were found to accelerate oestrus in the cow [Armstrong and Hansel (1958)]. In rats vasopressin enhanced the rate of I^{131} uptake in the thyroids. The effect was abolished by hypophysectomy and ascribed to TSH-releasing activity of vasopressin [Dubreuil and Martini (1956)]. On the other hand, Reichlin (1957) did not find any effect of vasopressin on the rates of thyroid iodine release in rats. Neither did naturally secreted antidiuretic hormone in dehydrated rats cause measurable release of TSH.

Attempts to stimulate TSH release from mouse hypophysis *in vitro* by means of hypothalamic tissue were negative [Florsheim et al (1957)].

Oxytocin has recently been ascribed a possible role in the release of prolactin from pars distalis. It was found that injections of synthetic oxytocin greatly retarded the involution of the mammary gland of lactating rats, which follows permanent removal of the litter. The effect was apparently due to prolactin that was released by the oxytocin injections [Benson and Folley (1958)]. It is still uncertain whether oxytocin acts directly on the pars distalis or indirectly via the central nervous system. The last alternative is rendered likely by the observation that the effect of oxytocin is greatly inhibited by dibenamine and atropin which are known to depress hypothalamic activity [Grosvenor and Turner (1958)]. Should oxytocin, however, turn out to stimulate prolactin release from the pars distalis directly, it remains to be shown that prolactin release is controlled by the hypothalamus in the intact animal and, if so, that oxytocin and not a chemically related polypeptide liberated from the median eminence acts as mediator between the hypothalamus and pars distalis.

Comparative anatomy of the hypothalamic-hypophysial region

The close chemical relationship that presumably exists between pars nervosa hormones and pars distalis-stimulating hypothalamic factors — at least in amphibians and mammals — leads naturally into an inquiry concerning anatomical and functional relationships between the structures that produce and regulate the release of these factors or hormones throughout the vertebrate system.

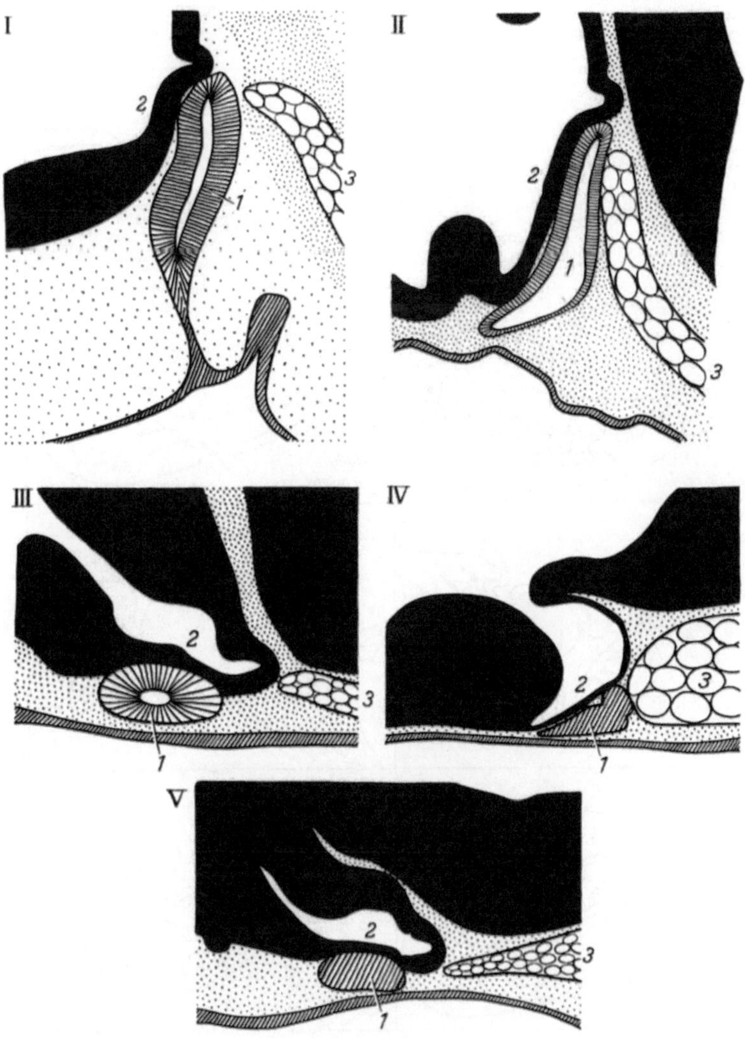

Fig. 1. Diagrams showing the early development of the vertebrate hypophysis. *I* Birds; *II* Elasmobranchs; *III* Ganoids (Amia); *IV* Amphibians; *V* Teleosts. *1* Adenohypophysial primordium; *2* Neurohypophysial primordium; *3* Chorda. Ectodermal structures are hatched and nervous structures black [WINGSTRAND (1959)]

The basic features in the embryonic development of the hypophysial region are similar in all vertebrates (Fig. 1). The adenohypophysis originates in the dorsal wall of the primitive oral cavity as an ectodermal invagination that at a very early stage makes contact with the neurohypophysial primordium. The importance of this contact for the further

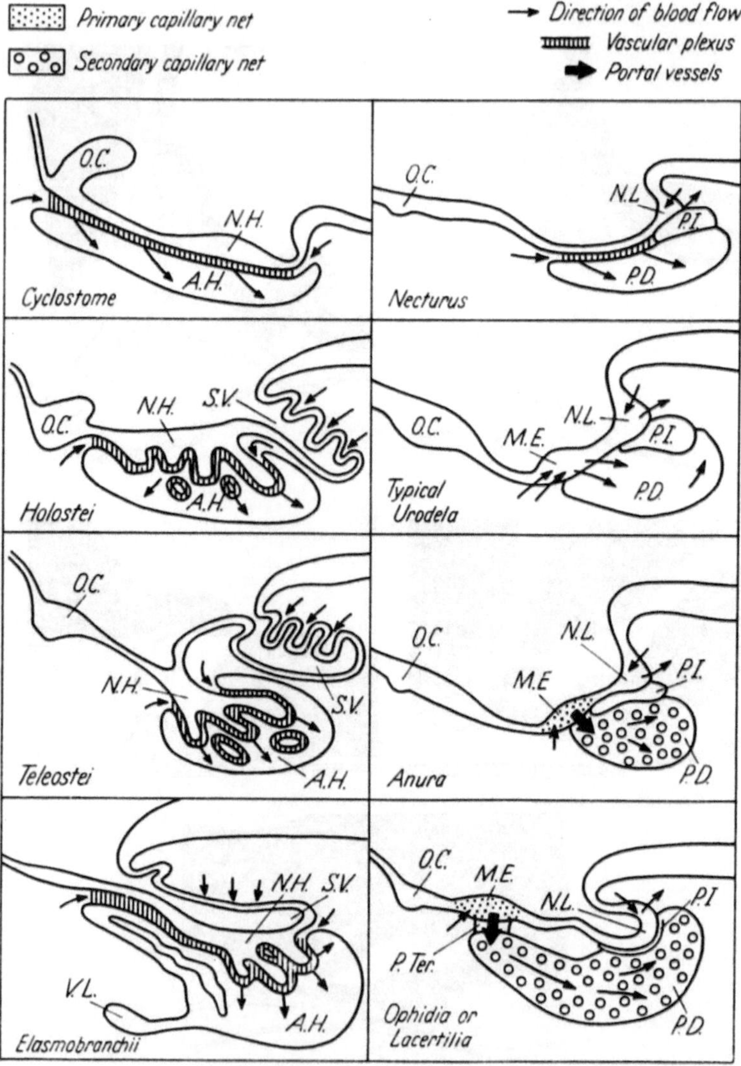

Fig. 2. Diagrams of sagittal section through the hypothalamic-hypophysial region of various vertebrates. *A.H.* Adenohypophysis; *M.E.* Median eminence; *N.H.* Neurohypophysis; *N.L.* Pars nervosa; *O.C.* Optic chiasma; *P.D.* Pars distalis; *P.I.* Pars intermedia; *P.T.* Pars terminalis; *S.V.* Saccus vasculosus; *V.L.* Ventral lobe. Arrows indicate direction of blood flow. Heavy arrows, portal vessels [Green (1951)]

development and differentiation of adenohypophysial structures has been investigated in amphibians and birds, but the results are inconclusive [see for instance BURCH (1946), ETKIN (1958), MAZZI and GUARDABASSI (1957), PASTEELS (1957), anurans and urodeles; HILLEMAN (1943), chicken]. It seems though that normal cellular differentiation of the pars distalis can depend upon induction from the neurohypophysial primordium. This of course agrees well with the previously mentioned observation that in adult vertebrates (p. 2 and 3) cellular dedifferentiation takes place in the pars distalis tissue transplanted to foreign sites, whereas redifferentiation occurs upon retransfer to the region of the median eminence [NIKITOWICH-WINER and EVERETT (1957), SIPERSTEIN and GREER (1956)]. It would be of interest to know whether the factors by which the neurohypophysial primordium stimulates cellular differentiation in the embryonic pars distalis are chemically related to the pars distalis-stimulating factors of the mature organism.

The neurohypophysial structures develop partly in a diverticulum arising from the diencephalic floor, partly in the floor itself between the diverticulum and the optic chiasma. The further differentiation within these areas is not identical in all vertebrate classes, and many problems of homology are still under discussion. However, the assumption seems well founded that the undivided neurohypophysis of lower vertebrates is homologous to the pars nervosa and median eminence of the higher vertebrates. The undivided neurohypophysis of fishes is derived from parts of both the diverticulum and the ventricular floor, whereas in tetrapods the pars nervosa develops in the diverticulum, the median eminence in the ventricular floor [WINGSTRAND (1951, 1959)].

The neurohypophysis varies much in size and differentiation in the various vertebrate classes. The cyclostome neurohypophysis is little developed, and is only a thickening of the ventricular wall. But in fishes, elasmobranchs and teleosts, it is a conspicuous, richly branched structure surrounded by adenohypophysial tissue (Fig. 2) [BARGMANN (1953), DA LAGE (1958), GREEN (1951), SCHARRER (1952)]. In one class of fishes, the dipnoans, the neurohypophysis is morphologically differentiated into a pars nervosa and a median eminence, the latter characterized by capillaries that communicate with the sinusoids of the adenohypophysis [WINGSTRAND (1956)]. In all tetrapods the neurohypophysis is divided into a median eminence, and a pars nervosa with a separate blood supply draining directly into the systemic circulation. The urodele median eminence is still of a simple type, similar to that of the dipnoans, whereas the median eminence of anurans and of all higher vertebrates possesses a fully developed portal system (Fig. 3) [GREEN (1951), WINGSTRAND (1951)].

The tissue of the vertebrate neurohypophysis consists to a large extent of nerve terminals of the hypothalamic-hypophysial tract. An important part of the nerve fibres of this tract originates from the nucleus preopticus of anamnians and from the homologous nucleus supraopticus and nucleus paraventricularis of amniotes, but several of the other hypothalamic nuclei are suspected of contributing fibres to the hypophysial tract. The preoptic, supraoptic and paraventricular nuclei are large and well defined and are found close to the optic chiasma in

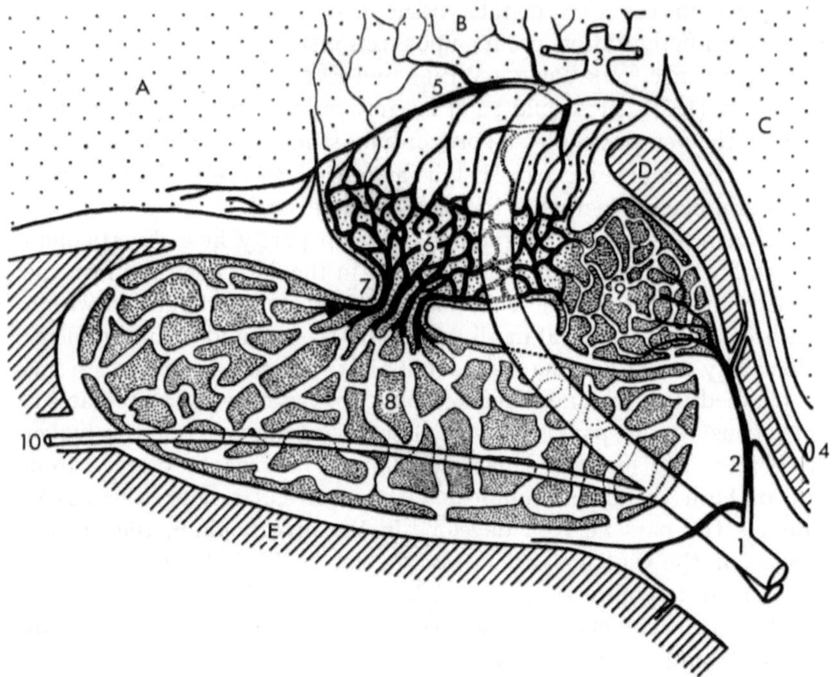

Fig. 3 Diagram of the vessels of the hypophysis of the pigeon. *A* Optic chiasma; *B* Hypothalamus; *C* Medulla oblongata; *D* Dorsum sellae; *E* Osseous floor of the sella. *1* Arteria carotis interna; *2* A. hypophysea inferior; *3* Anterior ramus of the a. carotis; 4 Posterior ramus of a. carotis; *5* A. infundibularis; *6* Primary capillary plexus in the median eminence; *7* Portal vessels; *8* Secondary plexus in the pars distalis; *9* Capillary bed of the pars nervosa [Wingstrand (1951)]

the anterior part of the hypothalamus. The nerve tracts from the nuclei terminate in the pars nervosa or, in fishes, in the undivided neurohypophysis.

Neurosecretion and site of formation of neurohypophysial hormones

The preoptic nucleus and the supraoptic and paraventricular nuclei have attracted special interest, because they probably produce the pars nervosa hormones and control their release. The preoptic, supraoptic and paraventricular nuclei are also the site of the so-called neurosecretion.

The products of neurosecretion consist of granules of varying size selectively stained by various staining procedures, especially by GOMORI's chrome alum hematoxylin and aldehyde fuchsin methods. An extensive literature deals with neurosecretory phenomena in the vertebrate hypothalamus [see SCHARRER and SCHARRER (1954a and b)]. Only the part of the literature dealing with the relationship between neurosecretion and neurohypophysial hormones is considered here. Such a relationship is especially evident from the observations that Gomori-positive material and hormone content of pars nervosa and hypothalamus varies in parallel under varying experimental conditions. Dehydration causes depletion of both neurosecretory material and antidiuretic hormone in the hypothalamus and pars nervosa; during recovery they reappear simultaneously. Transection of the hypothalamic-hypophysial tract leads to accumulation of Gomori-positive material as well as oxytocic and antidiuretic hormones proximal to the wound, whereas the pars nervosa is depleted of both Gomori-positive material and hormones. Destruction of supraoptic and paraventricular nuclei causes diabetes insipidus and disappearance of neurosecretory material from the nerve tract and pars nervosa. These and further observations led to the assumption that both neurosecretory material and hormones of the pars nervosa are produced in nerve cell bodies of the preoptic nucleus, transported along the axons, and eventually stored in the neurohypophysis or its pars nervosa [BACHRACH (1957), GREEN and VAN BREEMEN (1955), HILD (1956), OLIVECRONA (1957), SLOPER (1958), STUTINSKY (1957), TRAMEZZANI and URANGA (1954)]. The relation between the pars nervosa hormones and the neurosecretory material is still under discussion [ADAMS and SLOPER (1956), HILD and ZETLER (1955), SLOPER (1955)].

Neurosecretion in the hypothalamic nuclei has in recent years also been related to pars distalis-stimulating factors. In teleosts neurosecretory activity in the nucleus tuberis lateralis, but not in the nucleus preopticus, varies with the sexual cycle [POLENOV (1950) and ZAITZEV (1955), both cited from PICKFORD and ATZ (1957), p. 229; STAHL (1953), SCHIEBLER and v. BREHM (1958)]. It is noteworthy that chrome alum hematoxylin positive material in the teleost neurohypophysis is concentrated in the part that is surrounded by pars intermedia tissue, but largely absent from the part that branches into the pars distalis[1]. The latter part is perhaps innervated from nucleus tuberis lateralis whose neurosecretory products are not stainable by chrome alum hematoxylin, but by aldehyde fuchsin [BARGMANN (1953)].

---

[1] The question of homologies between the different parts of the adenohypophysis in fishes and tetrapods is not yet definitely settled.

Neurosecretory fibres have been observed in close association with the capillaries of the primary plexus of the hypophysial portal system in a great variety of tetrapods [ARKO and KIVALO (1958), ASSENMACHER and BENOIT (1958), DAWSON (1957), OKADA et al. (1955), STUTINSKY (1958), WINGSTRAND (1951)]. The neurosecretory granules found close to or even inside the capillaries of the median eminence have been related to the pars distalis-stimulating substances which are presumably released from the median eminence into the portal vessels. Of course, the mere presence of neurosecretory granules in the median eminence does not imply a direct relation to the pars distalis-stimulating factors, and it should be mentioned that neurosecretory granules may be present in the anuran median eminence, even when the pars distalis is not functioning. Destruction of the preoptic nucleus region in toads might cause cessation of the functioning of the pars distalis resulting in the death of the animals. In some of these toads at least part of the preoptic nucleus continued functioning as shown by the presence of neurosecretory material in the pars nervosa (see p. 8, footnote). In these toads normal amounts of neurosecretory granules may be present also in the median eminence.

Very little is known about the site and mode of formation of the pars distalis-stimulating factors. Investigations as to whether these factors are produced within or outside the median eminence have only been performed on the toad. The whole median eminence was extirpated and the pars distalis transplanted to the wound, where the gland presumably made contact with the cut nerve fibres which normally end in the median eminence. Most of the operated toads moulted and survived, whereas controls in which the pars distalis was prevented from making contact with the cut nerve fibres died from pars distalis deficiency. Normal hypothalamic control of the functioning of the pars distalis can therefore be exerted even in the absence of the median eminence. Consequently pars distalis-stimulating factors can originate outside the median eminence and be released from the end of the severed nerve tract. The factors must be formed either at the free ends of the cut axons or, more likely, in nerve cells of the nuclei from which the nerves arise and through which the hypothalamic control is mediated [JØRGENSEN et al. (1959)]. As in the case of the pars nervosa hormones, therefore, the factors stimulating the pars distalis are probably produced by nerve cells at a distance from their normal site of release and transported along nerve fibres from the nerve cell bodies. The formation of pars distalis-stimulating factors in the toad fulfill the essential requirements for neurosecretion. However, it remains to be decided whether the factors can be related to the neurosecretory granules demonstrated in the anuran median eminence, — neurosecretory material may exist which is not stained by the methods now in use [SCHARRER and SCHARRER (1954a)].

Regeneration of neurohypophysial structures

A close relationship between pars nervosa and median eminence is also evidenced from experiments on regeneration in these structures. Extirpation in the toad *Bufo bufo* of pars nervosa results in regeneration of a new pars nervosa in the tissue of the median eminence. The regenerate can obtain approximately normal size and hormone content. Extirpation of the median eminence is compatible with continued normal function of the pars distalis, presumably due to the reorganization of median eminence-like structures. However, if both median eminence and pars nervosa are extirpated pars distalis functions at most for some months, probably because permanent structures that can secure the proper hypothalamic control of the function of pars distalis do not develop. Moreover regeneration of pars nervosa is slight or missing after total extirpation of the neurohypophysis [JØRGENSEN et al. (1956a, b)]. Regeneration and reorganization of a functional pars nervosa has further been observed in rats [BILLENSTIEN and LEVEQUE (1955)].

Comparative physiology and evolution of the functions of the neurohypophysis

The evidence from the comparative anatomy, embryology and experimental morphology of neurohypophysial structures supports the theory that both the pars nervosa and the median eminence have evolved from an undivided neurohypophysis, still to be found in cyclostomes and most fishes. The close chemical relationship that has been demonstrated between pars nervosa and median eminence hormones, not only within the mammalian class, but also between hormones from different classes (amphibians and mammals) further indicates that the evolution of the functions of the neurohypophysis has proceeded with little chemical change of the original hormones. But apart from this little can be said about the early evolution of neurohypophysial functions. We do not even know whether the specific pars nervosa or the median eminence functions are phylogenetically the older. Before considering this problem, however, perhaps a survey should be given of the results of comparative studies on neurohypophysial functions. Studies in fishes especially, or even more likely in cyclostomes with a morphologically undifferentiated neurohypophysis, may provide decisive information about the function of the primitive neurohypophysis. Unfortunately, cyclostomes have not been investigated so far.

The physiology of the median eminence in higher vertebrates has been dealt with in previous sections. It was shown that its function as a mediator between the hypothalamus and the pars distalis is established in mammals, birds and amphibians. In fishes, a typical "median eminence function" of the neurohypophysis is not excluded, but neither is it proved. Light is thus found to stimulate gonads and thyroids in teleosts [BUSER-LAHAYE (1953), see also PICKFORD and ATZ (1957, p. 159)]. Such stimulation is likely to be exerted via the central nervous system and may accordingly be mediated by pars distalis-stimulating factors liberated in the neurohypophysis.

The functions of the pars nervosa in mammals are well known and will not be reviewed in detail here. It will suffice to recall that the antidiuretic hormone participates in the regulation of the osmotic concentration of the body fluids. Osmotic concentrations of the blood above normal values stimulate *via* the hypothalamic centres the release from the pars nervosa of antidiuretic hormone which increases reabsorption of water in the kidneys [THORN (1958), VERNEY (1947)].

Oxytocin, the only other hormone known with certainty to be released from the mammalian pars nervosa, has a known function only in females where it stimulates uterine contractions during labour and causes milk ejection during lactation. Suckling elicits a reflex release of oxytocic hormone that causes contraction of the myoepithelial cells in the secretory ducts of the mammary gland, in this way driving their contents into the milk cisterna. Without the action of oxytocic hormone the milk is not accessible to the young [CROSS (1958), FITZPATRICK (1956), FOLLEY (1956)].

The milk ejection effect of oxytocin is of course specific to mammals. But the uterus-stimulating effect is perhaps comparable to the premature oviposition which is caused by injection of mammalian pars nervosa extracts into birds (domestic fowl) [ROTHSCHILD and FRAPS (1946), MANUNTA (1951)]. However, it is uncertain whether the pars nervosa is of importance in normal egg laying, because neurohypophysectomy in the hen had no effect on the length of time spent by the egg in the various portions of the oviduct [SHIRLEY and NALBANDOW (1956)].

A peculiar effect of neurohypophysial factors has been observed in the teleost *Fundulus heteroclitus*. In hypophysectomized fish spawning reflexes were elicited by injection of the equally active synthetic vasopressin and oxytocin, as well as by preparations of fish hypophyses. The spawning reflex activity of hypophysial extracts of fish was approximately proportional to their oxytocic and vasopressor activity. Large doses, however, [about 1 mg hypophysial tissue (of pollock) to 1 g of *Fundulus*], had to be used in order to elicit a response [WILHELMI et al. (1955)]. The physiological significance of these findings is uncertain. They do not allow conclusions to be made about the rôle played by the neurohypophysis in the reproduction of fishes.

A special interest has been attached to the possible function of the neurohypophysis in the salt and water metabolism of vertebrates. Our knowledge of the comparative physiology of this subject is therefore relatively extensive.

In birds (domestic fowl) extirpation of the neurohypophysis has been found to cause diabetes insipidus. Injection of vasopressin produces antidiuresis [SHIRLEY and NALBANDOW (1956)]. Presumably the pars nervosa plays a similar rôle in the water metabolism of birds as it does in mammals.

The literature dealing with the effect of neurohypophysial extracts on water metabolism in amphibians has been reviewed by HELLER (1945, 1950), JØRGENSEN (1950) and SAWYER (1956). Injection of neurohypophysial extracts into amphibians kept in water generally enhances water uptake through the skin and inhibits urine flow. The total effect is an increase in body water. The response varies widely in strength from species to species. Apparently it is the stronger the more terrestrial are the habits of the species. In a purely aquatic amphibian, like the toad *Xenopus laevis,* injection of mammalian pars nervosa extracts has no effect on either water uptake through the skin or diuresis [EWER (1952)]. Dehydration of frogs and toads causes increased rates of secretion of water-balance hormone from the pars nervosa [JØRGENSEN et al. (1956b), SAWYER and ROTH (1953), TRAMEZZANI and URANGA (1954)]. Presumably the pars nervosa therefore also participates in the water economy in amphibians. It is noteworthy that the development of the pars nervosa, like the sensitivity to neurohypophysial extract, can be correlated with the availability of water in the environment. Phylogenetically an independently vascularized pars nervosa appears first in fishes with semi-terrestrial habits (dipnoans) and in the amphibians. In amphibians, and perhaps also in the rest of the vertebrates, the pars nervosa is larger the more independent the animal is of water in the environment [GREEN (1951)].

It is well known that amphibians are capable of taking up certain ions, especially sodium and chloride, through the skin from very dilute solutions. The active uptake of sodium is stimulated by the pars nervosa hormones, whether of fish, amphibian or mammalian origin. Synthetic oxytocin is also active. The effect has been demonstrated on intact animals [JØRGENSEN et al. (1946), axolotl; KALMAN and USSING (1955), toad)], as well as on the isolated skin of various anuran species [FUHRMAN and USSING (1951), MOREL et al. (1958)].

Attempts to trace definite neurohypophysial functions back through the vertebrate system again end with the amphibians. In fishes it has not been possible to attribute to the neurohypophysis any definite function in the water and salt metabolism. Injections of neurohypophysial extract had no effect on body water in the several species of fish investigated, even when extracts of fish hypophyses were used [FONTAIEN

(1956)]. Injection of mammalian pars nervosa extract did not change the salinity tolerance of freshwater trout [Smith (1956)]. Neither did extirpation of the hypophysis in the euryhaline eel. Hypophysectomized eels tolerated changes from fresh water to sea water and *vice versa*, and the osmotic concentration of the blood remained normal [Callamand et al. (1951)]. In the killifish *Fundulus heteroclitus* hypophysectomy did change the euryhalinity. In contrast to normal fish hypophysectomized killifish were unable to survive in fresh water. Injections of extracts of the hypophysis from killifish or from a fresh water fish, *Perca flavescens*, improved survival in fresh water. Extracts of the hypophysis from a marine fish *Pollachius virens*, of mammalian pars nervosa or of a number of pars distalis hormone preparations had no beneficial effects on survival time [Burden (1956)]. These experiments leave the question unanswered as to whether the neurohypophysis is of importance in the ability of the killifish to live in freshwater.

Interesting results have been obtained by investigating the effect of neurohypophysial extracts from a species of one vertebrate class on members of other classes. Examples of such inter-class effects of neurohypophysial substances have been mentioned on previous pages. Vasopressin has been found to stimulate pars distalis and enhance water uptake through the skin when injected into toads, and to accelerate egg laying when injected into the hen. Neurohypophysial extracts of lower vertebrates also act on higher vertebrates. When extracts from fishes, amphibians, reptiles and birds are tested on mammals they are found to possess all the known effects of the mammalian pars nervosa: antidiuretic, pressor and oxytocic [Heller (1950), Waring and Landgrebe (1950)]. These effects are also exerted by extracts of the brains of both fresh water and marine cyclostomes [Lanzing (1954), Sawyer (1955)].

It should, however, be emphasized that experiments of this kind can only inform us about neurohypophysial function in the species into which the extract is injected, not in the species from which the extract was obtained. For example, extract from an amphibian pars nervosa injected into a mammal, enhances tubular reabsorption of water, as does the mammalian antidiuretic hormone. However, when injected into amphibians the mammalian antidiuretic hormone acts, as the amphibian pars nervosa factor does, mainly by constricting the afferent arterioles. Therefore, when it is found that fishes or even cyclostomes produce substances which have antidiuretic or oxytocic effects when tested on mammals this does not imply that the substances exert similar — or even related — functions in the species in which they are produced. Interest attaches to these experiments mainly because they provide further evidence of close chemical relations between neurohypophysial factors in all vertebrate classes. But till now they have contributed little to answering the question as to which were the original functions of the vertebrate neurohypophysis.

Summary and conclusions

Release of hormones from the pars distalis of the vertebrate hypophysis is controlled by the central nervous system. Control of the release of gonadotropic hormone has been demonstrated in amphibians, birds and mammals, and of adrenocorticotropic and thyrotropic hormones in mammals. Lower vertebrates have not been investigated.

The nervous control of pars distalis is exerted via hypothalamic structures. However, the pars distalis is apparently not innervated by secretory nerves, but stimulated humorally. Factors that are capable of releasing hormones from the pars distalis are presumably liberated from the nerve endings of the tracts controlling the pars distalis. These tracts terminate in the median eminence and the hormone-releasing factors are carried to the pars distalis by special blood vessels connecting the median eminence and the pars distalis.

The pars distalis-stimulating factors are probably closely related chemically to the hormones of the mammalian pars nervosa: vasopressin and oxytocin. This chemical relationship extends to the neurohypophysial factors of all classes of vertebrates. Furthermore, both the pars distalis-stimulating factors of the median eminence and the pars nervosa hormones are most likely produced by neurosecretion in hypothalamic nuclei and transported along the nerves to their sites of release in the median eminence and pars nervosa, respectively.

Phylogenetically, both the pars nervosa and the median eminence seem to be derived from an undivided neurohypophysis, still to be found in cyclostomes and most fishes. This would explain the close relationship between the hormones responsible for the quite unrelated functions of the pars nervosa and the median eminence.

The function of the primitive undifferentiated neurohypophysis is not known. It is therefore not possible to decide which functions are phylogenetically oldest, whether those of the pars nervosa that in higher vertebrates is known to participate in water and salt metabolism and in reproduction, or those of the median eminence that mediates the hypothalamic control of the pars distalis. The latter alternative is of course favoured by the fact that the blood of both the undivided neurohypophysis of cyclostomes and fishes and of the median eminence of higher vertebrates is drained through the pars distalis, whereas the development of a pars nervosa is accompanied by the establishment of a separate blood circulation. Studies on lower vertebrates, and especially cyclostomes, may shed light on the early evolution of the function of the neurohypophysis. Until now such studies have mainly attempted to find out whether well known functions of the pars nervosa of higher vertebrates can also be demonstrated in fishes. In future work, more emphasis will probably be placed on the possible rôle of the neurohypophysis as a mediator of hypothalamic control of the function of the pars distalis.

Literature

1. Monographs and symposia

Bargmann, W.: Das Zwischenhirn-Hypophysensystem. Berlin-Göttingen-Heidelberg: Springer-Verlag 1954.

Curri, S. B., e L. Martini, Editors: Pathophysiologia Diencephalica. Wien: Springer-Verlag 1958.

Fields, W. S., R. Guillemin and C. A. Corton, Editors: Hypothalamic-hypophysial interrelationships. Springfield, Ill. U.S.A.: Charles C. Thomas 1956.

Harris, G. W.: Neural control of the pituitary gland. London: Edward Arnold 1955.

Heller, H., Editor: The neurohypophysis. London: Butterworth 1957.

Pickford, G. E., and J. W. Atz: The physiology of the pituitary gland of fishes. New York Zoological Society. 1957.

2. Literature cited in the text

Adams, C. W. M., and J. C. Sloper: The hypothalamic elaboration of posterior pituitary principles in man, the rat and dog. Histochemical evidence derived from a performic acid-alcian blue reaction for cystine. J. Endocr. **13**, 221—228 (1956).

d'Angelo, S. A.: Role of the hypothalamus in pituitary-thyroid interplay. J. Endocr. **17**, 286—299 (1958).

Arko, H., and E. Kivalo: Neurosecretory material and the pituitary portal vessels. Acta endocr. (Kbh.) **29**, 9—14 (1958).

Armstrong, D.T., and W. Hansel: Alteration of the bovine cycle with oxytocin. Fed. Proc. **17**, 6 (1958).

Assenmacher, I.: Répercussions de lésions hypothalamiques sur le conditionnement génital du canard domestique. C. r. Acad. Sci. (Paris) **245**, 210—213 (1957).

— et J. Benoit: Quelques aspects du contrôle hypothalamique de la fonction gonadotrope de la préhypophyse. Pathophysiol. Diencephal. Symp. Intern. Milano 1956. P. 401—427. Wien: Springer-Verlag 1958.

Bachrach, D.: Über einige Probleme der hypothalamischen Neurosekretion. 1. Beiträge zur Herkunft des Neurosekrets. Z. Zellforsch. **46**, 457—473 (1957).

Bargmann, W.: Zwischenhirn-Hypophysensystem von Fischen. Z. Zellforsch. **38**, 275—298 (1953).

Benoit, J., et I. Assenmacher: Rapport entre la stimulation sexuelle préhypophysaire et la neurosécrétion chez l'oiseau. Arch. Anat. micr. Morph. exp. **42**, 334—386 (1953).

Benson, G. K., and S. J. Folley: The effect of oxytocin on mammary gland involution in the rat. J. Endocr. **16**, 189—201 (1958).

Billenstien, D. C., and T. F. Leveque: The reorganization of the neurohypophyseal stalk following hypophysectomy in the rat. Endocrinology **56**, 704—717 (1955).

Bogdanove, E. M.: Selectivity of the effects of hypothalamic lesions on pituitary trophic hormone secretion in the rat. Endocrinology **60**, 689—697 (1957).

Brown-Grant, K., G. W. Harris and S. Reichlin: The effect of pituitary stalk section on thyroid function in the rabbit. J. Physiol. **136**, 364—379 (1957).

Burch, A. B.: An experimental study of the histological and functional differentiation of the epithelial hypophysis in *Hyla regilla*. Univ. Calif. Publ. Zool. **51**, 185—214 (1946).

Burden, C. E.: The failure of hypophysectomized *Fundulus heteroclitus* to survive in fresh water. Biol. Bull. **110**, 8—28 (1956).

Buser-Lahaye, J.: Étude expérimentale du déterminisme de la régéneration des nageoires chez les poissons téléostéens. Ann. Inst. Océanogr. Monaco **28**, 1—61 (1953).

Callamand, O., M. Fontaine, M. Olivereau et A. Raffy: Hypophyse et osmorégulation chez les poissons. Bull. Inst. Oceanogr. Monaco, no. 984 (1951).

Clayton, G. W., W. R. Bell and R. Guillemin: Stimulation of ACTH-release in humans by a non-pressor fraction from commercial extracts of posterior pituitary. Proc. Soc. exp. Biol. (N. Y.) **96**, 777—779 (1957).

Cross, B. A.: On the mechanism of labour in the rabbit. J. Endocr. **16**, 261—276 (1958).

Curri, S. B.: I lipidi semplici e complessi della regione diencefalica. Pathophysiol. Diencephal. Symp. Intern., Milano 1956. P. 443—534. Wien: Springer-Verlag 1958.

Da Lage, C.: L'innervation neurosécrétoire de l'adénohypophyse chez quelques Syngnathidés. Pathophysiol. Diencephal. Symp. Intern., Milano 1956. P. 118—121 Wien: Springer-Verlag. (1958).

Dawson, A. B.: Morphological evidence of a possible functional interrelationship between the median eminence and the pars distalis of the anuran hypophysis. Anat. Rec. **128**, 77—90 (1957).

De Wied, D., P. R. Bouman, and P. G. Smelik: The effect of a lipide extract from the posterior hypothalamus and of pitressin on the release of ACTH from the pituitary gland. Endocrinology **62**, 605—613 (1958).

Donovan, B. T., and G. W. Harris: Hypothalamic injections and ovulation in the rabbit. J. Physiol. **128**, 13—14 P (1955).

Dubreuil, R., and L. Martini: Possible mechanism of the hypothalamic control of thyrotrophic hormone secretion. Abstr. XX. int. physiol. Congr., Brussels **2**, 257 (1956).

Endröczi, E., J. Szalay und K. Lissák: Untersuchungen über die Entwicklung der Funktion des Hypothalamus-Hypophysen-Nebennierenrinden-Systems bei neugeborenen Ratten mit chronischen Tiefenelektroden. Endokrinologie **34**, 331—336 (1957).

Eser, S., et P. Tüzünkam: Mécanisme de l'excitation de l'antéhypophyse par les extraits post-hypophysaires. Pathophysiol. Diencephal. Symp. Intern., Milano 1956. P. 583—586. Wien: Springer-Verlag 1958.

Etkin, W.: Independant differentiation in components of the pituitary complex in the wood frog. Proc. Soc. exp. Biol. (N. Y.) **97**, 388—393 (1958).

Euler, C. von, and B. Holmgren: (a) The thyroxine "receptor" of the thyroid-pituitary system. J. Physiol. **131**, 125—136 (1956).

— (b) The role of hypothalamo-hypophysial connections in thyroid secretion. J. Physiol. **131**, 137—146 (1956).

Everett, J. W.: The time of release of ovulating hormone from the rat hypophysis. Endocrinology **59**, 580—585 (1956).

— and C. H. Sawyer: A 24-hour periodicity in the "LH-release apparatus" of female rats, disclosed by barbiturate sedation. Endocrinology **47**, 198—218 (1950).

— — Estimated duration of the spontaneous activation which causes release of ovulating hormone from the rat hypophysis. Endocrinology **52**, 83—92 (1953).

Ewer, R. F.: The effects of posterior pituitary extracts on water balance in *Bufo carens* and *Xenopus laevis* together with some general considerations of anuran water economy. J. exp. Biol. **29**, 429—439 (1952).

Fitzpatrick, R. J.: On oxytocin and uterine function. Proc. 8. Symp. Colston Res. Soc. **1956**, 203—217.

Flerkó, B., and J. Szentagothai: Oestrogen sensitive nervous structures in the hypothalamus. Acta endocr. (Kbh.) **26**, 121—127 (1957).

Florsheim, W. H.: The effect of anterior hypothalamic lesions on thyroid function and goiter development in the rat. Endocrinology **62**, 783—789 (1958).

— D. T. Imagawa and M. A. Greer: Failure of hypothalamic tissue to reinitiate thyrotropin production by mouse pituitaries in tissue culture. Proc. Soc. exp. Biol. (N. Y.) **95**, 664—667 (1957).

Folley, S. J.: Physiology and biochemistry of lactation. Edinburgh: Oliver and Boyd 1956.

Fontaine, M.: The hormonal control of water and salt-electrolyte metabolism in fish. Mem. Soc. Endocr. **5**, 69—81 (1956).

Fortier, C.: Observations on the socalled ACTH-releasing activity of neurohypophyseal extracts in Saffran and Schally's pituitary incubation system. Texas Rep. Biol. **16**, 68—78 (1958).

— G. W. Harris and I. R. McDonald: The effect of pituitary stalk section on the adrenocortical response to stress in the rabbit. J. Physiol. **136**, 344 (1957).

— and D. N. Ward: Limitations of the in vitro pituitary incubation system as an assay for ACTH-releasing activity. Canad. J. Biochem. **36**, 111—118 (1958).

Fraps, R. M.: Neural basis of diurnal periodicity in release of ovulation-inducing hormone in fowl. Proc. nat. Acad. Sci. (Wash.) **40**, 348—356 (1954).

— The varying effects of sex hormones in birds. Mem. Soc. Endocr. **4**, 205—219 (1955).

Fuche, J., and G. Kahlson: Histamine as a stimulant to the anterior pituitary gland as judged by the lymphopenic response in normal and hypophysectomized rabbits. Acta physiol. scand. **39**, 327—347 (1957).

Fuhrman, F. A., and H. H. Ussing: A characteristic response of the isolated frog skin potential to neurohypophysial principles and its relation to the transport of sodium and water. J. cell. comp. Physiol. **38**, 109—130 (1951).

Fulford, B. D., and S. M. McCann: Suppression of adrenal compensatory hypertrophy by hypothalamic lesions. Proc. Soc. exp. Biol. (N. Y.) **90**, 78—80 (1955).

Green, J. D.: Vessels and nerves of amphibian hypophysis. A study of the living circulation and of the histology of the hypophysial vessels and nerves. Anat. Rec. **99**, 21—53 (1947).

— The comparative anatomy of the hypophysis, with special reference to its blood supply and innervation. Amer. J. Anat. **88**, 225—311 (1951).

— and V. L. van Breemen: Electron microscopy of the pituitary and observations on neurosecretion. Amer. J. Anat. **97**, 177—227 (1955).

Greer, M. A.: Studies on the influence of the central nervous system on anterior pituitary function. Recent Progr. Hormone Res. **13**, 67—104 (1957).

Grosvenor, C. E., and C. W. Turner: Effects of oxytocin and blocking agents upon pituitary lactogen discharge in lactating rats. Proc. Soc. exp. Biol. (N. Y.) **97**, 463—465 (1958).

Guillemin, R.: Über die hypothalamische Kontrolle der ACTH-sekretion betrachtet an den Ergebnissen von in vitro-Versuchen. Endokrinologie **34**, 193—201 (1957)

— W. R. Hearn, W. R. Check and D. E. Housholder: Control of corticotrophin-release: Further studies with in vitro methods. Endocrinology **60**, 488—506 (1957).

HARRIS, G. W.: (a) Neural control of the pituitary gland. Physiol. Rev. **28**, 139—179 (1948).
— (b) Electrical stimulation of the hypothalamus and the mechanism of neural control of the adenohypophysis. J. Physiol **107**, 418—429 (1948).
— Neural control of the pituitary gland. London: Edward Arnold 1955.
— and D. JACOBSOHN: Functional grafts of the anterior pituitary gland. Proc. roy. Soc. B **139**, 263—276 (1952).
— and J. W. WOODS: The effect of electrical stimulation of the hypothalamus or pituitary gland on thyroid activity. J. Physiol. **143**, 246—274 (1958).
HELLER, H.: The effect of neurohypophysial extracts on the water balance of lower vertebrates. Biol. Rev. **20**, 147—157 (1945).
— The comparative physiology of the neurohypophysis. Experientia (Basel) **6**, 368—376 (1950).
HILD, W.: Neurosecretion in the central nervous system. Hypothalamic-Hypophysial Interrelationships. P. 17—26. Springfield, Ill. USA: Charles C. Thomas 1956.
— and G. ZETLER: Über die Funktion des Neurosekrets im Zwischenhirn-Neurohypophysensystem als Trägersubstanz für Vasopressin, Adiuretin und Oxytocin. Z. ges. exp. Med. **120**, 236—243 (1953).
HILLEMAN, H. H.: An experimental study of the development of the pituitary gland in chick embryos. J. exp. Zool. **93**, 347—373 (1943).
HOUSSAY, B. A., A. BIASOTTI, et R. SAMMARTINO: Modifications fonctionelles de l'hypophyse après les lésions infundibulo-tubériennes chez le crapaud. C. R. Soc. Biol. (Paris) **120**, 725—727 (1935).
HUME, D. M.: The method of hypothalamic regulation of pituitary and adrenal secretion in response to trauma. Pathophysiol. Diencephal. Symp. Intern., Milano 1956. P. 217—228. Wien: Springer-Verlag 1958.
JACOBSOHN, D., and C. B. JØRGENSEN: Survival and function of auto- and homografts of adenohypophysial tissue in the toad, *Bufo bufo* (L). Acta physiol. scand. **36**, 1—12 (1956).
JØRGENSEN, C. B.: The amphibian water economy, with special regard to the effect of neurohypophyseal extracts. Acta physiol. scand. **22**, suppl. **78** (1950).
— and LIS NIELSEN: Effect of synthetic lysine-vasopressin on adenohypophysial activity in toads. Proc. Soc. exp. Biol. (N. Y.) **98**, 393—395 (1958).
— H. LEVI and H. H. USSING: On the influence of the neurohypophyseal principles on the sodium metabolism in the axolotl (*Ambystoma mexicanum*). Acta physiol. scand. **12**. 350—371 (1946).
— P. ROSENKILDE and K. G. WINGSTRAND: (a) Regeneration of the neural lobe of the pituitary gland in the toad, *Bufo bufo* (L). Bertil Hanström, zool. papers in honour of his 65. birthday, p. 184—195, 1956.
— K. G. WINGSTRAND and P. ROSENKILDE: (b) Neurohypophysis and water metabolism in the toad *Bufo bufo* (L.). Endocrinology **59**, 601—610 (1956).
— LIS OLESEN LARSEN, P. ROSENKILDE and K. G. WINGSTRAND: Effect of extirpation of median eminence on function of pars distalis of the hypophysis of the toad *Bufo bufo* (L). J. comp. Biochem. Physiol. In the press.
KALMAN, S. M., and H. H. USSING: Active sodium uptake by the toad and its response to the antidiuretic hormone. J. gen. Physiol. **38**, 361—370 (1955).
KNIGGE, K. M., and S. M. BIERMAN: Evidence of central nervous system influence upon cold-induced acceleration of thyroidal I^{131} release. Amer. J. Physiol. **192**, 625—630 (1958).
LANZING, W. J. R.: The occurrence of a waterbalance-, a melanophore-expanding and an oxytocic principle in the pituitary gland of the river-lamprey (*Lampetra fluviatilis* L.). Acta endocr. (Kbh.) **16**, 277—284 (1954).

Markee, J. E., C. H. Sawyer and W. H. Hollinshead: Activation of the anterior hypophysis by electrical stimulation in the rabbit. Endocrinology **38**, 345—357 (1946).

Manunta, G.: Azione dell'estratto di lobo posteriore di ipofisi sull'utero di gallina. Arch. Sci biol. (Bologna) **35**, 228—232 (1951).

Martini, L.: Neurosecretion and stimulation of the adenohypophysis. 2. int. Symp. Neurosekretion. P. 52—54. Berlin: Springer-Verlag 1958.

— and A. de Poli: Neurohumoral control of the release of adrenocorticotrophic hormone. J. Endocr. **13**, 229—234 (1956).

Mason, J. W.: Plasma 17-hydroxycorticosteroid response to hypothalamic stimulation in the conscious Rhesus monkey. Endocrinology **63**, 403—411 (1958).

Mazzi, V.: Effetti di lesioni ipotalamiche sull'ipofisi e sul testiculo del Tritone crestato. Arch. di Anat. **77**, 1—26 (1952).

— e A. Guardabassi: Effetti del saccarosio sui fenomeni neurosecretori e sull'ipofisi di girini di rospo (*Bufo bufo bufo* L.). Arch. ital. Anat. Embriol. **62**, 172—196 (1957).

Mazzi, V., e A. Peyrot: La eminenza mediale della neuroipofisi negli anfibi urodeli. Monitore Zool. ital. **64**, 181—188 (1957).

McCann, S. M.: The ACTH-releasing activity of extracts of the posterior lobe of the pituitary. Endocrinology **60**, 664—676 (1957).

— and J. R. Brobeck: Evidence for a role of the supraopticohypophyseal system in regulation of adrenocorticotrophin secretion. Proc. Soc. exp. Biol. (N. Y.) **87**, 318—324 (1954).

— and A. Fruit: Effect of synthetic vasopressin on release of adrenocorticotrophin in rats with hypothalamic lesions. Proc. Soc. exp. Biol. (N. Y.) **96**, 566—567 (1957).

— — and B. D. Fulford: Studies on the loci of action of cortical hormones inhibiting the release of adrenocorticotrophin. Endocrinology **63**, 29—42 (1958).

McDonald, R. K., and V. K. Weise: Effect of oxytocin on adrenocortical activity in man. Proc. Soc. exp. Biol. (N. Y.) **92**, 109—110 (1956).

— — and R. W. Patrick: Effect of synthetic Lysine-vasopressin on plasma hydrocortisone levels in man. Proc. Soc. exp. Biol. (N. Y.) **93**, 348—349 (1956).

Metuzals, J.: The innervation of the adenohypophysis in the duck. J. Endocr. **14**, 87—95 (1956).

— The innervation of the anterior pituitary gland in the cat. Pathophysiol. Diencephal. Intern. Symp., Milano 1956. P. 148—158. Wien: Springer-Verlag 1958.

Morel, F., J. Maetz, et C. Lucarain: Action des deux peptides neurohypophysaires sur le transport actif de sodium et le flux net d'eau à travers la peau de diverses espèces de batraciens anoures. Biochim. biophys. Acta **28**, 619—626 (1958).

Nichols, Jr., B., R. Guillemin and R. A. Seibert: Endogenous and exogenous vasopressin on ACTH release. Fed. Proc. **17**, 398 (1958).

Nikitovitch-Winer, M., and J. W. Everett: Resumption of gonadotrophic function in pituitary grafts following retransplantation from kidney to median eminence. Nature (Lond.) **180**, 1434—1435 (1957).

Okada, M., T. Ban, and T. Kurotsu: Relation of the neurosecretory system to the third ventricle and the anterior pituitary gland. Med. J. Osaka Univ. **6**, 359—372 (1955).

Olivecrona, H.: Paraventricular nucleus and pituitary gland. Acta physiol. scand. **40**, Suppl. **136**, 1—178 (1957).

PASTEELS, J. L., Jr.: Recherches expérimentales sur le rôle de l'hypothalamus dans la différenciation cytologique de l'hypophyse, chez *Pleurodeles waltlii*. Arch. Biol. **68**, 65—114 (1957).

PICKFORD, G. E., and J. W. ATZ: The physiology of the pituitary gland of fishes. New York Zool. Soc. 1957.

POLENOV, A. L.: The morphology of the neurosecretory cells of the hypothalamus and the question of the relation of these cells to the gonadotropic function of the hypophysis of sazan and the mirror carp. Dokl. Akad. Nauk USSR **73**, 1025—1028 (1950) (In Russian).

POPA, G. T., and U. FIELDING: A portal circulation from the pituitary to the hypothalamic region. J. Anat. **65**, 88—91 (1930).

PORTER, J. C., and J. C. JONES: Effect of plasma from hypophysial-portal vessel blood on adrenal ascorbic acid. Endocrinology **58**, 62—67 (1956).

— and H. W. RUMSFELD: Effect of lyophilized plasma and plasma fractions from hypophyseal-portal vessel blood on adrenal ascorbic acid. Endocrinology **58**, 359—364 (1956).

REICHLIN, S.: The effect of dehydration, starvation, and Pitressin injections on thyroid activity in the rat. Endocrinology **60**, 470—487 (1957).

REZÁBEK, K.: The effect of ethyl alcohol on secretion of the adrenocorticotrophic hormone of the pituitary gland in rats. Notes on the mechanism of secretion of ACTH. Physiol. Bohemoslov. **6**, 516—522 (1957).

ROTHSCHILD, J., and R. M. FRAPS: Induced ovipositions in relation to age of oviducal egg in the domestic hen. Proc. Soc. exp. Biol. (N. Y.) **63**, 511—514 (1946).

ROYCE, P. C., and G. SAYERS: Corticotropin releasing activity of a pepsin labile factor in the hypothalamus. Proc. Soc. exp. Biol. (N. Y.) **98**, 677—680 (1958).

SAFFRAN, M., and A. V. SCHALLY: (a) The release of corticotrophin by anterior pituitary tissue in vitro. Canad. J. Biochem. **33**, 408—415 (1955).

— — (b) In vitro bioassay of corticotropin: Modification and statistical treatment. Endocrinology **56**, 523—532 (1955).

— — M. SEGAL, and B. ZIMMERMAN: Characterization of the corticotrophin releasing factor of the neurohypophysis. 2. Int. Symp. Neurosekretion. P. 55—59. Berlin: Springer-Verlag 1958.

SAWYER, W. H.: Oxytocic, antidiuretic and vasopressor activities in the neurohypophysis of the sea lamprey, *Petromyzon marinus*. Fed. Proc. **14**, 130 (1955).

— The antidiuretic action of neurohypophysial hormones in Amphibia. Colston Papers **8**, 171—182 (1956).

— and W. D. ROTH: Neurohypophyseal function in dehydrated and adrenalectomized rats as indicated by hormone assay and neurosecretory activity. Fed. Proc. **12**, 125 (1953).

SCHALLY, A. V., and M. SAFFRAN: Effect of histamine, hog vasopressin, and corticotropin-releasing factor (CRF) on ACTH release in vitro. Proc. Soc. exp. Biol. (N. Y.) **92**, 636—667 (1956).

— — and B. ZIMMERMANN: A corticotrophin-releasing factor: Partial purification und amino acid composition. Biochem. J. **70**, 97—103 (1958).

SCHAPIRO, S., J. MARMORSTON, and H. SOBEL: The steroid feedback mechanism. Amer. J. Physiol. **192**, 58—62 (1958).

SCHARRER, E.: Zwischenhirndrüse und Häutung bei der Erdkröte *Bufo vulgaris*. Verh. dtsch. Zool. Ges. **36**, 23—27 (1934).

— Das Hypophysen-Zwischenhirnsystem von *Scyllium stellare*. Z. Zellforsch. **37**, 196—203 (1952).

— und B. Scharrer: (a) Neurosekretion. Handbuch der mikroskopischen Anatomie des Menschen. p. 953—1066. Berlin-Göttingen-Heidelberg: Springer-Verlag 1954.

— — (b) Hormones produced by neurosecretory cells. Recent Progr. Hormone Res. **10**, 183—232 (1954).

Schiebler, T. H., und H. v. Brehm: Über jahreszyklische und altersbedingte Veränderungen in den neurosekretorischen Systemen von Teleostiern. Naturwissenschaften **45**, 450—451 (1958).

Schmid, R., L. Gonzalo, R. Blobel, E. Muschke, und E. Tonutti: Über die hypothalamische Steuerung der ACTH-Abgabe aus der Hypophyse bei Diphtheri-Toxin-Vergiftung. Endokrinologie **34**, 65—91 (1957).

Scow, R. O., and M. A. Greer: Effect on the thyroid gland of experimental alteration of the level of circulating thyroxine in mice with heterotopic pituitaries. Endocrinology **56**, 590—596 (1955).

Shirley, H. V., and A. V. Nalbandov: Effects of neurohypophysectomy in domestic chickens. Endocrinology **58**, 477—483 (1956).

Siperstein, E. R., and M. A. Greer: Observations on the morphology and histochemistry of the mouse pituitary implanted in the anterior eye chamber. J. nat. Cancer Inst. **17**, 569—599 (1956).

Sloper, J. C.: Hypothalamic neurosecretion in the dog and cat, with particular reference to the identification of neurosecretory material with posterior lobe hormone. J. Anat. **89**, 301—310 (1955).

— The application of newer histochemical and isotope techniques for the localisation of protein-bound cystine or cysteine to the study of hypothalamic neurosecretion in normal and pathological conditions. 2. Intern. Symp. Neurosekretion. P. 20—25. Berlin: Springer-Verlag 1958.

Slusher, M. A.: Dissociation of adrenal ascorbic acid and corticosterone responses to stress in rats with hypothalamic lesions. Endocrinology **63**, 412—419 (1958).

— and S. Roberts: Fractionation of hypothalamic tissue for pituitary-stimulating activity. Endocrinology **55**, 245—254 (1954).

Smelik, P. G., and D. de Wied: Corticotrophin-releasing action of adrenaline, serotonin and pitressin. Experientia (Basel) **14**, 17—18 (1958).

Smith, D. C. W.: The role of the endocrine organs in the salinity tolerance of trout. Mem. Soc. Endocr. **5**, 83—98 (1956).

Smith, R. N.: The presence of non-myelinated nerve fibres in the pars distalis of the pituitary gland of the ferret. J. Endocr. **14**, 279—283 (1956).

Sobel, H., R. S. Levy, J. Marmorston, S. Schapiro, and S. Rosenfeld: Increased excretion of urinary corticoid sby guinea pigs following administration of pitressin. Proc. Soc. exp. Biol. (N. Y.) **89**, 10—13 (1955).

Stahl, A.: La neurosécrétion chez les poissons téléostéens. Contribution à l'étude de la neurohypophyse chez les mugilidés. C. R. Soc. Biol. (Paris) **147**, 841—844 (1953).

Stutinsky, F. S.: Recherches expérimentales sur le complexe hypothalamo-neurohypophysaire. Arch. Anat. micr. Morph. exp. **46**, 93—158 (1957).

— Rapports du neurosécrétat hypothalamique avec l'adénohypophyse dans des conditions normales et expérimentales. Pathophysiol. Diencephal. Symp. Intern., Milano 1956. P. 78—103. Wien: Springer-Verlag 1958.

— J. Schneider et P. Denoyelle: Dosage de l'ACTH sur le rat normal et influence de la présence des principes posthypophysaires. Ann. Endocr. (Paris) **13**, 641—650 (1952).

Thorn, N. A.: Mammalian antidiuretic hormone. Physiol. Rev. **38**, 169—195 (1958).

TRAMEZZANI, J. H., y J. V. URANGA: Variaciones de la substancia Gomori-positiva y actividad antidiurética en la neurohipófisis de sapos hidratados y deshidratados. Rev. Soc. argent. Biol. **30**, 148—151 (1954).

VAZQUEZ-LOPEZ, E.: Innervation of the adenohypophysis. Nature (Lond.) **162**, 458 (1948).

VERNEY, E. B.: The antidiuretic hormone and the factors which determine its release. Proc. roy. Soc. **135**, 25—106 (1947).

VIVIEN, J. H., et J. SCHOTT: Contributions à l'étude des corrélations hypothalamo-pituitaires chez les Batraciens. Le contrôle de l'activité gonadotrope. J. de Physiol. **50**, 561—563 (1958).

WARING, H., and F. W. LANDGREBE: Hormones of the posterior pituitary. The Hormones II, Ed. G. Pincus, p. 427—514, (1950.).

WELLER, O.: Der Einfluß von Vasopressin und Oxytocin auf die Nebennierenrinden-funktion. Z. ges. inn. Med. **12**, 759—760 (1957).

WILHELMI, A. E., G. E. PICKFORD, and W. H. SAWYER. Initiation of the spawning reflex response in *Fundulus* by administration of fish and mammalian neuro-hypophysial preparations and synthetic oxytocin. Endocrinology **57**, 243—252 (1955).

WINGSTRAND, K. G.: The structure and development of the avian pituitary. Lund 1951.

— The structure of the pituitary in the african lungfish, *Protopterus annectens* (Owen). Vidensk. Medd. Dansk naturh. Foren. **118**, 193—210 (1956).

— Attempts at a comparison between the neurohypophysial region in fishes and tetrapods, with particular regard to amphibians. Symposium on Comparative Endocrinology. (In press.)

WINKLER, G., R. BLOBEL, und E. TONUTTI: 17-OH-Corticoid-Ausscheidung beim Meerschweinchen nach Ausschaltung der Nucl. hypothalamici ventro- und dorsomediales und nach Diphtherietoxin. Naturwissenschaften **45**, 13—14 (1958).

WISLOCKI, G. W.: The vascular supply of the hypophysis cerebri of the cat. Anat. Rec. **69**, 361—387 (1937).

ZAITZEV, A. V.: Questions of neurosecretory activity of the ganglion cells of the hypothalamic nuclei of pike and sazan in connection with seasonal manifestation of the gonadotropic function of the hypophysis. Dokl. Akad. Nauk USSR **101**, 351—354 (1955). (In Russian.)

Der Stoffwechsel der Trypanosomen*

Von THEODOR VON BRAND

U. S. Department of Health, Education, and Welfare, Public Health Service, National Institutes of Health, National Institute of Allergy and Infectious Diseases**, Bethesda, Maryland

Mit 3 Abbildungen

Der Erforschung des Stoffwechsels der Protozoen stellen sich Schwierigkeiten entgegen, denen man bei den meisten anderen Phyla nicht begegnet. Diese Schwierigkeiten sind im allgemeinen auf die Kleinheit der Versuchstiere zurückzuführen. Es ist häufig nicht leicht und manchmal unmöglich, genügend Material für chemische Bestimmungen herbeizuschaffen, für die keine spezifischen Mikromethoden zur Verfügung stehen. Weiter ist die Frage bakterieller Verunreinigungen nicht zu vernachlässigen. Es liegt auf der Hand, daß dieser Faktor bei einigermaßen sauberem Arbeiten mit größeren Wirbellosen, wie Schnekken, vielen freilebenden Würmern und Insekten, eine nur untergeordnete Rolle spielt, obwohl selbst hier in Spezialfällen und für besondere Fragestellungen Vorsicht am Platz ist. Wenn man die Literatur über den Stoffwechsel der Protozoen studiert, wird man sofort gewahr, daß wahrer Fortschritt nur oder, vielleicht richtiger gesagt, fast nur mit solchen Formen erzielt wurde, bei denen eine bakterienfreie Züchtung größerer Mengen möglich war. Das ist z. B. der Fall mit dem viel untersuchten Ciliaten *Tetrahymena*, und es ist bezeichnend, daß für keinen anderen Ciliaten auch nur annähernd so viel bekannt geworden ist als gerade für *Tetrahymena*.

Ähnlich steht es mit den *parasitischen* Protozoen. Wegen ihrer praktischen Bedeutung ist viel Mühe und Zeit darauf verwandt worden, Näheres über den Stoffwechsel von *Entamoeba histolytica* herauszufinden. Das Ergebnis ist durchaus enttäuschend. Die Arbeiten verschiedener Autoren sind voller Widersprüche, und man wird den Verdacht nicht los, daß sie die Bakterien nicht so vollständig eliminiert hatten, als sie glaubten annehmen zu dürfen.

---

* Nach einem im Dezember 1958 an den Universitäten Erlangen, München und Würzburg sowie dem Institut für Schiffs- und Tropenkrankheiten, Hamburg, gehaltenen Vortrag.

** Laboratory of Parasitic Diseases.

Die Trypanosomen auf der anderen Seite sind für stoffwechselphysiologische Untersuchungen sehr geeignet. Es ist leicht, die Blutform vieler Arten in großen Mengen steril zu erhalten, und sterile Kulturen vieler Species können ohne größere Schwierigkeiten gehalten werden. Es hat sich fernerhin herausgestellt, daß der Stoffwechsel dieser Parasiten ungewöhnlich hoch ist, so daß man selbst mit relativ geringen Mengen auskommen kann.

Im Lebenscyclus der meisten Säugetiertrypanosomen kommen eine Reihe verschiedener Stadien vor, die sich sowohl in morphologischen als auch physiologischen Merkmalen deutlich unterscheiden. In letzterer Hinsicht freilich sind bisher nur einzelne Formen zur Untersuchung gekommen. Es stehen zur Verfügung: einesteils im Endwirt die Blutform, bei *Trypanosoma cruzi* dazu noch die Gewebsform, und andernteils die verschiedenen in den Überträgern vorkommenden Stadien. Während man die Blutform für physiologische Studien direkt aus dem zirkulatorischen System geeigneter Wirte entnehmen kann, ist es nicht möglich, genügend Material aus infizierten Zwischenwirten (Insekten) zu gewinnen. Maß muß sich hier mit Kulturen begnügen, die, wie bekannt, zu Stadien führen, wie sie in den Überträgern vorkommen. Die Trypanosomen der *lewisi*-Gruppe, zu der *T. cruzi* gehört, durchlaufen in den Kulturen den ganzen Lebenscyclus, d. h. die Kulturen sind infektiös. Die Trypanosomen der *congolense*- und *brucei*-Gruppe dagegen entwickeln sich in den Kulturen nur zu den im Darm der Glossinen vorkommenden Stadien. Solche Kulturen sind nicht infektiös, da die Speicheldrüsenformen nicht gebildet werden. Neuerdings hat WEINMAN (1957) in einer kurzen Mitteilung angegeben, daß Kulturen von *T. rhodesiense* infektiös werden, wenn sie das Disaccharid Trehalose enthalten, das ja neuerdings als wichtiger Bestandteil des Insektenblutes erkannt wurde. Wir haben mit Trehalose kein analoges Ergebnis erzielt, und man wird WEINMANS ausführliche Arbeit abwarten müssen, bevor ein endgültiges Urteil möglich ist.

Alle bisher untersuchten Trypanosomen müssen als aerobe Gärer bezeichnet werden, und zwar sowohl ihre Blut- wie ihre Kulturformen; im Gegensatz zu der großen Mehrzahl freilebender Wirbelloser oxydieren sie ihre Nahrung nicht vollständig, vielmehr scheiden sie alle mehr oder weniger unvollständig oxydierte Stoffwechselendprodukte aus. Das hat nun einige wichtige Folgen.

Jede Gärung, sei sie nun aerob oder anaerob, ist unökonomisch, da ein großer Teil, oft der größte Teil, der in der Nahrung enthaltenen potentiellen Energie unausgenützt bleibt, eben weil sie in den ausgeschiedenen Stoffwechselprodukten enthalten ist (Tab. 1). Mit dieser Tatsache hängt es folgerichtig zusammen, daß von Gärungen lebende Organismen viel größere Mengen an Nahrungsmaterial umsetzen müssen

Tabelle 1. *Energie-Bilanzen*

1. Vollständige Oxydation
   $C_6H_{12}O_6 + 6\,O_2 \rightarrow 6\,CO_2 + 6\,H_2O + 674$ kg cal.
2. Unvollständige, zu Oxalsäure führende Gärung
   $2\,C_6H_{12}O_6 + 9\,O_2 \rightarrow 6\,C_2H_2O_4 + 6\,H_2O + 493$ kg cal.
3. Reine Gärungen
   A. Milchsäuregärung
   $C_6H_{12}O_6 \rightarrow 2\,C_3H_6O_3 + 22{,}5$ kg cal.
   B. Alkoholische Gärung
   $C_6H_{12}O_6 \rightarrow 2\,C_2H_6O + 2\,CO_2 + 22$ kg cal.
   C. Essigsäuregärung
   $C_6H_{12}O_6 \rightarrow 3\,C_2H_4O_2 + 15$ kg cal.

als die von vollständigen Oxydationen abhängenden Organismen, um die zur Aufrechterhaltung des Lebens nötige Energie zu gewinnen. Als Beispiel sei hier angeführt, daß die Blutform der afrikanischen pathogenen Trypanosomen in einer Stunde und bei einer Temperatur von 37° C eine Zuckermenge umsetzt, die zwischen 50 und 100% ihres Trockengewichtes liegt, ein wahrhaft erstaunlicher Umsatz. Es ist klar, daß ein derartiges Kohlenhydratbedürfnis nur befriedigt werden kann, wenn den Organismen eine dauernde Zuckerquelle zur Verfügung steht, im obigen Fall eben der Blutzucker des Wirtes. Eine Anhäufung intracellulärer Kohlenhydratreserven, wie sie z. B. in der Form von Glykogen in so vielen freilebenden und auch parasitischen Organismen vorkommt, würde den genannten Trypanosomen nur wenig nützen, da diese Reserven bestenfalls nur für sehr kurze Zeitspannen als Energiequellen ausreichen würden. In der Tat ist bisher in den Trypanosomen kein Glykogen nachgewiesen worden.

Es ist hervorzuheben, daß der geschilderte enorm hohe Zuckerumsatz nur für die Blutform der pathogenen afrikanischen Trypanosomen charakteristisch ist. Die Blutform von *Trypanosoma cruzi* verbraucht wesentlich weniger Zucker [RYLEY (1956)], und dasselbe trifft zu für die Kulturformen verschiedener Arten, im letzteren Falle auch wenn man berücksichtigt, daß die Kulturformen bei niedrigerer Temperatur untersucht werden als die Blutform (Beispiele in Tab. 2). Diese Unterschiede hängen wahrscheinlich mit der Tatsache zusammen, daß verschiedene

Tabelle 2. *Zuckerverbrauch verschiedener Trypanosomen in mg per 100 Millionen per 1 Std.*

Species	Form	Temp. °C	Glucose	Autor
T. gambiense	Blut	37	1,5	VON BRAND et al, 1955
T. rhodesiense	Blut	37	1,5	VON BRAND et al, 1948
T. congolense	Blut	37	0,5	AGOSIN et al, 1954
T. lewisi, jung	Blut	37	0,4	MOULDER, 1948
T. lewisi, alt	Blut	37	0,2	MOULDER, 1948
T. congolense	Kultur	30	0,2	VON BRAND et al, unveröff.

Trypanosomen das Zuckermolekül verschieden weit abbauen und so vermutlich verschiedene Prozentsätze der potentiellen Energie verwerten können. Vom biologischen Standpunkt aus ist der Unterschied zwischen Blut- und Kulturform der afrikanischen Species besonders interessant. Wie erwähnt, ist die Blutform vom Blutzucker des Wirtes völlig abhängig; schneidet man die Zuckerquelle ab, verenden diese Trypanosomen in wenigen Minuten. Wenn nun eine *Glossina* Blut saugt, nimmt sie nur eine beschränkte Zuckermenge auf, die wahrscheinlich bald geringer wird, weil die Darmgewebe Zucker im allgemeinen rasch resorbieren. Man muß demgemäß annehmen, daß sich die Blutform sehr rasch auf den weniger Zucker erfordernden Stoffwechsel der Kultur-, d. h. der Insekten-Darm Form umstellt, vielleicht sogar schneller, als der morphologische Umbau erfolgt.

Eine wichtige, im Moment schwer einwandfrei zu beantwortende Frage ist, ob die geschilderte Intensität des Stoffwechsels der Blutform der pathogenen afrikanischen Trypanosomen wirklich den Verhältnissen in der Natur entspricht. Es muß nämlich hervorgehoben werden, daß die bisherigen Untersuchungen fast ausschließlich mit alten Laboratoriumsstämmen durchgeführt wurden, die meist für viele Jahre durch Blutpassagen weitergeführt wurden. Es ist schon lange bekannt, vor allem durch REICHENOWs Studien, daß solche Stämme biologisch verändert sind, zum Beispiel in ihrer Unfähigkeit, sich in Glossinen zu entwickeln. Im Jahre 1956 haben JENKINS und GRAINGE mitgeteilt, daß solche alten Stämme einen merklich höheren Sauerstoffverbrauch haben als junge, frisch isolierte. Diese Beobachtung wurde 1957 im wesentlichen bestätigt durch FULTON und SPOONER. Diese Forscher fanden aber, daß die niedrigere Rate des jungen Stammes, zum Teil wenigstens, durch seine größere Empfindlichkeit gegenüber dem Schütteln im Warburg-Apparat bedingt war. Warum sie leichter einer Lysis unterlagen, ist nicht klar. Serum wirkte teilweise schützend. Qualitativ scheint der aerobe Zuckerstoffwechsel zwischen alten und jungen Stämmen identisch zu sein [GRANT und FULTON (1957)]. Es ist sehr wahrscheinlich, daß in den nächsten Jahren eine endgültige Klärung dieser Frage in Laboratorien erfolgen wird, die in der glücklichen Lage sind, Zugang zu alten und jungen Stämmen zu haben. Was der Ausgang solcher Studien auch immer ergeben mag, die Studien an alten Stämmen werden ihren Wert behalten, höchstens werden sich Unterschiede aufzeigen lassen, wie sie zwischen Haustieren und ihren wilden Verwandten vorkommen; selbst in diesem Falle aber werden die Versuche mit alten Stämmen ihre Wichtigkeit behalten, weil sie biochemische Fähigkeiten dauernd fortzüchtbarer Organismen demonstrieren.

Wenn wir uns nun den Stoffwechselendprodukten zuwenden, denen man im aeroben Kohlenhydratstoffwechsel begegnet, so findet man die größte Einheitlichkeit in der Blutform der *brucei*-Gruppe, welche den Zucker fast quantitativ in Brenztraubensäure verwandelt (Tab. 3). Dies ist gezeigt worden für *T. gambiense* [VON BRAND, TOBIE, MEHLMAN, WEINBACH (1953)], *T. hippicum* [HARVEY (1949)] und *T. equiperdum* [REINER, SMYTHE und PEDLOW (1936); AGOSIN und VON BRAND (1954)]. Es ist dies von einem formalen Standpunkt aus ein relativ einfacher Prozeß. Wenn er quantitativ vor sich geht, wird 1 Mol Zucker in 2 Mol Brenztraubensäure überführt, wobei 1 Mol Sauerstoff verbraucht wird. In den soeben genannten Formen kommt die Stoffwechselbilanz der theoretischen recht nahe; gewöhnlich findet sich jedoch ein geringes Defizit an Brenztraubensäure. Dieses ist manchmal so klein, daß es als innerhalb der Fehlerquellen liegend betrachtet werden könnte; es gibt aber doch Beobachtungen, die für die Realität der Erscheinung sprechen. Wenn Zucker quantitativ in Brenztraubensäure verwandelt wird, kann keine respiratorische Kohlensäure gebildet werden, weil alle Kohlenstoffatome des Zuckermoleküls in den zwei Brenztraubensäuremolekülen erscheinen. Der respiratorische Quotient des Prozesses ist also Null. In Wirklichkeit produzieren die Blutformen der genannten Trypanosomen kleine Mengen an Kohlensäure, die zu dem immerhin noch außergewöhnlich niedrigen respiratorischen Quotienten von ungefähr 0,08—0,09 führen. Die Quellen der Kohlensäure sind noch nicht endgültig bekannt geworden. Eine derselben mag die fehlende Brenztraubensäure sein oder wenigstens ein gewisser Prozentsatz derselben. Sicher ist dies allerdings nicht, da manche Forscher, z. B. neuerdings GRANT und FULTON (1957) für *T. rhodesiense*, angeben, daß neben Brenztraubensäure auch Glycerin gebildet wird. Es ist aber hervorzuheben, daß die Glycerinbildung vornehmlich gefunden wurde, wenn die Trypanosomen in einem Hämoglobin-freien, höchstens etwas Serum enthaltenden Medium untersucht werden. Seit längerer Zeit ist aber bereits bekannt, daß in derartigen Medien die Oxydationen dieser Tiere niedriger sind, als wenn man sie in Blut hält. Auf der anderen Seite ist gezeigt worden, daß unter anaeroben Bedingungen ein Mol Zucker in ein Mol Brenztraubensäure und ein Mol Glycerin verwandelt wird. Der Verdacht liegt nahe, daß in aeroben blutfreien Medien die Glycerinbildung vielleicht höher ist, als natürlichen Bedingungen entspricht.

Ich habe vorhin den für die Blutform der *brucei*-Gruppe charakteristischen Zucker-Brenztraubensäure-Prozeß als formal einfach bezeich-

Tabelle 3. *Zuckerabbau der Blutform der brucei-Gruppe*

$C_6H_{12}O_6 + O_2 \rightarrow 2\, C_3H_4O_3 + 2\, H_2O$	
1,0 Mol : 1,0 Mol : 2,0 Mol	Theoretisch
1,0 Mol : 0,9 Mol : 1,9 Mol	*T. equiperdum* in Blut
1,0 Mol : 1,0 Mol : 1,7 Mol	*T. gambiense* in Blut
1,0 Mol : 1,2 Mol : 1,7 Mol	*T. gambiense*, Arsenikfest in Blut
1,0 Mol : 1,2 Mol : 2,1 Mol	*T. gambiense* in Serum

net. Ich möchte hier nicht mißverstanden werden. Der Nachdruck liegt auf dem Worte formal, das andeuten soll, daß zwischen der Muttersubstanz und dem Endprodukt weniger Schritte liegen, als man sie findet, wenn eine Zelle den Zucker völlig zu Kohlensäure und Wasser oxydiert. Nicht gemeint ist etwa eine phylogenetische Primitivität des Prozesses. Ich bin im Gegenteil davon überzeugt, daß der Stoffwechseltyp der obigen Blutform sekundär erworben ist und daß seine Eigenartigkeit durch einen Verlust der Enzyme des Krebscyclus bedingt ist. Wie unten gezeigt wird, bauen die Kulturformen der Trypanosomen den Zucker etwas weiter ab als die Blutform der *brucei*-Gruppe, was gut mit der obigen Auffassung übereinstimmt. Es ist allgemein angenommen, daß die Wirbeltiertrypanosomen von Insektenformen abstammen.

Die Säugetiertrypanosomen können bekanntlich in verschiedene Gruppen eingeteilt werden, und zwar unterscheidet HOARE (1948) auf Grund morphologischer und entwicklungsgeschichtlicher Merkmale außer der bereits besprochenen *brucei*-Gruppe die *congolense*-, *vivax*- und *lewisi*-Gruppen. Während mehrere Vertreter der *brucei*-Gruppe untersucht sind, sind von der *congolense*- und *vivax*-Gruppe nur je eine und von der *lewisi*-Gruppe zwei Arten studiert worden, wenn wir hier nur die Blutformen im Auge haben. Diese Studien haben gezeigt, daß die Endprodukte des Kohlenhydratstoffwechsels bei diesen Gruppen mannigfaltige sind. Brenztraubensäure wird zwar gefunden, tritt aber quantitativ mehr in den Hintergrund, während Essigsäure und unter Umständen Bernsteinsäure in den Vordergrund treten; selbst etwas Milchsäure wird gebildet. Auf Grund der bisherigen Kenntnisse scheint jede Gruppe in dieser Beziehung einen speziellen Charakter zu besitzen. Um das Problem endgültig zu klären, bedarf es aber noch der Untersuchung zahlreicherer Arten. Man muß auch die Frage im Auge behalten, ob an den obigen Unterschieden nicht vielleicht zum Teil verschiedene Versuchsanordnungen der Autoren beteiligt sein könnten. Das ist entschieden möglich, wie am Beispiel der Blutform von *T. congolense* gezeigt werden soll: Wenn man sie bei der niedrigen Kohlensäurespannung untersucht, die in Warburg-Gefäßen herrscht, in denen die Kohlensäure durch Alkali absorbiert wird, findet man als Hauptendprodukt Essigsäure und ein wenig Brenztraubensäure, während Bernsteinsäure höchstens in Spuren gebildet wird [AGOSIN und VON BRAND (1954)]. Wenn man sie dagegen in Gefäßen hält, die mit einem kohlensäurehaltigen Gasgemisch begast wurden, findet man eine erhebliche Bernsteinsäurebildung [RYLEY (1956)]. Dieser Unterschied deutet natürlich darauf hin, daß eine Kohlensäurefixierung aus der Umwelt bei der Bernsteinsäurebildung eine Rolle spielt; er zeigt aber auch, daß in vitro-Versuche biochemische Fähigkeiten anzeigen, die zu Ergebnissen führen, die nicht immer ohne weiteres vollständig mit dem Stoffwechsel im normalen

Habitat übereinzustimmen brauchen. Es sei hier noch betont, daß die *congolense*-Trypanosomen sowohl in einer Kohlensäure-freien wie Kohlensäure-haltigen Umwelt gut überleben und infektiös bleiben, jedenfalls für die Dauer der hier angeführten Versuche, daß also die besprochenen Veränderungen nicht ohne weiteres als Absterbephänomene gedeutet werden dürfen.

Während, wie ausgeführt, Unterschiede in den Stoffwechselendprodukten der Blutformen auftreten, sind solche Unterschiede bei ihren Kulturformen viel weniger ausgeprägt. Von besonderer Bedeutung ist hier die Tatsache, daß die Kulturform von *Trypanosoma gambiense*, im

Tabelle 4. *Zuckerstoffwechsel der Blut- und Kulturform von T. gambiense.* Endprodukte in Mol per 1 Mol verbrauchter Glucose

	Blutform	Kulturform
Brenztraubensäure .	2,1—1,7	0,20
Milchsäure	0	0,17
Bernsteinsäure . . .	0	0,33
Essigsäure	0	0,83

Gegensatz zur Blutform (Tab. 4), nur kleine Mengen von Brenztraubensäure ausscheidet, dagegen größere Mengen von Essigsäure und Bernsteinsäure sowie kleinere von Milchsäure, alles Säuren, die von der Blutform überhaupt nicht gebildet werden [VON BRAND, WEINBACH und TOBIE (1955)]. In Zusammenhang mit diesem Unterschied steht auch die Tatsache, daß, wieder im Gegensatz zur Blutform, der respiratorische Quotient der Kulturform hoch ist. Es ist offensichtlich, daß der Zuckerabbau der beiden Formen erhebliche Unterschiede aufweist und daß die Blutform den Zucker weniger weit abbaut als die Kulturform.

Da die aeroben Stoffwechselendprodukte der Blutformen der *congolense*- und *lewisi*-Gruppen vielfältiger sind als jene der *brucei*-Gruppe, ist es nicht verwunderlich, daß bei ersteren keine so ausgesprochenen Unterschiede zwischen Blut- und Kulturformen bestehen. Die zur Beobachtung kommenden Unterschiede sind mehr quantitativ als qualitativ. Zum Beispiel scheidet die Kulturform von *T. congolense*, wie wir in bisher unveröffentlichten Versuchen gefunden haben, relativ mehr Brenztraubensäure aus als die Blutform, und sie bildet Bernsteinsäure auch in Warburg-Gefäßen, wenn die Kohlensäure durch Alkali absorbiert wird, d. h. im Gegensatz zur Blutform ist die Bernsteinsäurebildung nicht von einem höheren Kohlensäuredruck abhängig.

Bisher ist der Stoffwechsel nur sehr weniger Arten von Kulturtrypanosomen untersucht worden; es wäre deshalb verfrüht, mit Sicherheit entscheiden zu wollen, ob sich bei ihnen stoffwechselphysiologische Gruppenunterschiede vorfinden oder nicht. Ein Punkt soll aber besonders hervorgehoben werden. Versuche, eine Produktion von Äthylalkohol durch die Kulturformen der Säugetiertrypanosomen nachzuweisen, sind völlig negativ verlaufen. Dies steht in scharfem Gegensatz zur Kulturform der zwischen Pflanzen und Insekten wechselnden Trypanosomide, *Strigomonas oncopelti*, deren Kohlenhydratstoffwechsel zur Bildung großer Mengen Äthylalkohols führt [RYLEY (1955)]. Leider sind bisher noch keinerlei Daten bekannt geworden bezüglich der großen Schar von Trypanosomenarten, die Vögel, Amphibien oder Fische parasitieren. Vergleichende Untersuchungen würden sich sicherlich lohnen und mögen noch manche Überraschung bringen.

Tabelle 5. *In Trypanosomen nachgewiesene glykolytische Enzyme*

Enzym	Species		
	T. hippicum[1] Blutform	*T. equiperdum*[2] Blutform	*T. cruzi*[3] Kulturform
Hexokinase	+		
Isomerase			+
Aldolase	+	+	+
Triosephosphat Dehydrogenase	+	+	+
Phosphorbrenztraubensäure Dephosphorylase	+		

[1] Nach HARVEY (1949).
[2] Nach CHEN und GEILING (1946).
[3] Nach BAERNSTEIN (1953).

Es ist wohl bekannt, daß die Hauptlinien des Kohlenhydratabbaues typischer aerober Zellen in zwei Abschnitte zerfallen, die aufeinanderfolgen. Es sind dies der Embden-Meyerhof- und der Krebscyclus. Der erstere führt bekanntlich von Kohlenhydrat zu Milchsäure und verlangt keinen atmosphärischen Sauerstoff. Er besteht aus einer Reihe enzymatisch gesteuerter Reaktionen, die im allgemeinen in den verschiedensten Organismen recht gleichartig verlaufen. Die Trypanosomen bilden keine Ausnahme. Eine Reihe von Arten sind auf das Vorkommen verschiedener Enzyme des Embden-Meyerhof-Cyclus hin geprüft worden. Einschlägige Fermente wurden nachgewiesen (Tab. 5) in der Blutform von *T. hippicum* [HARVEY (1949)] und *T. equiperdum* [Chen und GEILING (1946)] oder in der Kulturform von *T. cruzi* [BAERNSTEIN (1953)]. Im Falle von *T. evansi* [Blutform, MARSHALL (1948)] wurde das Auftreten phosphorylierter Zwischenstufen der Glykolyse nachgewiesen, was natürlich dem Nachweis der zugehörigen Enzyme entspricht. Die gefundenen Verhältnisse entsprechen ganz jenen, die man in Hefe oder Leberzellen findet, allerdings mit einer Ausnahme: die Suche nach einer Milchsäuredehydrase in den genannten Blutformen ist negativ verlaufen, d. h. es fehlt ihnen jenes Ferment, das Brenztraubensäure in Milchsäure überführt. Wie oben erwähnt wurde, findet sich eine geringe

Milchsäurebildung in anderen Arten, besonders den Kulturformen. Aber selbst in diesen Fällen muß die Milchsäuredehydrase nur schwach sein, weil die Milchsäurebildung im Vergleich zu anderen Endprodukten hier ebenfalls in den Hintergrund tritt.

Das vollständige Fehlen der Milchsäuredehydrase in den genannten Blutformen erklärt sodann, warum sie anaerob keine Milchsäure bilden. Wenn eine typische aerobe Wirbeltierzelle anoxybiotischen Verhältnissen ausgesetzt wird, wird bekanntlich 1 Mol Zucker in 2 Mol Milchsäure verwandelt; eine Bilanz der verschiedenen Atome zeigt in diesem Fall, daß alle Kohlenstoff-, Wasserstoff- und Sauerstoffatome des Zuckermoleküls in den 2 Milchsäuremolekülen auftreten. Wenn nun die Milchsäuredehydrase fehlt, sind nur jene intracellulären Fermente vorhanden, die zur Brenztraubensäure führen. Würde anaerob der ganze Zucker in Brenztraubensäure verwandelt, so würden für jedes verbrauchte Zuckermolekül 2 überschüssige Wasserstoffmoleküle verbleiben. Die Trypanosomen scheiden jedoch keinen gasförmigen Wasserstoff ab, wohl aber bilden sie anaerob reichlich Glycerin [RYLEY (1956)]. Es ist sehr wahrscheinlich, daß hier eine der Triosen, nämlich Phosphoglycerinaldehyd, als Wasserstoffacceptor für die Wiederoxydation des reduzierten Coenzymes I dient. Die Reaktion liefert ein Molekül Phosphoglycerin für jedes Molekül Phosphoglycerinsäure, das gebildet wird. Wenn sodann eine Phosphatase den Phosphor vom Phosphoglycerin abspaltet, verbleibt Glycerin, das ausgeschieden wird.

Wie erwähnt, erfordert der Embden-Meyerhof-Cyclus in typisch aeroben Zellen keinen Sauerstoff, da die durch atmosphärischen Sauerstoff bedingten Oxydationen alle im Krebscyclus ablaufen. Jenen Trypanosomen nun, die Zucker praktisch quantitativ in Brenztraubensäure verwandeln, fehlen die Enzyme des Krebscyclus. Trotzdem verbrauchen sie, wie ebenfalls erwähnt, reichlich Sauerstoff. Dies hängt wieder mit dem Fehlen der Milchsäuredehydrase und dem dadurch bei der Brenztraubensäurebildung bedingten Freiwerden von Wasserstoffatomen zusammen. Gewöhnlich wird ja dieser Wasserstoff via reduziertes Coenzym I an die Brenztraubensäure angelagert, was eben zur Milchsäurebildung führt. Bei den obigen Trypanosomen kann Brenztraubensäure nicht in dieser Weise dienen; bei ihnen dient vielmehr atmosphärischer Sauerstoff als Wasserstoffacceptor. Dies kann in dem Glykolyseschema nur an einer Stelle stattfinden, nämlich wenn Phosphoglycerinaldehyd oxydiert wird.

Die Frage, ob Trypanosomen einen Krebscyclus besitzen, ist bedeutungsvoll, weil eine aerobe Zelle den größten Teil ihres Energiebedarfes durch ihn gedeckt bekommt, dann aber auch, weil er ein wichtiges Bindeglied für den Kohlenhydrat-, Lipoid- und Proteinstoffwechsel darstellt. Der Grund dafür ist, daß Brenztraubensäure oder, streng genommen, ein Zwei-Kohlenstoffabkömmling davon, Acetyl-Coenzym A, nicht nur aus Kohlenhydrat entstehen kann, sondern auch über Essigsäure aus Fett oder über Alanin aus Protein. Auch kann der Cyclus vor sich gehen, wenn irgendeine seiner Zwischenstufen, und nicht notwendigerweise nur Brenztraubensäure, zur Verfügung steht. Von besonderer Bedeutung sind hier Abkömmlinge des Proteinstoffwechsels. So kann Oxalessigsäure durch oxydative Deamination oder Transamination von Asparaginsäure oder α-Ketoglutarsäure durch Deamination oder Transamination von Glutaminsäure entstehen. Es ist in dieser Hinsicht von Interesse, daß verschiedene Transaminationsreaktionen für *T. cruzi* nachgewiesen worden sind [BASH-LEWINSON und GROSSOWICZ (1957)] und daß die Ausnützung von Glutamin durch die Blutform von *T. lewisi* beobachtet wurde, wobei Ammoniak frei gemacht wird [THURSTON (1958)].

Wie mit jeder anderen Zelle kann die Frage nach der An- oder Abwesenheit eines Krebscyclus in Trypanosomen auf zweierlei Weise angegangen werden. Erstens können ihnen an Stelle eines anderen Substrates Zwischenstufen des Krebscyclus angeboten werden; es ist in diesem Falle zu untersuchen, ob diese ausgenützt werden, entweder durch Beobachtung, ob die Sauerstoffaufnahme über die endogene Rate gesteigert ist, oder durch chemische Bestimmung des Verschwindens der zugesetzten Substanz. Der zweite Weg besteht darin, in den Trypanosomen nach den die Reaktionen des Cyclus katalysierenden Enzymen zu suchen. Beide Wege sind wiederholt mit der Blutform der *brucei*-Gruppe eingeschlagen worden [z. B. HARVEY (1949)], mit völlig negativem Resultat. Man muß daher, wie bereits oben erwähnt, zu dem Schlusse kommen, daß den Blutformen dieser Gruppe der Krebscyclus völlig fehlt.

Wenn auf der anderen Seite die Kulturform von *T. cruzi* untersucht wird, ergibt sich, daß sie alle Zwischenstufen des Krebscyclus ausnützen kann, und zwar manche besser und manche schlechter, immer aber vorausgesetzt, daß der p_H des Mediums so niedrig gehalten wird, daß die in Frage kommenden Säuren die celluläre Membran passieren können [VON BRAND und AGOSIN (1955)].

Von den Enzymen des Krebscyclus sind für die Kulturform von *T. cruzi* Äpfelsäuredehydrase und Fumarase von BAERNSTEIN (1953) beschrieben worden. In unserem Laboratorium wurde die schon vorher von SEAMAN (1953) gefundene Bernsteinsäuredehydrase [AGOSIN und VON BRAND (1955)] und die Isocitronensäuredehydrase [AGOSIN und WEINBACH (1956)] näher untersucht. Die Auffindung dieser Fermente zeigt deutlich, daß die Kulturform von *T. cruzi* entweder einen typischen Krebscyclus besitzt oder doch jedenfalls eine diesem sehr ähnliche Reaktionsfolge. Es ist recht wahrscheinlich, daß dasselbe zutrifft für andere Kulturformen und selbst vielleicht für die Blutform der *lewisi*-Gruppe, obwohl hier die Verhältnisse noch nicht genügend untersucht sind.

Es wurde bereits bemerkt, daß die Kulturform von *T. cruzi*, genau wie alle anderen Trypanosomen, ein aerober Gärer ist und als solcher selbst in Anwesenheit von Sauerstoff verschiedene organische Säuren ausscheidet. Darunter findet sich Bernsteinsäure, die ein Zwischenprodukt des Krebscyclus ist. Warum *T. cruzi* diese Säure ausscheidet, wenn es doch eine aktive Bernsteinsäuredehydrase hat und außerdem nach SEAMAN (1956) noch einen zweiten Bernsteinsäure ausnützenden Mechanismus besitzt, ist ein interessantes Problem. Es zeigt dies natürlich an, daß bei intakter cellulärer Struktur die zur Bernsteinsäurebildung führenden Enzyme tätiger sind als jene, welche zu ihrer Ausnützung dienen. Warum dies so ist, ist bisher nicht bekannt.

Von hoher theoretischer sowohl wie praktischer Bedeutung ist die Frage, ob die Enzyme der Parasiten in allen Einzelheiten jenen des Wirtes gleichen oder nicht. Es muß hervorgehoben werden, daß eine Bezeichnung, wie z. B. Bernsteinsäuredehydrase, nur etwas über die katalytische Funktion aussagt, aber nichts über die chemische Identität oder Verschiedenheit des von verschiedenen Tieren isolierten Enzymes. Das theoretische Interesse dieser Fragestellung liegt auf der Hand und braucht nicht näher erörtert zu werden. Ihre praktische Bedeutung beruht auf der sich immer mehr bahnbrechenden Erkenntnis, daß eine große Zahl von Medikamenten ihre antiparasitäre Wirkung durch Hemmung intracellulärer Enzyme ausübt. Besonders für die Entwicklung eines rationellen Weges zur Chemotherapie ist die Erkenntnis wichtig, daß etwaige chemische Differenzen zwischen den Enzymen des Parasiten und des Wirtes die praktische Brauchbarkeit einer Verbindung ermöglichen können, wenn sie eine Prozeßfolge unterbricht, die sowohl für den Parasiten wie den Wirt lebenswichtig ist. Die Frage ist bisher eingehender für Helminthen als für Protozoen studiert worden, was nicht weiter verwunderlich ist, da von den ersteren ja gewöhnlich viel größere Mengen zur Isolierung der Enzyme zur Verfügung stehen als von letzteren. Immerhin aber ist es uns gelungen, im Falle von *T. cruzi* einige einschlägige Beobachtungen zu machen, die auf solche Unterschiede hinweisen. So fanden wir, daß die Bernsteinsäuredehydrase viel empfindlicher ist gegen Malonsäurehemmung als das Säugetierenzym [AGOSIN und VON BRAND (1955)]. Die Isocitronensäure zeigte weniger Affinität zum Substrat, aber größere zu Mangan als das Säugetierenzym. Gereinigte Präparate des *cruzi*-Enzymes waren völlig von der Anwesenheit von Mangan abhängig, was anscheinend beim Säugetierenzym nicht der Fall ist [AGOSIN und WEINBACH (1956)]. Diese Studien haben auf jeden Fall gezeigt, daß chemische Unterschiede vorkommen, und es darf zuversichtlich erwartet werden, daß die Zukunft solche Beobachtungen in verstärkter Zahl bringen wird.

Die Frage, ob Trypanosomen außer Kohlenhydrat andere Substanzen für Energiegewinnung ausnützen können, hat bisher relativ wenig Beachtung gefunden. Die Blutform der *brucei*-Gruppe kann dazu offenbar keine Aminosäuren benützen, wohl aber Glycerin, und zwar fast ebenso gut wie Glucose, und zu einem sehr geringen Grade Hefe-Extrakt, Difco Bacto Pepton und Casein Hydrolysat [THURSTON (1958)]. Daß Glycerin gut auswertbar ist, ist nicht sehr verwunderlich, da es ja leicht in die Reaktionen des glykolytischen Abbaues einbezogen werden kann. Die Blutform von *T. lewisi* ist entschieden vielseitiger als jene von *T. equiperdum*. Außer den für diese Art genannten Substanzen kann es auch Glucosamin, Glutaminsäure und, wenn auch nur in geringem Umfang, einige andere, aber nicht alle Aminosäuren ausnützen [MOULDER (1948); RYLEY (1951); THURSTON (1958)]. Gründlich untersucht ist der Stickstoffwechsel allerdings noch nicht; es ist aber bekannt, daß Ammoniak freigesetzt wird.

Es ist bereits erwähnt worden, daß Trypanosomen Sauerstoff verbrauchen und der respiratorische Quotient der Blutform verschiedener Gruppen verschieden ist und daß auch verschiedene Quotienten bei der Blutform und der Kulturform einer Art auftreten können. Es seien hier noch einige andere Seiten des Atmungsproblems kurz angeführt. Man kann manche Einzelheit der Atmung und der Atmungskette durch Anwendung verschiedener Hemmer klarlegen. Zahlreiche derartige Versuche sind mit Trypanosomen ausgeführt worden; hier können nur einige ihrer wichtigsten Ergebnisse erwähnt werden. Einleitend sei betont, daß in vergleichenden Untersuchungen gleichartige Bedingungen eingehalten werden müssen, da die Atmung selbst und ihre Beeinflußbarkeit durch chemische Verbindungen stark von dem Medium abhängt, in dem die Parasiten gehalten werden. So ist relativ häufig gefunden worden, daß die Atmung der Blutform am höchsten ist in Blut, während sie in Serum und besonders in Salzlösungen, wie Ringer-Phosphat, niedriger ist, selbst wenn die letzteren genügend Glucose enthalten. Die neueste einschlägige Untersuchung verdanken wir THURSTON (1958). Sie hat auch die neuesten Untersuchungen über die Beeinflußbarkeit der Atmungshemmung durchgeführt. Als Beispiel sei angeführt, daß die Atmung von *T. equiperdum* unter dem Einfluß von Jodessigsäure am wenigsten gehemmt ist, wenn das Substrat Glycerin ist, und am meisten, wenn es Glucose ist. Mit dem gleichen Hemmer war auch die Atmung von *T. lewisi* am meisten gehemmt, wenn das Substrat Glucose war, aber in diesem Falle am wenigsten, wenn die Tiere Glutamin als Substrat hatten. Es ist offensichtlich, daß eine verschiedene Empfindlichkeit eines oder mehrerer intracellulärer Enzyme, welche bei der Ausnützung verschiedener Substrate ja nicht immer die gleichen sind, diesen Erscheinungen zugrunde liegen muß.

Mit Trypanosomen sind so zahlreiche einschlägige Versuche durchgeführt worden, daß trotz der oben genannten Schwierigkeiten einige Verallgemeinerungen möglich sind. So ist einwandfrei festgestellt worden. daß die Atmung der Blutform der *brucei*-Gruppe allgemein gegen Blausäure völlig unempfindlich ist, ja durch dieses Gift sogar stimuliert werden kann (Abb. 1). Die Atmung der Blutform der *lewisi*-Gruppe dagegen ist blausäureempfindlich, was auch bei der Atmung aller bisher

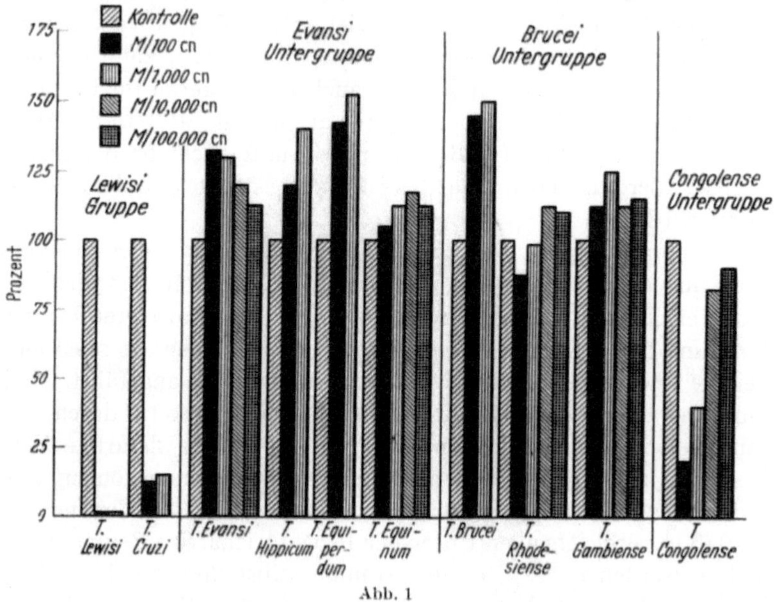

Abb. 1

untersuchter Kulturformen der Fall ist. Die Atmung der Blutform der *congolense*- und *vivax*-Gruppen ist bisher weniger oft untersucht worden; sie ist im Falle von *T. congolense* leicht empfindlich, in jenem von *T. vivax* unempfindlich [Literatur in VON BRAND (1956)].

Die Tatsache, daß die Atmung mancher Trypanosomen völlig blausäure-unempfindlich ist, zeigt an, daß sie offenbar nicht auf Schwermetallkatalyse beruhen kann. So ist es denn nicht verwunderlich, daß man in diesen Arten vergeblich nach den eisenhaltigen Cytochromen gesucht hat [z. B. HARVEY (1949); RYLEY (1956)]. Da nun, soweit untersucht, die Atmung der Kulturformen jener Arten, die blausäureunempfindliche Blutformen haben, immer blausäureempfindlich ist, kann man annehmen, daß die Blutformen diese Pigmente sekundär verloren haben. MOULDER (1950) hat mit Recht darauf hingewiesen, daß hier ein Rätsel vorliegt. Denn es muß betont werden, daß die cytochromfreien Trypanosomen einen sehr lebhaften Sauerstoffverbrauch haben, der tatsächlich höher ist als jener der blausäureempfindlichen Arten. Der Verlust des Schwermetallsystems muß also begleitet gewesen sein von dem Erscheinen eines anderen respiratorischen Enzymes, das eine höhere katalytische Wirksamkeit entfaltet. Seine Natur ist bisher unbekannt geblieben.

Ein anscheinend vollständiges Cytochromsystem ist von RYLEY (1951) in der Blutform von *T. lewisi* nachgewiesen worden. Er beobachtete die charakteristischen Absorptionsbanden der verschiedenen Cytochrome und fand, daß die Atmung durch Kohlenoxyd nennenswert gehemmt wurde, daß diese Hemmung aber durch Belichtung aufgehoben wurde. Dies ist eine wichtige Beobachtung, da andere Trypanosomen

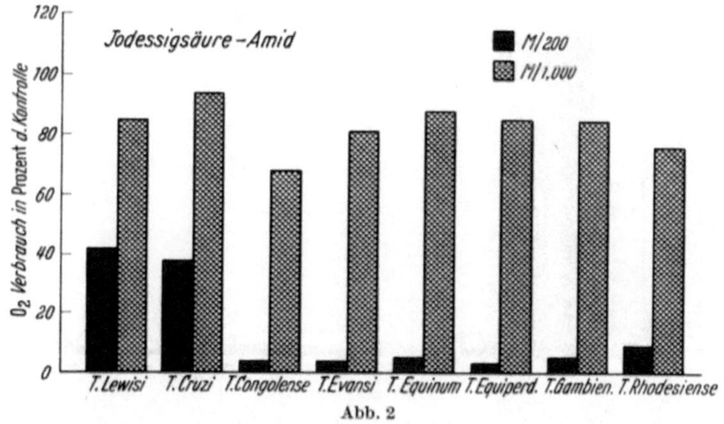

Abb. 2

sich anders verhalten. Die Kulturform von *T. cruzi* enthält kein Cytochrom c, wohl aber Cytochrom a und b. Ihre Atmung wird von Kohlenoxyd nicht gehemmt, wobei Belichtung keinen Unterschied macht [BAERNSTEIN (1953)]. Auf Grund dieser Beobachtungen muß geschlossen werden, daß *T. cruzi*, im Gegensatz zu *T. lewisi*, keine typische Cytochromoxydase besitzt. Die Aufklärung der ohne jeden Zweifel vorhandenen terminalen Oxydase bleibt ein offenes Problem.

Die Empfindlichkeit verschiedener Trypanosomen typischen Sulfhydrylhemmern (Abb. 2) gegenüber wie Jodessigsäure oder Quecksilberbenzoesäure, ist genau umgekehrt wie die eben gegenüber der Blausäure beschriebene. Dies will sagen, daß die Blutform der *lewisi*-Gruppe relativ, aber natürlich nicht vollständig, unempfindlich ist, während die Blutform der *brucei*-Gruppe ungewöhnlich empfindlich ist, und zwar kann hier eine nennenswerte Atmungshemmung selbst mit so schwachen Sulfhydrylhemmern wie Bromessigsäure (Abb. 3) und Chloressigsäure [VON BRAND, TOBIE und MEHLMAN (1950)] erzielt werden. Daß es sich in solchen Experimenten wirklich um eine Hemmung Sulfhydrylgruppen enthaltender Enzyme handelt, kann gezeigt werden durch Aufhebung der Hemmung durch Zugabe von Sulfhydrylsubstanzen, wie Dimercaptopropanol, Glutathion oder Cystein. Solche Umkehrversuche glücken ohne größere Schwierigkeit, wenn die Trypanosomen im Blut untersucht werden. FULTON und SPOONER (1956) haben neuerdings gezeigt, daß man dem Medium Katalase zusetzen muß, wenn man die Parasiten in Erythrocyten-freien Medien untersucht, da die genannten Sulfhydrylsubstanzen durch Autooxydation Peroxyd produzieren, das für die Trypanosomen sehr toxisch ist und das sie wegen fehlender Katalase nicht selbst zersetzen können.

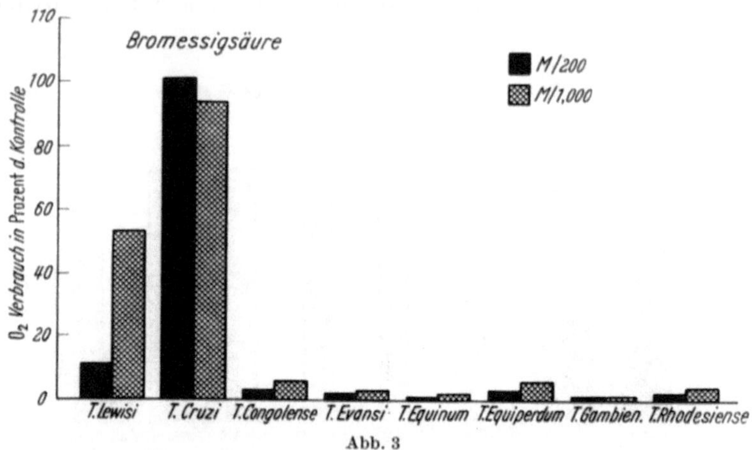

Abb. 3

Die eben erwähnte unterschiedliche Empfindlichkeit der Blutformen der *lewisi*- und *brucei*-Gruppen den Sulfhydrylhemmern gegenüber ist nicht nur von theoretischem Interesse; die Beobachtung erklärt auch, warum Arsenikalien, die ja bei der Bekämpfung der Schlafkrankheit von großer Bedeutung sind, gegen Chagas-Krankheit im wesentlichen versagen. Arsenikalien sind typische Sulfhydrylhemmer und entfalten ihre Wirkung sehr wahrscheinlich durch Unterbindung der Wirkung eines oder mehrerer intracellulärer Enzyme, deren Wirksamkeit von essentiellen Sulfhydrylgruppen abhängt. Ein typisches Beispiel ist im Glykolyseschema die Triosephosphatdehydrogenase. Es läßt sich nun sehr leicht im Experiment zeigen, daß die Blutform der *brucei*-Gruppe viel leichter durch organische Arsenikalien wie Mapharsen und selbst das anorganische Natriumarsenit gehemmt wird als die Blutform der *lewisi*-Gruppe.

Wir haben uns bisher im wesentlichen mit Abbauprozessen beschäftigt. Ein Blick in die zeitgenössische biochemische Literatur zeigt, daß mehr und mehr Aufmerksamkeit synthetischen Prozessen zugewandt wird. Mit Parasiten im allgemeinen und Trypanosomen im besonderen steht man hier noch ganz in den Anfängen. Es kann aber gesagt werden, daß mit großer Wahrscheinlichkeit hier ein ausbeutungsfähiges und interessantes Feld vorliegt. Es wäre jedoch verfrüht, auf diesem Teil-gebiet jetzt schon eine Zusammenfassung allgemeiner Art versuchen zu wollen. An deren Stelle seien von den wenigen vorliegenden Angaben nur zwei als Beispiele dieser Arbeitsrichtung angeführt.

Die Notwendigkeit, den in vitro gehaltenen Trypanosomen komplexe Nährböden zu geben, erschwert die Analyse, deutet aber auf der anderen Seite darauf hin, daß ihre synthetischen Fähigkeiten von denen der freilebenden Tiere in mancher Beziehung abweichen. Von besonderer Bedeutung für sich rasch vermehrende Zellen ist die Nucleinsäuresynthese. Zwei neuere Untersuchungen liegen hier für Trypanosomen vor, jene von BONÉ und STEINER (1956) an *T. mega* und jene von FERNANDES und CASTELLANI (1958) an *T. cruzi*. Beide stimmen darin überein, daß einfache Vorstufen wie Glycin oder Ameisensäure zwar rasch in die Proteine eingebaut werden, aber entweder nicht oder nur in geringem Grade zur Synthese von Purinbasen dienen. Adenin wurde dagegen von beiden Formen rasch in die Adeninnucleotide und die Nucleinsäurepurine eingebaut.

Literatur

Die ältere Literatur ist zusammengefaßt in: T. VON BRAND: Metabolism of Trypanosomidae and Bodonidae. A. LWOFF: Biochemistry and Physiology of Protozoa. Vol. **2**, p. 177. New York: Academic Press 1951.

AGOSIN, M., and T. VON BRAND: Studies on the carbohydrate metabolism of *Trypanosoma congolense*. Exp. Parasitol. **3**, 517 (1954).

— — The influence of Puromycin on the carbohydrate metabolism of *Trypanosoma equiperdum*. Antibiotics and Chemotherapy **4**, 624 (1954).

— — Characterization and intracellular distribution of the succinic dehydrogenase of *Trypanosoma curzi*. Exp. Parasitol. **4**, 548 (1955).

— and E. C. WEINBACH: Partial purification and characterization of the isocitric dehydrogenase from *Trypanosoma cruzi*. Biochim. biophys. Acta **21**, 117 (1956).

BAERNSTEIN, H. D.: The enzyme systems of the culture form of *Trypanosoma cruzi*. Ann. N. Y. Acad. Sci. **56**, 982 (1953).

BASH-LEWINSON, D., and N. GROSSOWICZ: Transaminases of *Trypanosoma cruzi*. Bull. Res. Counc. Israel **6** E, 91 (1957).

BONÊ, G. J., and M. STEINERT: Isotopes incorporated in the nucleic acids of *Trypanosoma mega*. Nature (Lond.) **178**, 308 (1956).

BRAND, T. VON: Beziehungen zwischen Stoffwechsel und taxonomischer Einteilung der Säugetiertrypanosomen. Zool. Anz. **157**, 119 (1956).

— and M. AGOSIN: The utilization of Krebs cycle intermediates by the culture forms of *Trypanosoma cruzi* and *Leishmania tropica*. J. inf. Dis. **97**, 274 (1956).

— and E. J. TOBIE: Further observations on the influence of cyanide on some trypanosomes. J. cell. comp. Physiol. **31**, 49 (1948).

— — and B. MEHLMAN: The influence of some sulfhydryl inhibitors and of fluoroacetate on the oxygen consumption of some trypanosomes. J. cell. comp. Physiol. **35**, 273 (1950).

— — — and E. C. WEINBACH: Observations on the metabolism of normal and arsenic-resistant *Trypanosoma gambiense*. J. cell. comp. Physiol. **41**, 1 (1953).

— E. C. WEINBACH, and E. J. TOBIE: Comparative studies on the metabolism of the culture form and the bloodstream form of *Trypanosoma gambiense*. J. cell. comp. Physiol. **45**, 421 (1955).

CHEN, G., and E. M. K. GEILING: Glycolysis in *Trypanosoma equiperdum*. Proc. Soc. exp. Biol. (N. Y.) **63**, 486 (1946).

FERNANDES, J. F., and O. CASTELLANI: Nucleotide and polynucleotide synthesis in *Trypanosoma cruzi*. I. Precursors of purine compounds. Exp. Parasitol. **7**, 224 (1958).

FULTON, J. D., and D. F. SPOONER: Inhibition of the respiration of *Trypanosoma rhodesiense* by thiols. Biochem. J. **63**, 475 (1956).

— — Comparison of the respiratory activity of an old and of a freshly isolated strain of *Trypanosoma rhodesiense*. Ann. Trop. Med. Parasitol. **51**, 417 (1957).

GRANT, P. T., and J. D. FULTON: The catabolism of glucose by strains of *Trypanosoma rhodesiense*. Biochem. J. **66**, 242 (1957).

HARVEY, S. C.: The carbohydrate metabolism of *Trypanosoma hippicum*. J. biol. Chem. **179**, 435 (1949).

HOARE, C. A.: The relationship of the haemoflagellates. Proc. Fourth Internat. Congr. Trop. Med. a. Malaria, Washington, D. C. **2**, 1110 (1948).

JENKINS, A. R., and E. B. GRAINGE: The oxidative metabolism of African pathogenic trypanosomes. I. Observations on *Trypanosoma rhodesiense* maintained by sub-inoculation and cyclical tsetse-fly transmission. Trans. roy. Soc. trop. Med. Hyg. **50**, 481 (1956).

MARSHALL, P. B.: The glucose metabolism of *Trypanosoma evansi* and the action of trypanocides. Brit. J. Pharmacol. Chemother. **3**, 8 (1948).

MOULDER, J. W.: The oxidative metabolism of *Trypanosoma lewisi* in a phosphate-saline medium. J. inf. Dis. **83**, 33 (1948).

— The oxygen requirements of parasites. J. Parasitol. **36**, 193 (1950).

REINER, L., C. V. SMYTHE, and J. T. PEDLOW: On the glucose metabolism of trypanosomes (*Trypanosoma equiperdum* and *Trypanosoma lewisi*). J. Biol. Chem. **113**, 75 (1936).

RYLEY, J. F.: Studies on the metabolism of the protozoa. 1. Metabolism of the parasitic flagellate, *Trypanosoma lewisi*. Biochem. J. **49**, 577 (1951).

— Studies on the metabolism of the protozoa. 4. Metabolism of the parasitic flagellate, *Strigomonas oncopelti*. Biochem. J. **59**, 353 (1955).

— Studies on the metabolism of the protozoa. 7. Comparative carbohydrate metabolism of eleven species of trypanosome. Biochem. J. **62**, 215 (1956).

SEAMAN, R. C.: The succinic dehydrogenase of *Trypanosoma cruzi*. Exp. Parasitol. **2**, 236 (1953).

— Succinate metabolism of haemoflagellates. Exp. Parasitol. **5**, 138 (1956).

THURSTON, J. P.: The oxygen uptake of *Trypanosoma lewisi* and *Trypanosoma equiperdum*, with special reference to oxygen consumption in the presence of amino-acids. Parasitology **48**, 149 (1958).

— The effect of some metabolic inhibitors on the oxygen uptake of *Trypanosoma lewisi* and *Trypanosoma equiperdum*. Parasitology **48**, 165 (1958).

WEINMAN, D.: Cultivation of trypanosomes. Trans. roy. Soc. trop. Med. Hyg. **51**, 560 (1957).

Centrifugal mechanisms of sensory control

By K. E. Hagbarth

Department of Clinical Neurophysiology, Akademiska Sjukhuset, Uppsala, Sweden

Contents

It is often assumed that the central nervous system (C.N.S.) is organized on the basis of more or less complex reflex arcs by which the receptors control the activity of the effector organs. The fundamental notion that sensory input controls motor output must not, however, obscure the fact that there also exist output-input relations. Recent studies have shown that sensory structures at various levels of the nervous system are exposed to centrifugal control and the general thesis that the C.N.S. can control its own input has since developed into a promising field of research. Output-input relations are evidently established in various ways, in the periphery as well as within the C.N.S., and there is a growing conviction that such relations are indispensable for the operation of a great number of neural mechanisms at various levels of the neuraxis. The present review summarizes some of the data which have led up to these concepts and, even though it might seem premature, attempts are made to emphasize some functional similarities in the various neural mechanisms concerned.

Ordinary motor action controlling sensory input

Intero- or proprioceptive input

It has been known for a century that the body contains a considerable number of intero-or proprioceptors, sense organs which are not primarily directed towards the environment but which give information concerning events within the body itself. Secretion of a gland or contraction of a muscle affects, in a mediate or immediate way, internal receptors which in turn affect the activity of the effector organs by various reflex arcs. It is a common notion that the C.N.S. in this way receives information about actively induced peripheral events, so that it can send appropriate messages back as motor impulses to the effector organs. Such feedback control of motor output was wellknown to the physiologists long before the modern concepts of cybernetics were introduced (cf. Wiener 1948), but since then biological servo-mechanisms have attracted a greater interest. Most workers, however, still attend to the problem of what such servo-mechanisms mean in terms of motor control. In the following, special attention is paid to the sensory aspects of these functional circuits.

A great deal of experimental work has been devoted to the problem of how stretch and contraction of a muscle affect the various muscle receptors, and how these receptors in turn (via the spinal cord) affect those motoneurones which govern the state of contraction in the muscle. The reader is referred to Granit's Silliman lectures (1955, a) for pertinent literature and an extensive review of this field [cf. also Burkhardt(1958)]. For the present purpose it will be sufficient to emphasize a few functional aspects of the autogenetic reflexes induced by the tension-recording Golgi organs and the length-recording annulo-spiral endings.

The Golgi organs are easily stimulated by contraction of the muscle (especially if the initial muscle tension is high) and their sensory discharges exert via spinal interneurones an inhibitory action on the motoneurones supplying the same muscle (autogenetic inhibition as demonstrated in extensor muscles of spinal and decerebrate cats). As suggested by Granit, this functional loop through the Golgi organs may serve to make motor outflow self-inhibitory. The Golgi organs, however, can also be stimulated by stretching the muscle and, providing the autogenetic inhibition in this case has a pre-existing motor outflow to operate upon, a relaxation of the muscle may occur and this in turn tends to unload the Golgi organs (the lengthening reaction described by Sherrington 1913). It seems, in other words, as if an alteration induced anywhere in this circuit, either on the motor or the sensory side, will induce a reaction which strives to counteract the alteration.

The same principle holds true for the circuit involving the annulo-spiral endings. These receptors tend to become unloaded when the muscle shortens during contraction but they are stimulated when the muscle is elongated by passive stretch; and their sensory discharges exert a monosynaptic excitatory action on the motoneurones supplying the muscle (the myotatic reflex). Providing there is a pre-existing sensory discharge from these receptors, their unloading during muscle contraction tends to deprive the motoneurones of an excitatory inflow; in such a way this loop may serve to make motor outflow self-inhibitory (cf. Merton 1951). When, on the other hand the annulo-spirals are stimulated, the resulting stretch reflex tends to unload the receptors and this means automatic suppression of the sensory input.

It is hazardeous to speculate about the "biological meaning" of these servo-mechanisms until it is better known to what extent and under which conditions they are free to operate in the intact organism. We will later attend to the efferent gamma control of the annulo-spiral endings (p. 50) and the supra-spinal control of the interneurones mediating the effect from the Golgi organs (p. 9). The relevant point is, however, that the circuits described may serve to check not only motor output but also sensory input. All kinds of elaborate mechanisms require for their operation that certain variables are kept within appropriate limits, and within the organism such variables are to be found both within the motor and the sensory sphere (cf. ASHBY 1952).

It is well known that the autonomic nervous system to a large extent is organized according to the principle of feedback control. A peripheral event, such as a change in pulse rate or blood-pressure, induces reflexes which change the activity of the effector organs in such a way that the peripheral event is counteracted. In this way the servo-mechanisms provide a constant internal environment, and in terms of output-input relations this means that the neural output varies in such a way that the input can be kept more or less constant.

The proprio- or interoceptive loops may also serve other purposes, however, than to induce self-regulatory output and/or input. They may, for instance, enable chain-reactions to occur in which one reflex automatically induces another until a complex behaviour pattern is established. It may be enough to refer to the act of swallowing, or to the act of locomotion, or the following elucidative example [cf. FULTON (1947)]: "If a cat walking in a straight line hears a mouse to the right of it, mere turning of the head to the right causes the extremities on that side to become extended and the animal is automatically prepared for a quick takeoff with its left foot — it has only to decide whether to go for the mouse. The neck reflexes arrange the rest." In this fashion complex motor coordination may become automatic by means of proprioceptive loops.

Exteroceptive input

Exteroceptors are also involved in functional motor-sensory loops. Movements change the relation of the organism to the environment and to such changes the exteroceptors react. ASHBY (1952) emphasizes that the organism and its environment should be regarded as a functional unit in which "the receptors . . . affect the muscles (by effects transmitted through the nervous system) and the muscles affect the receptors (by effects transmitted through the environment") and he makes the pertinent remark: . . . "most physiological experiments are deliberately arranged to avoid this feedback. Thus in experiments with spinal reflexes, a stimulus is applied and the resulting movement recorded; but the movement is not allowed to influence the nature or duration of the stimulus". Also when these exteroceptive functional loops are concerned, it may be misleading to assume that the movements in themselves represent the biological purpose of the motor output and that the altered input merely serves to inform the centers about the peripheral event. Many motor acts are undoubtedly initiated for the purpose of modifying sensory inflow: we turn our eyes to change (or retain) the visual image on the retina, by withdrawal movements we put an end to painful stimuli and when going to sleep we make motor efforts to avoid peripheral inflow [cf. ADRIAN (1954)].

In this connexion it is of interest to consider the organization of the spinal skin reflexes. The motor patterns involved in the scratch reflex vary according to the localization of the skin stimulus in such a way that the ensuing movements always tend to remove the offending object ("local sign" described by SHERRINGTON, see CREED, DENNY-BROWN et al. 1932); and spinal withdrawal movements are in a similar striking manner specifically orientated in relation to the position of the noxious stimulus [HAGBARTH (1952), KUGELBERG and HAGBARTH (1958)]. The reflex centers concerned are evidently organized to initiate an outflow which effectively interrupts noxious or annoying skin stimuli wherever they strike the body. In other words, this kind of inflow is "self-inhibitory".

Centrifugal paths specifically engaged in sense organ control

Many sense organs are controlled by specific neural mechanisms of their own. We may consider how illumination on the retina becomes self-regulatory by means of the pupillary reflexes and how auditory reflexes govern the state of contraction of the muscles in the middle ear. In recent years a special interest has been devoted to the centrifugal control of the mammalian muscle spindles and a short description of the mechanisms involved in this regulation may serve to elucidate some general notions about output-input relations [for literature see GRANIT (1955, a)].

The annulo-spiral receptors within the muscle spindles lie in the middle, non-contractile portion of the small intrafusal muscle fibres, and these intrafusal structures, in turn, lie in parallel with the ordinary extrafusal muscle fibres which are responsible for the actual muscle contraction and determine the total length of the muscle [for further data see MATTHEWS (1933) and BARKER (1948)]. The mechanical arrangements are such that the load on the receptors tends to decrease when the muscle shortens during contraction and it tends to increase when the muscle is passively extended or when the intrafusal fibres contract. The ability of the receptors to respond to variations in muscle length (due to contraction or passive extension) depends upon the initial length-relations between the intrafusal and extrafusal fibres, and these relations can be varied by means of separate motor supplies: the extrafusal fibres are innervated by the α motoneurones, the intrafusal fibres by the γ motoneurones [LEKSELL (1945); KUFFLER, HUNT and QUILLIAM (1951)].

The spinal and supraspinal control of the γ outflow has been intensively studied by HUNT (1951, 1952), by KOBAYASHI et al. (1952) and by GRANIT and co-workers [see GRANIT (1955a)]. All agree that a typical feature of this outflow is that tonic activity goes on all the time in a good preparation; in this respect it differs from the α outflow. Furthermore, the α and γ motoneurones are functionally linked in such a way that (even though the thresholds may be different) they are co-excited or co-inhibited by most peripheral and central stimuli. There is, however, one exception to this rule: muscle stretch excites the α motoneurones (the myotatic stretch reflex) while it inhibits the γ motoneurones [HUNT (1951)].

KUFFLER and HUNT (1952) conclude that the γ outflow serves to correlate intrafusal fibre length to the length of the extrafusal fibres. According to this notion, the functional linkage between the α and the γ neurones is such that, during a muscle contraction, the intrafusal fibres contract more or less in parallel with the extrafusal ones and this would prevent mechanical unloading of the sensory structures. The γ-inhibition due to passive extension of the muscle would in principle serve a similar purpose, namely to prevent excess stimulation of the spindle end-organs.

According to GRANIT and co-workers, however, the essential task of the γ system is to control movement or muscular tone [cf. ELDRED, GRANIT and MERTON (1953)]. In summary, their experiments have led to the following conclusions. Since the γ motoneurones often respond more promptly than the α (to various central and peripheral stimuli), an increase of the γ outflow may occur without concomitant α activation and, providing the muscle is sufficiently stretched, "this leads to pull on the nuclear bag containing the spindle organ which in turn discharge impulses to the spinal cord for the α motoneurones controlling the length of the extrafusal fibres. The muscles contract until the extrafusal fibres have assumed the length of the intrafusal ones and so the self-regulated system is balanced". These notions imply that the length of the intrafusal fibres determine the length of the muscle; essentially, the γ system is regarded as "a new motor system, indirect because it traverses the spindle loop but nevertheless ... singularly potent" [GRANIT (1959)]. There is a great amount of evidence that this is how the system works in decerebrate preparations, but there is still no conclusive evidence, that the system works in a similar way also in the intact organism, and that e.g. volontary movements are normally initiated by the indirect route via the spindles [as suggested by MERTON (1951, 1953)]. In any case, it may be asked whether it is appropriate to define the whole γ loop as a motor system or whether it is more revealing to say that the γ outflow sensitizes the muscle spindles so that the actual contraction can be brought about by the stretch reflex. In general, centrifugal control of sensory structures implies control also of those reflexes which are elicitable from (or conveyed by) these sensory elements, but for this reason there is probably no need to change our definitions as to what constitutes the sensory and the motor limb of the reflex arcs.

Another relevant problem is how the tonic activity in the γ system is maintained. The experiments by GRANIT and KAADA (1952) indicate that the reticular system in the brain stem serves an important role in the maintainance of the sustained spindle discharge. They stimulated various structures in the brain and recorded both excitatory and inhibitory effects on the γ system. Powerful excitatory effects were regularly obtained from that portion of the brain stem reticular formation which has been designated the reticular activating system and which in a non-specific fashion receives connexions from practically all sensory paths [STARTZL, TAYLOR and MAGOUN (1951), FRENCH, VERZEANO and MAGOUN (1953)]. It had earlier been demonstrated that electrical stimulation within these areas brings about cortical arousal as defined by the electroencephalogram and also generalized excitation of subcortical motor centers [NIEMER and MAGOUN (1947), MORUZZI and MAGOUN (1949), MAGOUN (1952)]. Now it was shown that such stimuli also bring about "arousal" of the muscle spindles [cf. also VON EULER and SÖDERBERG (1957), ELDRED and FUJIMORI (1958)]. The general notion is that the reticular arousal system provides the spindles with a certain tonus which strives to maintain a sustained sensory inflow. And this inflow in turn may serve not only to "energize" neural centers; it also supplies a necessary background for sensory inhibitory processes [cf. GRANIT (1955a)].

So far, we have only considered motor-sensory loops of various kinds, but there is evidence that output-input relations can be established also by direct neural effects upon sensory elements.

CAJAL (1911) described centrifugal fibres projecting to the retina and to the olfactory bulb [cf. ALLISON (1953)], and other histological studies show that efferent fibres project also to the hair cells in the cochlea [HELD (1893), RASMUSSEN (1946, 1953, 1955)] and to the vestibular sensory epithelia [ENGSTRÖM and WERSÄLL (1958)].

Although the olfactory bulb and the neural layers of the retina properly belong to the C.N.S., they may for didactic reasons be included under this heading. GRANIT (1955b) recorded with micro-electrodes from the cat's retina and stimulated through needle electrodes inserted into the mesencephalic reticular substance. Reticular stimulation frequently increased the firing of the retinal units, but also inhibitory effects were seen; and he concluded that the effects probably were induced by the centrifugal fibre system. Centrifugal impulses to the rabbit's retina have recently been described by DODT (1956).

KERR and HAGBARTH (1955) noted that stimulation of the basal rhinencephalic area or the anterior commisure caused a depression of both intrinsic and induced waves in the olfactory bulb; and these effects were thought to be mediated by the centrifugal fibres of the anterior commisure which are known to terminate upon the granule cells in the bulb [ALLISON (1953)].

GALAMBOS (1956) recorded the synchronized auditory nerve discharge to a click stimulus and found that this response was reduced or abolished when electrical stimulation was applied to the medulla at the site of the decussation of the olivo-cochlear pathway. The phenomenon persisted after removal of the bones and muscles in the middle ear and he concluded that the neural inflow from the cochlea can be suppressed by centrifugal impulses in the olivo-cochlear tract which terminates near or on the hair cells.

Neural effects upon receptor elements have also been demonstrated in invertebrates and in the frog. KUFFLER and EYZAGUIRRE (1955) showed that stimulation of the afferent fibres to a stretch receptor in the crayfish prevents the appearance of afferent impulses, and they concluded that the effect was due to inhibition occurring in the dendritic processes of the receptor nerve cell. LOEWENSTEIN's experiments (1956) on an isolated skin preparation (from a frog) indicate that sympathetic outflow liberates an adrenaline-like substance in the skin which has a facilitatory effect upon cutaneous mechano-receptors. He supports the view that this receptor facilitation is part of a more generalized facilitatory action of the sympathetic system upon somatic neural elements.

Sensory relays exposed to centrifugal control

Recent experimental work has led to the conclusion that the C.N.S. may regulate not only the initiation of afferent impulses but also their transmission at the sensory synapses. Earlier anatomical studies had shown that centrifugal fibres or recurrent collaterals may project to sensory nuclei [cf. CAJAL (1909, 1911), LORENTO DE NÒ (1933)] and clinical observers had concluded that such paths may exert an inhibitory [HEAD and HOLMES (1911)] or excitatory [WALLENBERG (1928), BROUWER (1933)] effect upon afferent transmission.

As will be shown, recent experiments indicate that dorsal root inflow, as soon as it reaches the initial synapses can be modified by centrifugal influences from the brain. It may be asked what this means in terms of spinal inflow-outflow relations and what it means in terms of sensory inflow to the brain. For the sake of clarity, these two problems will be treated separately.

Centrifugal effects on spinal reflex patterns

It is an old concept that spinal interneurones, conveying polysynaptic reflexes, are exposed to excitatory or inhibitory influences which strive to affect transmission along the reflex arcs. When discussing inhibitory phenomena, SHERRINGTON concluded: "The motor-neurone itself seems not to be the actual seat of the inhibition, for if so, it would be inhibited for all reflexes; unless the motor-neurone is functionally divisable, and one part of it e.g. one set of dendrites, can be inhibited at a time when another is not. The seat of inhibition appears, therefore, with some likehood, to lie neither in the afferent neurone proper nor in the efferent neurone proper, but in an internuncial mechanism — synapse or neurone — between them" [cf. SKOGLUND (1955)].

It is a well known clinical fact that intracranial lesions may profoundly alter the thresholds and the patterns of the human skin reflexes. DUENSING (1940, 1952) concluded, based on the conviction that most of these reflexes are spinal that the effects must be due to alterations of a cerebral outflow which normally checks the excitability of the spinal interneurones. Physiologists are equally aware of the fact that spinal reflexes in many respects differ in decerebrate and in spinal preparations. Some of these discrepancies can easily be explained in terms of altered motoneurone excitability but this explanation is not sufficient to account for threshold changes of individual reflexes which converge upon the "final common pathway". As early as 1926 FULTON suggested that internuncial neurones participating in spinal polysynaptic reflexes may be under an inhibitory influence of descending pathways, and he proposed that these influences might serve to regulate the distribution of incoming afferent impulses at the spinal level, suppressing some and deflecting others rostrally for higher integration.

During the last decades, many workers have approached these problems, and various techniques have been used to study cerebral governing of internuncial relays in the cord. LLOYD showed in 1941, by recording from intraspinal structures, that the pyramidal system in cats impinges upon spinal interneurones rather than directly on the motoneurones themselves. Other workers have applied electrical stimuli to various stations in the brain (including the reticular formation) and studied the effects upon various spinal reflex responses; and the results have led them to conclude that the effects are exerted not only on the "final common pathway" but also on the interneurones mediating these reflexes [BACH (1950), AUSTIN (1952), KLEYNTJENS, KOISUMI and BROOKS (1955)]. LINDBLOM and OTTOSON (1953, 1956, 1957) arrived at a similar conclusion during their experiments on slow cord potentials. In this connexion reference should also be made to the experiments of HUGELIN

(1955) on the nociceptive linguo-maxillar reflex, in which he demonstrated that reticular activation inhibits the reflex even though the excitability of the motoneurones involved is not lowered.

A different approach was taken by JOB (1952) and ECCLES and LUNDBERG (1958) who studied the spinal reflexes from the muscle end organs in decerebrate cats before and after a high transection of the spinal cord. In the spinal preparation they readily observed the individual effects of three kinds of muscle afferents (from the annulo-spiral, flower spray and GOLGI endings) upon motoneurones, but in the decerebrate animal the effects from the GOLGI and the flower spray afferents were difficult to demonstrate. They concluded that in the decerebrate preparation descending impulses from suprasegmental structures inhibit the spinal interneurones involved in these reflex arcs.

Microelectrode studies of spinal interneurone activity have shown that quite often a single unit may be activated or inhibited from a great variety of peripheral sources [KOLMODIN and SKOGLUND (1954), KOLMODIN (1957), cf. also OSCARSSON (1957)]. When such results are obtained in spinal, decerebrate or anaesthetized preparations, however, they merely signify that a great number of converging functional connexions do exist. It must not be presupposed that all these potential functional connexions are simultaneously open to transmission in the intact organism. HAGBARTH and FEX (1959) recently demonstrated (in decerebrate cats) that spinal interneurones in the dorsal horn react not only to various kinds of peripheral stimuli but also to electrical stimulation in the cerebellum or in the brain stem; and additional results indicated that the "spontaneous" activity of the interneurones was sometimes maintained by tonic excitatory outflow from suprasegmental structures.

It should be added that, in recent years, histologists have been able to demonstrate corticospinal fibres projecting to sensory structures in the spinal cord [CHAMBERS and LIU (1957), KUYPERS (1958), cf. also SCHÄFER (1899) and HOFF (1932)].

In summary then, animal experiments yield evidence that supraspinal centers can control the excitability of at least some of those spinal interneurones which serve to establish the reflex patterns of the spinal cord. This raises the question: how are these mechanisms put into operation under physiological circumstances? We may suggest that by such interference from higher centers the motoneurones can be made available for those particular spinal patterns which "fit" the total sensory situation at any given time. We recognize in the spinal preparation a number of more or less well defined motor patterns which are specifically correlated to certain types of receptors (cf. the reflexes from the various muscle receptors) or at least to certain kinds of peripheral stimuli (cf. the scratch reflex) but, again, it must not be presupposed that all these converging functional connexions to the motoneurones are simultaneously open to transmission in the intact organism; and neither can it be excluded that in the intact organism temporary spinal reflex patterns may arise, which are not to be found in a spinal or decerebrate preparation.

These notions led HAGBARTH and KUGELBERG (1958) to study the skin reflexes of the trunk in unrestrained human beings. As judged by latency measurements, (on electromyographic records) these reflexes proved to be spinal; and it was found that the motor patterns induced varied according to the localisation of the skin stimulus in such a way that the ensuing movements always tended to cause a true withdrawal from the offending object wherever it struck the trunk [KUGELBERG and HAGBARTH (1958)]. The strength of the motor response, however, and the amount of motor radiation (to distant muscle groups) depended to a large extent upon the subject's "state of expectancy" in regard to the intensity of the skin stimulus. Reflexes to skin stimuli of low intensity could be exaggerated into forceful jumps (involving also the muscles of the extremities) simply by a preceding false statement that the stimulus intensity should be raised; and the reflexes usually diminished when the subject was convinced that no painful stimuli would occur. Repetition of a non-painful stimulus regularily caused a decline and often an enduring total extinction of the reflex response (habituation) even though the motor units involved in this response reacted as promptly as before to other kinds of peripheral stimuli. Furthermore, a single intervening pain-stimulus was often sufficient to cause a lasting "sensitization" to succeeding weak stimuli within the same skin area. The effects of habituation and sensitization often persisted for several hours.

Information derived from other senses (auditory in case of verbal suggestion) or past experiences (habituation and sensitization procedures) evidently help to determine whether a light skin stimulus on the trunk will not affect the spinal motoneurones at all, or whether it will give rise to a forceful spinal withdrawal reaction which, like a startle reflex, involves a great number of muscles in the body. There is a striking discrepancy between the relative "stability" of the spinal withdrawal reactions in spinal and decerebrate cats and the adaptive behaviour of these spinal protective reflexes in the intact human organism. The results are hard to explain unless we regard the network of spinal interneurones as a substrate for an immense number of potential sensory-motor connections which selectively can be closed or opened by effects emerging from those levels of the nervous system where the total sensory situation can be properly analysed.

It may be revealing to compare the centrifugal control of the muscle spindles with the descending effects upon the spinal interneurones. In both cases the centrifugal paths evidently serve to control the sensory signals impinging upon the spinal motoneurones. It is not wholly unlikely that the fundamental principles involved are similar either the centrifugal effect is exerted at the receptor level (in the case of the monosynaptic myotatic reflex) or at the level of the internuncial relays in the cord (in the case of polysynaptic reflexes). SOMMER (1940) and HOFFMANN (1951) suggested that enhancement of the tendon jerk in man by JENDRASSIK's well known method is due to spindle activation, and MATTHEWS (1956) made the interesting observation that the tendon reflexes are suppressed during free fall, but he draws no conclusions as to whether this effect is due to the γ system.

Centrifugal effects on ascending sensory inflow

HAGBARTH and KERR (1954) demonstrated, in experiments on curarized cats, how electrical stimulation of various stations in the brain affected the size of post-synaptic sensory volleys, which appeared in ascending spinal tracts in response to single shock stimulation of a lumbar dorsal root. Whether the central test stimuli were applied in the midbrain reticular formation, in cerebellum or in sensory-motor cortex the result was always a depression of the synchronized sensory response; no facilitatory effects were demonstrated. The effects were totally abolished by moderate barbiturate anaesthesia or by high cord transection which also caused a marked increase in the control size of the sensory responses. These effects of anaesthesia or cord transection were interpreted as release phenomena; release from tonic descending influences which normally, during wakefulness, check the afferent conduction in the spinal cord. LINDBLOM and OTTOSON (1953) had earlier demonstrated similar release phenomena on spinal dorsum potentials.

It remained to find out if similar results could be obtained from other afferent pathways subserving other sensory modalities. Postsynaptic evoked potentials were recorded from the spinal sensory trigeminal nucleus [HERNÁNDEZ-PEÓN and HAGBARTH (1955)], the gracilis nucleus [HERNÁNDEZ-PEÓN, SCHERRER and VELASCO (1956)] and the cochlear nucleus [JOUVET and DESMEDT (1956)] and a common finding was that the responses were depressed during electrical stimulation in the midbrain reticular formation. Conversely, barbiturate anaesthesia or lesions within the reticular formation caused an enhancement of the evoked sensory potentials [HERNÁNDEZ-PEÓN and SCHERRER (1955), HERNÁNDEZ-PEÓN, JOUVET and SCHERRER (1957)]. With regard to the visual pathway HERNÁNDEZ-PEÓN, SCHERRER and VELASCO (1956) reported that stimulation of the reticular formation usually decreased the size of photically evoked potentials in the lateral geniculate body and in the visual cortex, while the responses recorded from the optic tract were often enhanced.

Moreover, KING, NAQUET and MAGOUN (1957) demonstrated that reticular stimulation also affects sensory transmission at thalamic levels. When reviewing these results, HERNÁNDEZ-PEÓN (1955) concludes that the brain stem reticular formation, either as a source or as a passage, plays an essential role in checking transmission along the sensory paths. The experiments by JOUVET and DESMEDT (1956) indicate, however, that a distinction should be made between those reticular regions which yield electrographic arousal and those which yield depression of sensory responses [cf. also DESMEDT and MECHELSE (1958)]. DAWSON recently contributed to this subject by showing that, in rats under trichlorethylene anaesthesia, a single electrical stimulus to the primary receiving cortical area markedly reduces the size of the post-synaptic potential in the cuneate nucleus (1958, a and b).

In all these experiments macroelectrode recordings of synchronized sensory volleys were used as an index of the afferent inflow. This technique however, is subject to limitations which may be relevant for the interpretation of the results. Microelectrode recordings of single unit activity represent an alternative method of investigation which may allow more definite conclusions about excitatory and inhibitory processes and may help to evaluate the results obtained with the technique of synchronized evoked potentials.

HAGBARTH and FEX (1959) recently studied centrifugal effects on single unit activity in spinal sensory paths. Most experiments were performed on decerebrate, curarized cats; the sensory records were usually obtained from the distal stump of the dorsal spinocerebellar tract while the central test stimuli were applied to the exposed cerebellar cortex or to the cut surface of the brain stem (at the intracollicular level). The curarization served a two-fold purpose: to miminize movement artefacts during the recording and to paralyze the muscle spindles so that the peripheral γ loop was blocked. Units responding to natural peripheral stimuli (touch, pressure, pinch) were identified and attempts were then made to modify the activity of these units by electrical stimulation in the supra-spinal structures.

It was found that most of the units were "spontaneously" active and they responded to adequate peripheral stimuli either with an increase or with a decrease of frequency [cf. HOLMQVIST, LUNDBERG and OSCARSSON (1956) and OSCARSSON (1957)]. Similar effects, however, could also be induced by electrical stimulation applied to the anterior lobe of vermis, the brain stem or (in a chloralose cat) the sensory-motor cortex. In this way, some sensory units were inhibited while others were excited; but the effect upon a certain unit could often be reversed by changing the parameters of the central stimuli.

The next step was to analyze how central and peripheral stimuli interacted upon a single sensory unit. It was frequently observed that e.g. a cerebellar stimulus inhibited a sensory unit both when it was "spontaneously" active and when its activity was maintained by peripheral stimulation. Occasionally, however, more complex results were obtained which indicated that sometimes a distinction must be made between "spontaneous" activity and activity induced by peripheral stimuli: the background activity could by inhibited at a time when the responses to peripheral stimuli were unaffected or even exaggerated.

The results indicate that these post-synaptic sensory elements in the cord may be exposed to excitatory as well as inhibitory influences from the brain, and furthermore it seems as if their "spontaneous" activity is not always maintained from the periphery; it may probably be maintained by descending impulses acting on the initial sensory relays.

Recent observations by SEGUNDO and LARRANAGA (1959) are of interest in this connexion. They studied striatal influence upon unit activity in gracilis and cuneatus nuclei, and they found that striatal stimulation fires units in these posterior column nuclei (very rarely it inhibits them) and interacts with regular sensory influences.

Reference should also be made to recent experiments by ARDEN and SÖDERBERG (1959). They recorded single unit activity in the lateral geniculate body (in rabbits, encéphale isolé) and found that the resting discharge often was increased when the animal showed EEG arousal in response to e.g. whistling or other noises. On the other hand, the average firing rate of the geniculate resting discharge was but little affected by retinal stimulation (flickering light) or by complete blocking of the optic nerve; but such a blocking often caused a reversal of the geniculate response to arousal, so that auditory stimuli now caused a decrease in firing rate.

The simple conception that the sensory pathways deliver receptor messages relatively unchanged to the cortex is certainly no longer tenable. WALSH (1957) expresses the notion that "probably there is available at the periphery much more sensory data than can be transmitted centrally. It is almost to be expected that according to the conditions at any given time the sensory systems can be readjusted by means of activity in descending pathways to listen in to specific aspects of the information that is available". This suggestion is now supported by the following experimental data.

HERNÁNDEZ-PEÓN and co-workers studied afferent transmission in unanaesthetized unrestrained cats with implanted electrodes in various sensory relays [see HERNÁNDEZ-PEÓN (1955)] and they developed their experiments around the ideas of "habituation" and "attention".

The neural mechanisms underlying attention were studied by HERNÁNDEZ-PEÓN, SCHERRER and JOUVET (1956) who recorded evoked synchronized responses to sound clicks of a certain intensity from the dorsal cochlear nucleus in unrestrained cats. They noted that these responses were greatly reduced when certain visual (showing a mouse to the cat), olfactory (fish odour) or nociceptive (shock to the forepaw) stimuli were introduced and the sensory blockade lasted as long as there was behavioural evidence that the diverting stimuli attracted the cat's attention. In a similar manner, photically evoked responses in the visual pathway were reduced during attentive behaviour elicited by acoustic or olfactory stimuli [Hernández-Peón, Guzmán-Flores et al. (1957)].

Habituation, as defined by Dodge (1923) "is a process whereby certain sensory stimuli by repeated application lose significance to be individual, and result in a decline and the final disappearance of the corresponding response". Hernández-Peôn and co-workers showed that repeated application of auditory [Hernández-Peón, Jouvet and Scherrer (1957)], visual [Hernández-Peôn, Guzmán-Flores et al. (1957)] or olfactory stimuli [Hernández-Peón, Alcocer-Cuarón et al. (1957)] resulted in a clear-cut decrease of the sensory synchronized responses as they appeared in, respectively, the cochlear nucleus, the lateral geniculate body and the olfactory bulb. Furthermore, they showed that the process of habituation specifically affected the response to that particular stimulus which had been repeated, e.g. auditory responses to tones of a high frequency were unaffected by habituation to low frequency tones. Sharpless and Jasper (1956) demonstrated that cortical synchronized responses to auditory clicks were not reduced in size, at a time when the arousal effect (on EEG) of the stimuli had diminished as a consequence of habituation.

An increase of auditory cochlear responses has been reported to occur after repeated presentations of the acoustic stimulus together with a nociceptive stimulus [Galambos, Sheatz and Vernier (1956)]. It is doubtful whether this effect should be defined as conditioning, sensitization or dishabituation, and it remains to find out whether the changes occur at the receptor level or at the level of the initial sensory synapses.

It is still uncertain to what extent and in which way the brain stem reticular formation is involved in the control of sensory transmission [see the reviews on this subject by Rossi and Zanchetti (1957) and Livingston (1958)]. It is true that evoked potentials in sensory paths are often depressed during reticular stimulation while they appear to be released during nembutal anaesthesia [cf. also Hagbarth and Höjeberg (1957)] or following injury to the brain stem tegmentum. It should be emphasized, however, that there is often a sustained background activity in the sensory neurones which escapes detection by a macro-electrode. It is not wholly unlikely that such background activity can be maintained by reticular outflow acting upon the sensory relays [cf. Hagbarth and Fex

(1959)]. According to this hypothesis the activating reticular system serves to "energize" these relays in a similar way as it "energizes" the muscle spindles; and the depression of the synchronized potentials during reticular stimulation should be an occlusion phenomenon due to excess reticular excitation of individual sensory neurones.

The reticular formation, in its capacity as a diffuse activating system, can hardly be held responsible for the habituation process which in a specific fashion affects the response to that particular stimulus which has been repeated. This phenomenon is probably more easily explained on the basis of specific feed-back systems acting upon the sensory neurones. There is anatomical evidence that corticofugal fibres project to various sensory relays [BRODAL, SZABO and TORVIK (1956), KUYPERS (1958a)] and, in the case of the auditory system, a specific neural loop involving the cochlea and the auditory relay neurones has been demonstrated anatomically [RASMUSSEN (1955)] and physiologically [GALAMBOS (1956)]. It should also be recalled that in the case of the muscle spindles the inflow is exposed not only to reticular influences, it is also governed by specific spinal feedback systems.

On their centripetal course to the thalamic and cortical relays most sensory paths send collaterals to the midbrain reticular formation where sensory convergence on single neural elements occurs [for ref. see ROSSI and ZANCHETTI (1957)]. It is also well established that corticofugal fibres project to these reticular relays, and it was recently shown that the cephalically directed conduction within the reticular formation itself could be either suppressed or facilitated by excitation applied to the cortical foci concerned [ADEY, SEGUNDO and LIVINGSTON (1957)]. It is not within the scope of this review to discuss in which way these cortico-reticular paths are involved in the attention process [see FRENCH (1958), cf. also LARSSON (1959)] or if they play any role for the establishment of temporary reticular connections [see MORRELL (1958), GASTAUT (1958)]. There is no valid reason to suppose, however, that these events at the midbrain reticular level are wholly different from those which occur within the spinal internuncial pool. Centrifugal paths may, in both cases, be engaged in the establishment, of temporary internuncial connections.

In the case of lower reflex centers, adjustment of sensory inflow implies modulation of motor outflow, but what does adjustment of sensory transmission mean in terms of perception? In an attempt to analyze this intricate problem HERNÁNDEZ-PEÓN and DONOSO were able to record photically evoked potentials (to flash stimuli of constant intensity) by electrodes deeply implanted in the white matter of the occipital lobes in five human beings. The records were made in awake subjects and the results showed that the magnitude of the evoked potentials varied with the degree of attention. Furthermore, they noted that, in a suggestible patient, they could depress or enhance the response merely by suggesting a decrease or an increase of the flash intensity; and in this case there was a close correlation between the size of the evoked responses and the perceptual intensity of the light, as reported by the subject.

BROUWER said prophetically in 1933: "We accept that there is also a centrifugal side in the process of sensation, of vision, of hearing and so on. I believe that a further analysis of these descending tracts to pure sensory centers will also help physiologists and psychologists to understand some of their experiences". His notions are now supported by experimental data but much work remains to be done before any final conclusions can be drawn about the many intricate problems involved in this stimulating field of research.

References

ADEY, W. R., J. P. SEGUNDO and R. B. LIVINGSTON: Corticofugal influences on intrinsic brainstem conduction in cat and monkey. J. Neurophysiol. **20**, 1 (1957).

ADRIAN, E. D.: The physiological basis of perception. In "Brain mechanisms and conciousness". Oxford: Blackwell scient. Publ. 1954.

ALLISON, A. C.: The structure of the olfactory bulb and its relationship to the olfactory pathways in the rabbit and the rat. J. comp. Neurol. **98**, 309 (1953).

ARDEN, G. B., and U. SÖDERBERG: The relationship of lateral geniculate activity to the electrocorticogram in the presence or absence of the optic tract input. In course of publication (1959).

ASHBY, W. R.: Design for a brain. London: Chapman & Hall Ltd. See p. 37, 1952.

AUSTIN, G. M.: Suprabulbar mechanisms of facilitation and inhibition of cord reflexes. Rec. Publ. Ass. nerv. ment. Dis. **30**, 196 (1952).

BACH, L. M. N.: Effect of bulbar facilitation and inhibition on peripheral reflex inhibition. J. Neurophysiol. **13**, 259 (1950).

BARKER, D.: The innervation of the muscle spindles. Quart. J. micr. Sci. **89**, 143 (1948).

BRODAL, A., T. SZABO and A. TORVIK: Corticofugal fibers to sensory trigeminal nuclei and nucleus of solitary tract. J. comp. Neurol. **106**, 527 (1956).

BROUWER, B.: Centrifugal influence on centripetal systems in the brain. J. nerv. ment. Dis. **77**, 621 (1933).

BURKHARDT, D.: Die Sinnesorgane des Skeletmuskels und die nervöse Steuerung der Muskeltätigkeit. Erg. Biol. **20**, 27 (1958).

CAJAL, S. RAMÓN Y: Histologie du système nerveux de l'homme et des vertébrés. Paris: Maloine. Vol. 1 (1909); Vol. 2 (1911).

CHAMBERS, W. W., and C.-N. LIU: Cortico-spinal tract of the cat. J. comp. Neurol. **108**, 23 (1957).

CREED, R. S., D. DENNY-BROWN, J. C. ECCLES, E. G. T. LIDDELL and C. S. SHERRINGTON: Reflex activity of the spinal cord. Oxford: Clarendon Press 1932.

DAWSON, G. D.: The effect of cortical stimulation on transmission through the cuneate nucleus in the anaesthetized rat. J. Physiol. **142**, 2—3 p (1958, a).

— The central control of sensory inflow. Proc. roy. Soc. Med. **51**, 531 (1958, b).

DESMEDT, J. E., and K. MECHELSE: Suppression of acoustic input by thalamic stimulation. Proc. Soc. exp. Biol. (N. Y.) **99**, 772 (1958).

DODGE, R.: Habituation to rotation. J. exp. Psychol. **6**, 1 (1923).

DODT, E.: Centrifugal impulses in rabbit's retina. J. Neurophysiol. **19**, 301 (1956).

DUENSING, F.: Zur normalen und pathologischen Physiologie der Bauchdeckenreflexe. Z. ges. Neurol Psychiat. **168**, **171** (1940).

— Zur Pathologie der exteroceptiven Reflexe des Menschen. J. nerv. ment. Dis. **116**, 973 (1952).

ECCLES, R. M., and A. LUNDBERG: Significance of supraspinal control of reflex actions by impulses in muscle afferents. Experientia (Basel) **14**, 197 (1958).

ELDRED, E., and B. FUJIMORI: Relations of the reticular formation to muscle spindle activation. In "Reticular Formation of the Brain", p. 275. Henry Ford Hospital, Internat. symp. (1958).

—, R. GRANIT and P. A. MERTON: Supraspinal control of the muscle spindles and its significance. J. Physiol. **122**, 498 (1953).

ENGSTRÖM, H., and J. WERSÄLL: The ultrastructural organization of the organ of corti and of the vestibular sensory epithelia. Exp. Cell Res. Suppl. **5**, 460 (1958).

EULER, C. VON, and U. SÖDERBERG: The influence of hypothalamic thermoceptive structures on the electroencephalogram and motor activity. EEG. clin. Neurophysiol. **9**, 391 (1957).

FRENCH, J. D.: Corticofugal connections with the reticular formation. In "Reticular Formation of the Brain". Henry Ford Hospital. Internat. symp. p. 177 (1958

—, M. VERZEANO and H. W. MAGOUN: An extralemniscal sensory system in the brain. A. M. A. Arch. Neurol. Psychiat. **69**, 505 (1953).

FULTON, J. F.: Muscular contraction and the reflex control of movement. Baltimore: Williams and Wilkins Co. 1926. See pp. 531—537.

— In Howell's textbook of physiology. Ed. Fulton, J. F. 15th Ed. See p. 194, 1947.

GALAMBOS, R.: Suppression of auditory nerve activity by stimulation of efferent fibers to cochlea. J. Neurophysiol. **19**, 424 (1956).

—, G. SHEATZ and V. G. VERNIER: Electrophysiological correlates of a conditioned response in cats. Science **123**, 376 (1956).

GASTAUT, H.: The role of the reticular formation in establishing conditioned reactions. In "Reticular Formation of the Brain". Henry Ford Hospital. Internat. symp. p. 561 (1958).

GRANIT, R.: Receptors and sensory perception. New Haven: Yale Univ. Press **1955**, a.

— Centrifugal and antidromic effects on ganglion cells of retina. J. Neurophysiol, **18**, 388 (1955, b).

— Descending effects of the reticular formation with special reference to the gamma neurones. To be published in: Medizinische Grundlagenforschung. Stuttgart: Georg Thieme 1959.

—, and B. R. KAADA: Influence of stimulation of central nervous structures on muscle spindles in cat. Acta physiol. scand. **27**, 130 (1952).

HAGBARTH, K.-E.: Excitatory and inhibitory skin areas for flexor and extensor motoneurones. Acta physiol. scand. **26**, suppl. 94 (1952).

—, and J. FEX: Centrifugal influences on single unit activity in spinal sensory paths. J. Neurophysiol. In Press (1959).

—, S. HÖJEBERG: Evidence for subcortical regulation of the afferent discharge to the somatic sensory cortex in man. Nature (Lond.) **179**, 526 (1957).

—, and D. I. B. KERR: Central influences on spinal afferent conduction. J. Neurophysiol. **17**, 295 (1954).

—, and E. KUGELBERG: Plasticity of the human abdominal skin reflex. Brain **81**, 305 (1958).

HEAD, H., and G. HOLMES: Sensory disturbances from cerebral lesions. Brain **34**, 102 (1911).

Held, H.: Die centrale Gehörleitung. Arch. Anat. Physiol., Anat. Abt. 201 (1893).

Hernández-Peón, R.: Central mechanisms controlling conduction along central sensory pathways. Acta neurol. Latinoamer. **1**, 256 (1955).

—, C. Alcocer-Cuarón, A. Lavin y G. Santi-Báñez: Regulación centrifuga de la actividad eléctrica del bulbo olfactorio. Primera Reunión Cientifica de Ciencias Fisiol. Montevideo, p. 192 (1957).

—, and C. M. Donoso: Subcortical photically evoked electric activity in the human waking brain. Abstr. IV int. Congr. EEG clin. Neurophysiol. VIII Meeting int League against Epilepsy, Bruxelles, p. 155 (1957).

—, C. Guzmán-Flores, M. Alcaraz and A. Fernández-Guardiola: Sensory transmission in visual pathway during "attention" in unanaesthetized cats Acta neurol. Latinoamer. **3**, 1 (1957).

—, and K.-E. Hagbarth: Interaction between afferent and cortically induced reticular responses. J. Neurophysiol. **18**, 44 (1955).

—, M. Jouvet and H. Scherrer: Auditory potentials at cochlear nucleus during acoustic habituation. Acta neurol. Latinoamer. **3**, 144 (1957).

—, and H. Scherrer: Inhibitory influence of brain stem reticular formation upon synaptic transmission in trigeminal nucleus. Fed. Proc. **14**, 71 (1955).

— — and M. Jouvet: Modification of electric activity in cochlear nucleus during "attention" in unanaesthetized cats. Science **123**, 331 (1956).

— — and M. Velasco: Central influences on afferent conduction in the somatic and visual pathways. Acta neurol. Latinoamer. **2**, 8 (1956).

Hoff, E. C.: Distribution of spinal terminals (boutons) of pyramidal tract, determined by experimental degeneration. Proc. roy. Soc. B, **111**, 226 (1932).

Hoffmann, P.: Die Aufklärung der Wirkung des Jendrassikschen Handgriffs durch die Arbeiten von Sommer und Kuffler. Dtsch. Z. Nervenheilk. **166**, 60 (1951).

Holmqvist, B., A. Lundberg and O. Oscarsson: Functional organization of the dorsal spino-cerebellar tract in the cat. Acta physiol. scand. **38**, 76 (1956).

Hunt, C. C.: The reflex activity of mammalian small-nerve fibers. J. Physiol. **115**, 456 (1951).

— The effect of stretch receptors from muscle on the discharge of motoneurones. J. Physiol. **117**, 359 (1952).

Hugelin, A.: Analyse de l'inhibition d'un réflexe nociceptif (réflexe linguomaxillaire) lors de l'activation du système réticulo-spinal dit "facilitateur". C. R. Soc. Biol. (Paris) **149**, 1893 (1955).

Job, C.: Über autogene Inhibition und Reflexumkehr bei spinalisierten und decerebrierten Katzen. Pflügers Arch. ges. Physiol. **256**, 406 (1953).

Jouvet, M., and J. E. Desmedt: Contrôle central des messages acoustiques afférents. C. R. Acad. Sci. (Paris) **243**, 1916 (1956).

Kerr, D. I. B., and K.-E. Hagabarth: An investigation of olfactory centrifugal fiber system. J. Neurophysiol. **18**, 362 (1955).

King, E. E., R. Naquet and H. W. Magoun: Alterations in somatic afferent transmission through the thalamus by central mechanisms and barbiturates. J. Pharmacol. **119**, 48 (1957).

Kleyntjens, F., K. Koizumi and C. McC. Brooks: Stimulation of suprabulbar reticular formation. Arch. Neurol. Psychiat. (Chicago) **73**, 425 (1955).

Kobayashi, Y., K. Oshima and I. Tasaki: Analysis of afferent and efferent systems in the muscle nerve of the toad and cat. J. Physiol. **117**, (1952).

Kolmodin, G. M.: Integrative processes in single spinal interneurones with proprioceptive connections. Acta physiol. scand. **40**, suppl. 139, 1—89 (1957).

—, and C. R. Skoglund: Interneuron activity in the cat's spinal cord studied with intracellular electrodes. Acta physiol. scand. **31**, suppl. 114, 32 (1954).

KUGELBERG, E., and K.-E. HAGBARTH: Spinal mechanisms of the abdominal and erector spinae skin reflexes. Brain **81**, 290 (1958).

KUFFLER, S. W., and C. EYZAGUIRRE: Synaptic inhibition in an isolated nerve cell. J. gen. Physiol. **39**, **155** (1955).

—, and C. C. HUNT: The mammalian small-nerve fibers: a system for efferent nervous regulation of muscle spindle discharge. Rec. Publ. Ass. nerv. ment. Dis. **30**, 24 (1952).

— — and J. P. QUILLIAM: Function of medullated small-nerve fibers in mammalian ventral roots: efferent muscle spindle innervation. J. Neurophysiol. **14**, 29 (1951).

KUYPERS, H.: An anatomical analysis of cortico-bulbar connexions to the pons and lower brain stem in the cat. J. Anat. **92**, **198** (1958a).

— Pericentral cortical projections to motor and sensory nuclei. Science **128**, 662 (1958b).

LARSSON, L.-E.: The effect of variations in the psychological significance of stimuli on the magnitude of the startle blink reaction and evoked potentials in the human EEG. In course of publication (1959).

LEKSELL, L.: The action potential and excitatory effects of the small ventral root fibres to skeletal muscle. Acta physiol. scand. **10**, suppl. 31 (1945).

LINDBLOM, U. F., and J. O. OTTOSSON: Effects of spinal sections on the spinal cord potentials elicited by stimulation of low threshold cutaneous fibres. Acta physiol. scand. **29**, suppl. 106, 191 (1953).

— — Bulbar influence on spinal cord dorsum potentials and ventral root reflexes. Acta physiol. scand. **35**, 203 (1956).

— — Influence of pyramidal stimulation upon the relay of coarse cutaneous afferents in the dorsal horn. Acta physiol. scand. **38**, 309 (1957).

LIVINGSTON, R. B.: Central control of afferent activity. In "Reticular Formation of the Brain", Henry Ford Hospital. Internat. symp. p. **177** (1958).

LLOYD, D. P. C.: The spinal mechanism of the pyramidal system in cats. J. Neurophysiol. **4**, **525** (1941).

LOEWENSTEIN, W. R.: Modulation of cutaneous mechanoreceptors by sympathetic stimulation. J. Physiol. **132**, 40 (1956).

LORENTO DE NÓ, R.: Vestibulo-ocular reflex arc. Arch. Neurol. Psychiat. (Chicago) **30**, 245 (1933).

MAGOUN, H. W.: An ascending reticular activating system in the brain stem. Arch. Neurol. Psychiat. (Chicago) **67**, **145** (1952).

MATTHEWS, B. H. C.: Nerve endings in mammalian muscle. J. Physiol. **78**, 1 (1933).

— Tendon reflexes in free fall. J. Physiol. **133**, 31 P (1956).

MERTON, P. A.: Speculations on the servo-control of movement. The spinal cord. Ciba Foundation. Boston, p. 247 (1953).

— The silent period in a muscle of the human hand. J. Physiol. **114**, 183 (1951).

MORRELL, F.: Some electrical events involved in the formation of temporary connections. In "Reticular Formation of the Brain". Henry Ford Hospital. Internat. Symp. p. **545** (1958).

MORUZZI, G., and H. W. MAGOUN: Brain stem reticular formation and activation of the EEG. EEG clin. Neurophysiol. **1**, **455** (1949).

NIEMER, W. T., and H. W. MAGOUN: Reticulospinal tracts influencing motor activity. J. comp. Neurol **87**, **367** (1947).

OSCARSSON, O.: Functional organization of the ventral spino-cerebellar tract in the cat. Acta physiol. scand. **42**, suppl. 146, 1—107 (1957).

RASMUSSEN, G. L.: The olivary peduncle and other fiber projections of the superior olivary complex. J. comp. Neurol. **84**, 141 (1946).

RASMUSSEN, G. L.: Further observations of the efferent cochlear bundle. J. comp. Neurol. **99**, 61 (1953).

— Descending or "feedback" connections of auditory system of the cat. Amer. J. Physiol. **183**, 653 (1955).

ROSSI, G. F., and A. ZANCHETTI: The brain stem reticular formation. Arch. ital. Biol. **95**, 199 (1957).

SCHÄFER, E. A.: Some results of partial transverse section of the spinal cord. J. Physiol. **24**, Proc. 22 (1899).

SEGUNDO, J. P., and W. LARRAÑAGA: Striatal influence upon units in posterior column nuclei. In course of publication. Personal Communication (1959).

SHARPLESS, S., and H. JASPER: Habituation of the arousal reaction. Brain **79**, 655 (1956).

SHERRINGTON, C. S.: Reflex inhibition as a factor in the co-ordination of movements and postures. Quart. J. exp. Physiol. **6**, 251 (1913).

SKOGLUND, C. R.: The functional organization of spinal interneurones. Special University Lectures in Physiology. Univ. of London **1955**.

SOMMER, J.: Periphere Bahnung von Muskeleigenreflexen als Wesen des Jendrassikschen Phänomens. Dtsch. Z. Nervenheilk. **150**, 249 (1940).

STARZL, T. E., C. W. TAYLOR and H. W. MAGOUN: Collateral afferent excitation of the reticular formation of the brain stem. J. Neurophysiol. **14**, 479 (1951).

WALLENBERG, A.: Das sensible System (anatomischer Teil). Dtsch. Z. Nervenheilk. **101**, 111 (1928).

WALSH, E. G.: Physiology of the nervous system. London: Longmans, Green & Co. **1957**.

WIENER, N.: Cybernetics. Cambridge, Mass.: Technology Press **1948**.

Photomorphogenetische Reaktionssysteme in Pflanzen

1. Teil: Das reversible Hellrot-Dunkelrot-Reaktionssystem und das Blau-Dunkelrot-Reaktionssystem[1]

Von HANS MOHR, Tübingen

Aus dem Botanischen Institut der Universität Tübingen

Mit 16 Abbildungen

Inhaltsübersicht

Vorbemerkung

Als „Licht" bezeichnen wir jene Strahlung, die beim Menschen die Lichtempfindung verursacht, also nur Strahlung zwischen etwa 400 und 750 mμ Wellenlänge. Daß dieses „Licht" beim Menschen die „Lichtempfindung" verursacht, ist für die Phänomene, die hier zur Diskussion stehen, belanglos. Es wäre wohl besser, bei allen physikalischen und pflanzen- bzw. tierphysiologischen Untersuchungen den Begriff „Licht" zu vermeiden. Man sollte generell von Strahlung sprechen und diese durch die Wellenlänge näher definieren. Um der Kürze willen werden jedoch

[1] Der 2. Teil dieses Artikels, in dem Reaktionssysteme besprochen werden, bei denen ausschließlich kurzwelliges Licht und nahes Ultraviolett wirksam sind, folgt im nächsten Band.

auch wir den Begriff im Sinn obiger Definition verwenden. Den an das langwellige Rot anschließenden Spektralbereich nennen wir „kurzwelliges = nahes Infrarot", den an das Violett anschließenden Bereich nennen wir „langwelliges = nahes Ultraviolett". Wir behandeln experimentelle Resultate, die durch Bestrahlung mit Strahlung zwischen 320 und 800 (— 1000) mμ gewonnen wurden. Die Grenze im nahen Infrarot ist dadurch gegeben, daß es bisher nicht möglich war, jenseits dieser Grenze mit Sicherheit spezifische physiologische Strahlungseffekte nachzuweisen. Die Absorption von Quanten im Bereich des Infrarot scheint lediglich in unspezifischen thermischen Effekten zu resultieren. Die alleinige Beeinflussung der Schwingungs- bzw. Rotationsenergie der vielatomigen organischen Moleküle führt offenbar nicht zu spezifischen Stoffwechseländerungen. Im infraroten Bereich würde auch, falls spezifische Strahlungseffekte nachweisbar wären, die zwischen 800 und 1000 mμ einsetzende, starke Absorption des Wassers eine Interpretation von Wirkungsspektren sehr erschweren.

In dem für uns wichtigen Bereich ($\lambda < 800 — 1000$ mμ) sind, falls ein Quant absorbiert wird, elektronische Anregungen möglich. Die Energie der Quanten dieses Spektralbereichs ist so groß, daß, falls das Molekül seine Anregungsenergie nicht anderweitig verliert, eine photochemische Reaktion erfolgen kann. [Der Energiegehalt der Quanten in dem fraglichen Bereich liegt etwa zwischen 35 kcal/Einstein[1] (800 mμ) und 90 kcal/Einstein (320 mμ). Dies ist z. B. die Größenordnung der thermochemischen Bindungsenergie gewisser Verbindungstypen in organischen Molekülen.]

Im kurzwelligen Bereich wollen wir prinzipiell den Bereich bis etwa 320 mμ betrachten; wir wollen also das nahe Ultraviolett noch einschließen, da für diesen Bereich grundsätzlich noch dieselben Verhältnisse gelten wie für den Bereich der sichtbaren Strahlung (elektronische Anregung der Moleküle, keine Ionisation). Um 320—300 mμ beginnen zahlreiche in der Zelle vorkommende Stoffe (z. B. solche, die aromatische Aminosäuren enthalten) stark zu absorbieren. Damit ist die Grenze des Bereichs erreicht, wo für die Interpretation von Wirkungsspektren nur wenige Molekültypen in Frage kommen. (Es können nur relativ wenige organische Molekültypen im sichtbaren Bereich absorbieren.) Dazu kommt noch, daß das Spektrum der natürlichen Sonnenstrahlung um 300 mμ endet. Dieser Punkt ist biologisch sehr bedeutsam. Denn es zeigte sich, daß bei Verwendung von Strahlung kürzerer Wellenlängen ($\lambda < 300$ mμ) plötzlich sehr starke abiotische Effekte auftreten, was offenbar damit zusammenhängt, daß die Organismen an diese im natürlichen Biotop nicht vorkommende Strahlung nicht angepaßt sind [vgl. LOOFBOUROW (1948)].

1. Einleitung

a) Man wird, wenn von einem Einfluß des Lichts auf das Wachstum und die Entwicklung der Pflanzen die Rede sein soll, zunächst an die Bedeutung der *Photosynthese* für den Lebencyclus der Pflanze denken. Es hat sich jedoch gezeigt, daß das Licht auch *direkt*, d. h. ohne den „Umweg" über die Photosynthese, einen mächtigen Einfluß auf die Entwicklung und das Wachstum vieler Pflanzen ausüben kann („formative Lichtwirkung", „Photomorphogenese"). Besonders augenfällig ist dieser von der Photosynthese unabhängige Lichteinfluß auf Wachstum und Ent-

[1] 1 Einstein = 1 Mol Quanten = N Quanten, N = Loschmidtsche Zahl = $6{,}025 \cdot 10^{23}$. 1 Einstein repräsentiert den Energiebetrag $E = N \cdot h \cdot \nu$.

wicklung bei heterotrophen Pflanzen, z. B. bei vielen Pilzen. Wir wollen den Begriff „Photomorphogenese" (= Beeinflussung der pflanzlichen Gestaltbildung durch Licht) im weitesten Sinn verstehen, also auch solche Phänomene einschließen, die nicht unmittelbar die „Gestaltbildung" der Pflanzen betreffen. Wir wollen (wenigstens vorläufig) zu den „Photomorphosen" (= durch formative Lichteinflüsse entstandene, unmittelbar beobachtbare Veränderungen) alle mit morphologisch-anatomisch-histologischen, biochemischen oder biophysikalischen Methoden erfaßbaren lichtinduzierten Reaktionen der Pflanze rechnen, die nicht auf die Photosynthese zurückgehen. Die Registrierung anatomisch-morphologischer Veränderungen steht heute noch bei weitem im Vordergrund. — Freilich übt die Photosynthese indirekt über die Stoff- und Energieproduktion einen großen Einfluß auf Wachstum und Entwicklung der Pflanzen aus. Andererseits wird natürlich auch die photosynthetische Leistungsfähigkeit der Pflanze durch die Photomorphogenese stark beeinflußt. Es bleibt dem experimentellen Geschick des Forschers überlassen, auch bei autotrophen Pflanzen den Einfluß der Photosynthese von dem direkten Lichteinfluß auf das Wachstums- und Entwicklungsgeschehen zu trennen. Während früher diese Trennung meist nicht befriedigend erfolgte, ist es den Pflanzenphysiologen in jüngster Zeit gelungen, das photomorphogenetische Geschehen auch bei potentiell autotrophen Pflanzen relativ frei von den Einflüssen der Photosynthese zu untersuchen. Es muß dabei beachtet werden, daß bei den untersuchten Pflanzen die Versorgung mit Nährstoffen nicht der limitierende Faktor für die photomorphogenetische Reaktionsfähigkeit wird (z. B. Downs u. Mitarb. 1957). Beliebte Objekte sind daher frühe Keimlingsstadien, wo noch ausreichende Mengen von Nährstoffen zur Verfügung stehen und keinerlei Abhängigkeit von der Photosynthese besteht. Bei genauen Studien mit monochromatischer Strahlung (kurze Belichtungszeit!) ist die Verwendung junger, chlorophyllfreier Dunkelkeimlinge auch deshalb besonders günstig, weil unter diesen Bedingungen die Fluorescenz des Chlorophylls die spektrale Reinheit (z. B. des eingestrahlten Blaulichts) nicht beeinflußt.

b) Es ist gut bekannt, daß die Absorption der Strahlungsenergie, die bei der Photosynthese in chemische Energie umgesetzt wird, hauptsächlich durch die Chlorophylle erfolgt (wegen der in neuerer Zeit entdeckten Komplikationen vgl. z. B. Duysens 1956). Die für uns wichtigen Fragen sind, welche Pigmente die Strahlung im Fall der Photomorphogenese absorbieren und welche Reaktionen diese absorbierte Energie auslöst. Wir müssen uns folgende, allgemeine Vorstellung machen: Durch ein in der Zelle vorhandenes Pigment wird Strahlungsenergie absorbiert. Durch die Aufnahme dieser Energie in die Zelle können nun chemische Reaktionen ablaufen, die ohne diese Energiezufuhr nicht oder nicht in ausreichendem Maße stattfinden konnten. Dadurch kommt es im Stoffwechsel der Zelle zu Umsteuerungen, die letztlich zu den photomorphogenetischen Veränderungen führen. Meist wurden die morphologisch-anatomisch-histologischen Veränderungen untersucht (z. B. das lichtinduzierte Blattwachstum oder die lichtinduzierte Etiolementverhinderung des Hypokotyls), zuweilen auch mit physikalisch-chemischen Mitteln registrierbare Veränderungen (z. B. lichtabhängige Pigmentsynthese); nur in seltenen Fällen setzte die Registrierung des photomorphogenetischen Lichteinflusses noch früher ein (z. B. lichtabhängige Änderungen der Atmungsintensität bzw. von Enzymaktivitäten). Unser Wunsch ist, die Kausalverhältnisse von dem Moment der primären Energieabsorption bis hin zu den morphologisch-anatomischen Veränderungen kennenzulernen. Zur Zeit ist unser Wissen darüber noch recht beschränkt. Die experimentelle Arbeit an dieser Problematik kann von zwei Seiten her erfolgen. Einmal kann man den Kausalverhältnissen vom Endeffekt her nachgehen, also z. B. versuchen, von den anatomisch-morphologisch erfaßbaren Veränderungen zu den zugrunde liegenden Stoffwechseländerungen vorzudringen. Man versucht z. B. immer wieder, die anatomisch-morphologisch erfaßbaren Photomorphosen mit histologischen bzw. cytologisch-zellphysiologischen Veränderungen in Beziehung zu bringen [z. B. BURKHOLDER (1936), AVERY u. Mitarb. (1937), GOODWIN (1941), THOMSON (1951), WILLIAMS u. Mitarb. (1955), WILLIAMS (1956), WETMORE (1956), WILLIAMS u. Mitarb. (1957), REIMERS (1957), IKUMA und THIMANN (1958)]. Andererseits aber kann man versuchen, sich direkt oder indirekt Aufschluß über die an der Energieabsorption beteiligten Substanzen und über die ersten durch die absorbierte Strahlungsenergie ermöglichten photochemischen Reaktionen zu verschaffen. Neben reaktionskinetischen Studien ist eine korrekte Aufstellung des Wirkungsspektrums in diesem Falle von sehr großem Wert, denn man kann vom Wirkungsspektrum einer Reaktion unter Umständen auf das die wirksame Energieabsorption durchführende Pigment schließen.

c) Wirkungsspektren zeigen quantitativ die Abhängigkeit des untersuchten Effekts (z. B. Blattgröße bei lichtinduziertem Blattwachstum) von der Wellenlänge der verabreichten Strahlung. Im allgemeinen wird bestimmt, wieviel Energie/Flächeneinheit, oder (besser) wieviel Einstein/Flächeneinheit bei jeder Wellenlänge zur Erzielung ein und derselben Reaktionsgröße benötigt werden. Die reziproken Werte werden in Abhängigkeit von der Wellenlänge aufgetragen. Die Aufstellung und Interpretation von Wirkungsspektren ist nicht einfach und manche diesbezüglichen Versuche müssen als mißglückt bezeichnet werden. Es ist hier nicht der Ort, die theoretischen Grundlagen der Aufstellung und Interpretation von Wirkungsspektren darzustellen [vgl. hierzu z. B. LOOFBOUROW (1948)]. Es soll lediglich auf die Problematik einiger Voraussetzungen hingewiesen werden, die meist stillschweigend gemacht werden, wenn man ein Wirkungsspektrum aufstellt und interpretiert. Ist z. B. die Quantenausbeute des zugrunde liegenden, photochemischen Prozesses innerhalb des untersuchten Spektralbereichs konstant? Ist die Beeinflussung der Strahlung durch Streuung, Reflexion und Absorption durch nicht effektive Pigmente für alle untersuchten Wellenlängen annähernd dieselbe, bevor die Strahlung den Ort der wirksamen Absorption erreicht? Ist der Selbstbeschattungseffekt des die wirksame Strahlung absorbierenden Pigments ausreichend gering? Ist das Absorptionsspektrum des wirksamen Pigments (auf das man schließen will!) in vivo wenigstens näherungsweise identisch mit dem (meist nur bekannten) Absorptionsspektrum in vitro? Hier ist besonders zu beachten, daß die Absorptionsspektren der Pigmente von den physikalisch-chemischen Bedingungen in der Zelle (z. B. Art der Bindung, Art der Lösung bzw. Verteilung, p_H, r_H) stark modifiziert werden können. Weitere, unbedingt notwendige Voraussetzungen für die Ausarbeitung sinnvoller Wirkungsspektren sind adäquate reaktionskinetische Vorstudien und eine physikalisch einwandfreie Ausrüstung zur Erzeugung und Messung monochromatischer Strahlung. Die früher häufig verwendeten, relativ einfachen und weitgehend unzulänglichen Hilfsmittel zur Erzeugung und Messung „monochromatischer" Strahlung müssen durch neue, physikalisch korrekte Konstruktionen ersetzt werden [z. B. PARKER u. Mitarb. (1946) (Spektrograph), MONK und EHRET (1956) (Spektrograph), WITHROW (1957) (Interferenzfiltermonochromatorsystem), MOHR und SCHOSER (1959) (Interferenzfiltermonochromatorsystem)]. Man versucht nun, von dem einwandfreien Wirkungsspektrum auf das die Energieabsorption durchführende Pigment zu schließen. Dies ist möglich, weil, unter bestimmten Voraussetzungen (vgl. oben), das Wirkungsspektrum das Absorptionsspektrum des die Strahlungsabsorption durchführenden Pigments repräsentiert. Ein schönes Beispiel dafür, daß unter günstigen Bedingungen das absorbierende Pigment, dessen Absorptionsspektrum bekannt ist, durch das Wirkungsspektrum identifiziert werden kann, findet man bei KOSKI u. Mitarb. (1951). Sie stellten fest, daß die für die photochemische Umwandlung von Protochlorophyll zu Chlorophyll a notwendige Strahlungsenergie vom Protochlorophyll selber absorbiert wird.

Auch wenn das Wirkungsspektrum einer Photomorphose zunächst keine eindeutige Identifizierung des absorbierenden Pigments zuläßt, kann es doch äußerst wertvoll sein, indem es die Möglichkeit schafft, durch Vergleich mit den Wirkungsspektren anderer Photomorphosen festzustellen, ob mehrere, phänomenmäßig zunächst getrennte und unabhängige Photomorphosen nicht letztlich auf die Energieabsorption durch dasselbe Pigmentsystem zurückgehen.

d) Der nun folgende Bericht soll einen knappen Überblick über unser derzeitiges Wissen und unsere gegenwärtigen Vorstellungen hinsichtlich einiger photomorphogenetisch besonders wichtiger Reaktionssysteme geben. Der wesentliche Teil unseres derzeitigen Wissens ist in den letzten 6—8 Jahren erarbeitet worden. Dies hängt besonders mit den methodischen Fortschritten während des vergangenen Jahrzehnts zusammen, und zwar einmal mit der verbesserten physikalischen Technik im Bereich der Photobiologie [vgl. Withrow und Withrow (1956)], andererseits aber auch mit entscheidenden Verbesserungen in der Versuchsplanung selber.

2. Historische Notiz

Der morphogenetische Einfluß des Lichts hat die Pflanzenphysiologen seit jeher interessiert. Vor allem am Beispiel gewisser Pilze hat man bald gemerkt, daß die im Zusammenhang mit der durch Licht erfolgenden „Etiolementverhinderung" beobachteten formativen Lichteinflüsse mit der Photosynthese und mit dem Angebot an Nährstoffen nichts zu tun haben. Ein instruktives Beispiel sei kurz angeführt [nach Boysen-Jensen (1939)]: Aus der Kartoffelknolle entstehende Schößlinge etiolieren im Dunkeln trotz des Überflusses an Nährstoffen. Werden die Schößlinge im CO_2-freien Raum belichtet, so treten die „normalen" morphogenetischen Lichtwirkungen ein (Etiolementverhinderung). Schon frühzeitig ist die auf zu einfache Versuche gestützte, summarische Behauptung aufgestellt worden, es sei vor allem das kurzwellige Licht, das etiolementverhindernd wirke. Das Blaulicht wurde als der Bereich des sichtbaren Spektrums betrachtet, der notwendig sei, damit der „normale" Habitus der Pflanzen, z. B. der Habitus im Weißlicht hoher Intensität, entstehe. Die formative Wirkung des Weißlichts wurde ausschließlich dem im Weißlicht vorhandenen Blaulicht zugeschrieben. Im Gegensatz dazu fand man, daß die Photosynthese im Blau- und im Rotlicht abläuft. Der Befund hinsichtlich der Photosynthese ist korrekt, die hinsichtlich der morphogenetischen, etiolementverhindernden Lichtwirkung vertretene Auffassung ist dagegen *nicht* richtig, obgleich diese Auffassung seit Sachs (1874) und Pfeffer (1904) bis in die neueste Zeit in den meisten Lehrbüchern der Pflanzenphysiologie vertreten wird. Der dominierende Einfluß des Blaulichts auf das pflanzliche Wachstum (im Sinn von: Weißlicht ist nur durch Blaulicht ersetzbar) ist nur *ein* Aspekt der Photomorphogenese und läßt sich nur bei einer bestimmten Versuchsanordnung nachweisen [Strahlung relativ hoher Intensität, lange Dauer der Bestrahlung, z. B. längere Kultur älterer Pflanzen im ausschließlich „monochromatischen" Licht hoher Intensität; vgl. z. B. Popp und Brown (1936), Crocker (1948), Wassink (1954), Wassink und Stolwijk (1956)].

Man weiß heute, auf Grund der analytischen Arbeit der vergangenen Jahre, daß bei den potentiell grünen Pflanzen mehrere Reaktionssysteme mit ganz verschiedenen Wirkungsspektren an der Photomorphogenese beteiligt sind. Es hängt von dem Zeitpunkt und der Dauer der Belichtung, von der Intensität des Lichts und von der Vorbehandlung der Pflanzen ab, welcher Spektralbereich wirkungsmäßig in den Vordergrund tritt. Es ist notwendig gewesen, recht komplizierte Bestrahlungsprogramme anzuwenden, um die einzelnen Systeme physiologisch voneinander zu trennen. Dabei mußte bei potentiell grünen Pflanzen große Sorgfalt darauf verwendet werden, die störenden Einflüsse der in den verschiedenen Spektralbereichen verschieden starken Photosynthese weitgehend auszuschalten. Besonders bei länger dauernden Bestrahlungen mit ausschließlich monochromatischer Strahlung höherer Intensität kann der indirekte Einfluß der Photosynthese auf das Wachstum der Pflanzen leicht zu falschen Schlüssen hinsichtlich der spezifisch formativen Lichteinflüsse führen.

Ein wichtiges Prinzip der neueren Forschung ist, im Gegensatz zu vielen älteren Arbeiten, die für die Induktion der Photomorphosen notwendige Belichtungszeit so kurz wie möglich zu halten. Nur so lassen sich solche photomorphogenetischen Reaktionssysteme, die schon bei Einstrahlung geringer Energiemengen praktisch saturiert werden, von solchen trennen, die erst dann sich photomorphogenetisch auswirken, wenn hohe Energiemengen eingestrahlt werden. Daß schon durch sehr kurzfristige Belichtungen, z. B. ein paar Sekunden, starke formative Effekte induziert werden können, ist ein weiterer Beleg dafür, daß die Photomorphosen nicht auf indirekten Beeinflussungen via Photosynthese beruhen können.

3. Das reversible Hellrot-Dunkelrot Reaktionssystem (reversible red far-red reaction system)[1]

a) Einführung und Übersicht

Es ist seit langem bekannt [CASPARY (1860)], daß die Keimung mancher Samen durch Licht gefördert werden kann. Dieser markante Lichteinfluß ist sehr häufig untersucht worden. Man hat schon früh gelernt, „Lichtkeimer" (= Samen, deren Keimung gefördert wird, wenn sie im gequollenen Zustand belichtet werden, z. B. *Lythrum salicaria*) und „Dunkelkeimer" (= Samen, deren Keimung durch Belichtung gehemmt wird, z. B. *Phacelia tanacetifolia*) zu unterscheiden. Trotz der zahlreichen Arbeiten [vgl. z. B. die Zusammenfassungen bei KINZEL (1926), LEHMANN und AICHELE (1931), CROCKER (1936), EVENARI (1956) und TOOLE u. Mitarb. (1956)] war es lange Zeit nicht möglich, die Lichtwirkungen näher zu analysieren.

Diese Analyse wurde erst durch die Verwendung monochromatischer Strahlung und durch neue Wege der Versuchsanstellung ermöglicht.

[1] Folgende Abkürzungen werden benutzt: Hellrot = HR, Dunkelrot = DR.

FLINT und MCALISTER (1935, 1937) haben als erste ein Wirkungsspektrum der Keimung von *Lactuca*-Achänen (Lichtkeimer) aufgestellt. Sie fanden, daß Hellrot = HR (Hauptwirkung zwischen 600 und 700 mμ) die Keimung der Achänen fördert, daß jedoch Dunkelrot = DR ($\lambda >$ 700 mμ, etwa 700—800 mμ) und auch Blaulicht (400—500 mμ) die durch HR induzierte Keimung hemmen, falls nach dem HR DR bzw. Blau gegeben werden. Nach MEISCHKE (1936) und RESÜHR (1939) werden Licht- und Dunkelkeimer durch dieselben Spektralbereiche gefördert bzw. gehemmt. Bei Einstrahlung von „Weißlicht" überwiegt in dem einen Fall der Förder-, im andern Fall der Hemmeffekt. Nach JONES und BAILEY (1956) ist bei den Samen von *Lamium amplexicaule* (Dunkelkeimer) die Empfindlichkeit für DR sehr hoch, verglichen mit der Empfindlichkeit für HR. Das Resultat ist, daß die Samen von Sonnen- oder ungefiltertem Glühlampenlicht in ihrer Keimung gehemmt werden. Ähnliche Verhältnisse hatten schon BORTHWICK u. Mitarb. (1954) für gewisse, in bestimmter Weise vorbehandelte *Lactuca*-Achänen beschrieben. (In manchen Fällen kann der Lichteinfluß auf die Keimung komplizierter sein. Darauf wird später hingewiesen.)

Die erst halb-quantitativen Resultate von FLINT und MCALISTER wurden von BORTHWICK u. Mitarb. (1952a, 1954, *Lactuca sativa*-Achänen) und TOOLE u. Mitarb. (1955, *Lepidium*-Samen) prinzipiell bestätigt und verfeinert. Bei *Lactuca* genügt schon ein einmaliger, relativ kurzer Lichtstoß (wenige Minuten HR von relativ niedriger Intensität), um bei einem hohen Prozentsatz der gequollenen Samen die Keimung zu induzieren. Die aufgestellten Wirkungsspektren (Abb. 1 und Abb. 2) geben die für einen bestimmten Prozentsatz an Förderung oder Hemmung benötigte Energie in Abhängigkeit von der Wellenlänge an. Der Gipfel des Förderspektrums liegt bei 660 mμ, der Gipfel des Hemmspektrums bei 735 mμ. Während also die von FLINT und MCALISTER gefundenen Verhältnisse im HR-DR-Bereich perfekt reproduziert werden konnten, waren die Angaben für den Blaubereich nicht entsprechend sicher verifizierbar. Darauf kommen wir später zurück. — Das Charakteristische dieses an der Samenkeimung beteiligten Reaktionssystems ist, daß der Induktionseffekt einer bestimmten HR-Dosis durch nachfolgende Bestrahlung mit DR wieder annulliert werden kann („Reversibles HR-DR-Reaktionssystem"). In den letzten Jahren ist nun bei sehr zahlreichen anderen lichtabhängigen pflanzlichen Reaktionen gefunden worden, daß ihnen ebenfalls dieses Reaktionssystem, das durch den HR-DR-Antagonismus gekennzeichnet ist, zugrunde liegt. In allen diesen Fällen scheinen dieselben photoreceptorischen Pigmente und derselbe photochemische Mechanismus vorzuliegen. Diese Pigmente sind offenbar unter den potentiell grünen Pflanzen weit, vielleicht universell, verbreitet (bei Pilzen konnte das Reaktionssystem bisher nicht nachgewiesen werden) und scheinen für die Photomorphogenese der Pflanzen eine ähnliche Bedeutung zu haben wie das Chlorophyll für die Photosynthese.

Das Reaktionssystem ist z. B. bei folgenden lichtbedingten Reaktionen der Pflanzen nachgewiesen worden (Induktion des Effekts durch HR, Annullierung der Induktion durch nachfolgende DR-Bestrahlung):

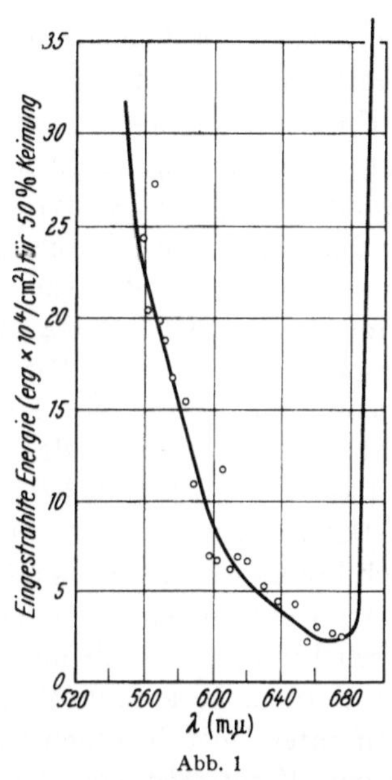

Abb. 1

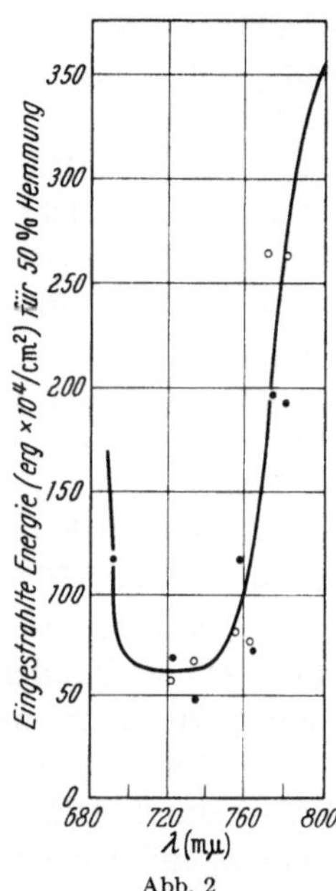

Abb. 2

Abb. 1. Das Wirkungsspektrum für die Förderung der Keimung von *Lactuca*-Achänen (Grand Rapids) nach 16 stündiger Quellung. [Nach BORTHWICK u. Mitarb. (1954)]

Abb. 2. Das Wirkungsspektrum für die Hemmung der durch Hellrot induzierten Keimung von *Lactuca*-Achänen (Grand Rapids) nach 16stündiger Quellung. [Nach BORTHWICK u. Mitarb. (1954)]

Samen- bzw. Fruchtkeimung bei Licht- und Dunkelkeimern [BORTHWICK u. Mitarb. (1952a, 1954), TOOLE u. Mitarb. (1955a, b), JONES und BAILEY (1956)], Beeinflussung der Atmung von *Lactuca*-Früchten und anderen Geweben [LEOPOLD und GUERNSEY (1954), EVENARI u. Mitarb. (1955)], Hemmung des Internodienwachstums bei Dunkelkeimlingen von mono- und dikotylen Pflanzen [WENT (1941), PARKER u. Mitarb. (1949), BORTHWICK u. Mitarb. (1951); in diesem Fall ist bisher nur der HR-Effekt bearbeitet], Hemmung des Hypokotylwachstums bei Dunkelkeimlingen [DOWNS (1955)], Beeinflussung des Wachstums von Internodien und Petiolen bei älteren, grünen Pflanzen [WASSINK und STOLWIJK (1956), DOWNS u. Mitarb. (1957), VAN DER VEEN und MEIJER (1958)], Beeinflussung des Längenwachstums von Segmenten der *Avena*-Koleoptile [LIVERMAN und BONNER (1953)], Förderung der Öffnungsbewegung des „Plumulahakens" (plumular hook) bei Bohnenkeimlingen [KLEIN u. Mitarb. (1956, 1957b), WITHROW u. Mitarb. (1957)], Förderung des Blattwachstums bei Dunkelkeimlingen [DOWNS (1955), LIVERMAN u. Mitarb. (1955), MOHR (1959b)], Auslösung der Blattbewegung bei *Phaseolus multiflorus* [LÖRCHER (1958)], Förderung des Dunkelwachstums von *Lemna minor* [HILLMAN (1957)], Beeinflussung der Blütenbildung bei Lang- und Kurztagpflanzen durch Zusatzlicht in der Mitte der auf einen Weißlicht-Kurztag folgenden Dunkelperiode [PARKER u. Mitarb. (1945, 1946, 1950), BORTHWICK u. Mitarb. (1948, 1952b), DOWNS (1956)], Auslösung der Haarbildung aus Epidermiszellen des Hypokotyls bei Dunkelkeimlingen von *Sinapis alba* [MOHR (1959a)], Bildung eines gelben Pigments in der Tomatenepidermis [PIRINGER und HEINZE (1954)], Bildung von Anthocyan in gewissen Keimlingen [SIEGELMAN und HENDRICKS (1957), MOHR (1957)], Bildung von Protochlorophyll und damit von Chlorophyll a [WITHROW u. Mitarb. (1956), VIRGIN (1958)], Keimung von Farnsporen [MOHR (1956)], Drehung des Chloroplasten bei *Mougeotia* [HAUPT (1958)].

Es scheint, daß durch das HR der Stoffwechsel der Zellen fundamental beeinflußt wird. Alle die genannten Reaktionen von Pflanzen auf eine Bestrahlung mit HR dürften auf *dieselbe* Umsteuerung des Stoffwechsels zurückgehen. Ob eine solche fundamentale Umsteuerung direkt mit der in gewissen Geweben beobachteten Beeinflussung der Aktivität des IES[1]-Oxydase-Systems durch das reversible HR-DR-Reaktionssystem zusammenhängt [HILLMAN und GALSTON (1957)], muß noch weiter geklärt werden. Auch der Einfluß des reversiblen HR-DR-Reaktionssystems auf die Rate der oxydativen Phosphorylierung durch Mitochondrien (aus Rattenleber und *Avena*-Koleoptilen) könnte ein Hinweis dafür sein, wie man sich eine grundsätzliche Beeinflussung des Stoffwechsels vorstellen darf [GORDON und SURREY (1958)]. Gleichzeitig ist damit gezeigt, daß das reversible HR-DR-Reaktionssystem auch in tierischem Gewebe vorhanden ist. — Nach LOCKHART (1958) hängt auch die Reaktion gewisser Pflanzen (Dunkelpflanzen) auf eine Behandlung mit Gibberellinsäure von einer HR-Vorbelichtung ab. Auch hier kann der HR-Effekt durch DR reversibilisiert werden. MOH und WITHROW (1957) berichten, daß der Effekt einer bestimmten Dosis Röntgenstrahlen auf die Chromosomen in Keimwurzeln von *Vicia faba* von der Vorbehandlung der Wurzeln mit HR bzw. DR abhängt. Diese Angaben zeigen, daß sich die durch das HR im Stoffwechsel der Zelle verursachten Änderungen auch darin äußern, daß sich die Reaktionsfähigkeit der Organismen gegenüber anderen exogenen Faktoren ändert.

In der Literatur sind noch andere Rotlichteffekte beschrieben [z. B. JOHNSTON (1937), BIEBEL (1942), WEINTRAUB und MCALISTER (1942), WEINTRAUB und PRICE (1947), GOODWIN und OWENS (1948, 1951), TORREY (1952)], bei denen wohl die weitere Untersuchung erweisen würde, daß auch sie auf die Wirkung des reversiblen HR-DR-Reaktionssystems zurückgehen.

b) Das Wirkungsspektrum der HR-Induktion und der DR-Reversibilisierung

Mehrere Male ist das Wirkungsspektrum der Induktion und der Reversibilisierung des Induktionseffekts genau bestimmt worden [BORTHWICK u. Mitarb. (1954), MOHR (1956), WITHROW u. Mitarb. (1957)]. Die Ausarbeitung erfolgte an ganz verschiedenen Objekten: *Lactuca-*

[1] IES = β-Indolylessigsäure (Heteroauxin).

Achänen, *Dryopteris*-Sporen, „Plumula-Haken" von *Phaseolus*. Auch die eingesetzte physikalische Technik, die verwendete physiologische Versuchsanstellung und die Darstellung der Wirkungsspektren waren verschiedenartig. Trotzdem stimmen die erarbeiteten Wirkungsspektren weitgehend überein (vgl. z. B. die Abb. 1—4). Die genauesten Wirkungsspektren haben Withrow u. Mitarb. (1957) publiziert (Abb. 3 und 4). Aus den Wirkungsspektren geht hervor, daß sich die (aus den Wirkungsspektren erschlossenen!) Absorptionskurven des HR- und des DR-Pigments[1] überschneiden. Es wird also oft zutreffen, daß Licht einer bestimmten spektralen Zusammensetzung sowohl von dem HR- wie auch von dem DR-Pigment absorbiert wird und daß entsprechend sowohl die Induktionswirkung wie auch der Reversibilisierungseffekt auftreten. Bei Einstrahlung einer bestimmten Wellenlänge in diesem Bereich (z. B.

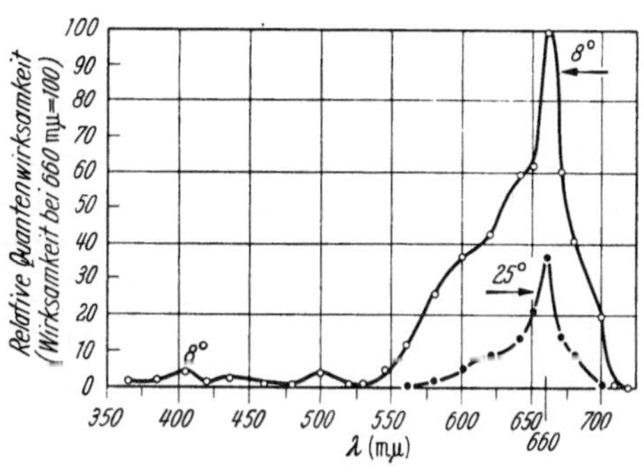

Abb. 3. Das Wirkungsspektrum der lichtgeförderten Öffnungsbewegung des isolierten „Plumula-Hakens" der Dunkelkeimlinge von *Phaseolus vulgaris*. Auf der Ordinate ist die relative Quantenwirksamkeit aufgetragen, berechnet für einen Öffnungswinkel von 8° bzw. 25°. [Nach Withrow u. Mitarb. (1957)].

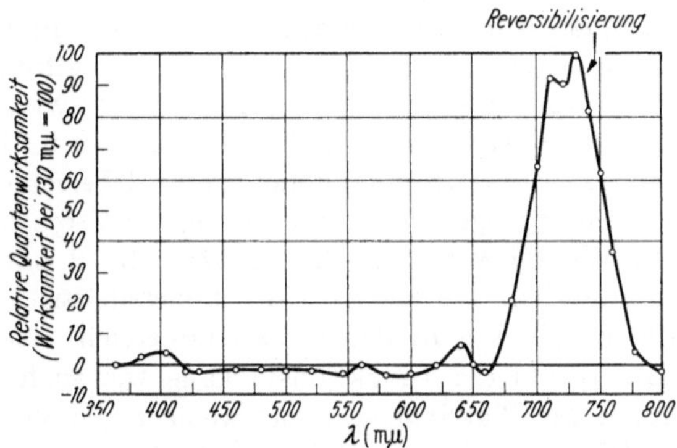

Abb. 4. Das Wirkungsspektrum der Reversibilisierung der Hellrot-Induktion (Isolierte „Plumula-Haken" von Dunkelkeimlingen von *Phaseolus vulgaris*). Mit einer bestimmten Quantität Hellrot (600—700 mμ) wird ein bestimmter Öffnungswinkel des Plumula-Hakens induziert. Anschließend werden die Plumula-Haken mit monochromatischem Licht bestrahlt, um die Reversibilisierung der Hellrotinduktion zu messen. Auf der Ordinate ist die relative Quantenwirksamkeit hinsichtlich der Reversibilisierung aufgetragen [Nach Withrow u. Mitarb. (1957)]

---

[1] Als HR-Pigment (= P_{HR}) bezeichnen wir jenes Pigment, dessen (erschlossenes) Absorptionsmaximum im hellroten Spektralbereich liegt, als DR-Pigment (= P_{DR}) jenes, dessen (ebenfalls erschlossenes) Absorptionsmaximum im dunkelroten Bereich des Spektrums liegt.

700 mμ) wird der physiologische Effekt von dem Wirkungsgleichgewicht abhängen [vgl. BORTHWICK u. Mitarb. (1954)]. TOOLE u. Mitarb. (1955a) geben an, daß bei der Keimung der Samen von *Lepidium virginicum* (Lichtkeimer, 20° C, 0,2 % KNO_3) derjenige Wellenlängenbereich, bei dem auch bei sehr hoher Energieeinstrahlung (konstante Intensität) nicht mehr als 50% der Samen keimen, zwischen 687,5 und 690 mμ liegt. Bei der Einstrahlung von „Weißlicht" konstanter Intensität wird der Endeffekt von der spektralen Verteilung des Lichts und von der relativen Empfindlichkeit des Induktions- bzw. Reversibilisierungsprozesses abhängen. Man sieht, daß die Verwendung von nicht genau definierten und konstanten Strahlungsquellen für quantitative Versuche zwecklos und irreführend sein kann! Der Verlauf der Wirkungsspektren wird durch die Überschneidung der Induktions- und Reversibilisierungsspektren sicher stark beeinflußt. Dies muß besonders beachtet werden, wenn man von den Wirkungsspektren auf die Photoreceptoren schließen will. So ist es durchaus möglich, daß der Gipfel des Absorptionsspektrums des HR-Pigments bei etwas längerer Wellenlänge als 660 mμ liegt, wahrscheinlich zwischen 660 und 680 mμ. Das Verhältnis der beiden Gipfel des Reversibilisierungsspektrums (710 bzw. 730 mμ) ist nicht konstant. Bei niedriger eingestrahlter Energie dominiert der 710 mμ-Gipfel, bei höherer Energie der 730 mμ-Gipfel. Es ist möglich, daß der Hauptabsorptionsgipfel des Pigments bei 710 mμ liegt und daß bei höherer Energie dieser Gipfel im Wirkungsspektrum wegen der Interferenz mit dem Induktionsprozeß erniedrigt wird [WITHROW u. Mitarb. (1957)]. — Aus Abb. 3 und 4 ersieht man, daß Blaulicht ($\lambda < 500$ mμ) nur von sehr geringer Bedeutung bei diesen Reaktionen ist. Die Verhältnisse im Blaubereich sind indessen nicht ganz eindeutig. Nach BORTHWICK u. Mitarb. (1954) sollen sowohl das HR- wie auch das DR-Pigment im Blaubereich absorbieren und es soll dort eine komplizierte Überschneidung der Absorptionsspektren vorliegen, was die Deutung der Versuchsergebnisse erschwert. Auch WITHROW u. Mitarb. [(1957), Abb. 3 und 4] finden sowohl bei der Induktion wie auch bei der Reversibilisierung gewisse, allerdings sehr kleine Effekte im blauen Spektralbereich. Sicher ist, daß, wenn überhaupt, sowohl der Induktionsvorgang wie auch der Reversibilisierungsprozeß für Blaulicht sehr viel weniger „empfindlich" sind als für HR bzw. DR [vgl. hierzu auch VIRGIN (1958)]. In manchen Fällen, in denen starke Blaulichteffekte auftreten, z. B. im Falle der Hemmung der mit HR induzierten Keimung von Farnsporen [MOHR (1956)] oder gewisser Samen [z. B. WAREING und BLACK (1958)] dürften die beobachteten Blaulichteffekte auf der Strahlungsabsorption in einem anderen Pigmentsystem beruhen. (Man kann eine Energieübertragung von diesem Pigmentsystem auf das P_{DR} vermuten). Man muß auch stets damit rechnen, daß, besonders bei Einstrahlung hoher Energie, neben

dem reversiblen HR-DR-Reaktionssystem auch noch andere Reaktionssysteme wirksam werden.

c) Beobachtungen zur Kinetik des HR-DR-Reaktionssystems

Die an verschiedenen Objekten gewonnenen Resultate sind hier nicht so einheitlich wie im Falle der Wirkungsspektren. Dies spricht jedoch nicht dagegen, daß es sich stets um dieselben Pigmente und um dieselben primären Vorgänge handelt, denn man darf nicht erwarten, daß z. B. die quantitativen Beziehungen zwischen dem Produkt der primären, von der absorbierten Energie bewirkten photochemischen Reaktion (von der wir annehmen, daß sie stets dieselbe ist) und dem Ausmaß der beobachtbaren Photomorphose in allen Fällen dieselben sind.

Die Quantität der durch HR induzierten Reaktion ist bei verschiedenen Photomorphosen [vgl. z. B. PARKER u. Mitarb. (1949), BORTHWICK u. Mitarb. (1951), DOWNS (1955), KLEIN u. Mitarb. (1957a), WITHROW u. Mitarb. (1957)] linear proportional dem log der eingestrahlten Energie; das Ausmaß der DR-Reaktion (Reversibilisierung) hängt dagegen, wenigstens bei der Öffnungsbewegung des Plumula-Hakens von *Phaseolus*, linear von der eingestrahlten Energie ab (Abb. 5). BORTHWICK u. Mitarb. (1954) und TOOLE u. Mitarb. (1955) geben dagegen für *Lactuca*-Früchte und *Lepidium*-Samen an, daß sowohl bei der HR-Induktion wie auch bei der DR-Reversibilisierung die probits für die Keimung eine einheitliche, lineare Funktion des log der eingestrahlten Energie sind. Zuweilen [z. B. PARKER u. Mitarb. (1949), BORTHWICK u. Mitarb. (1951), DOWNS (1955)] zeigte sich bei höheren Energien ein Bruch in der Steigung der log-Energie-Effekt-Kurve, so daß sich zwei Gerade mit etwas verschiedener Steigung ergeben (Abb. 6). Dies dürfte daher rühren, daß das verwendete Material aus zwei Populationen mit etwas verschiedener Strahlungsempfindlichkeit bestand. — Das „Reizmengengesetz" ist über lange Zeiten hinweg nicht exakt gültig [WITHROW u. Mitarb. (1957)], es wurde jedoch bei relativ kurzen Belichtungszeiten (immerhin bis 100 min) als gültig befunden [BORTHWICK u. Mitarb. (1951)]. In Übereinstimmung damit wird gefunden, daß die Intensitätsabhängigkeit der HR-Reaktion logarithmisch ist [KLEIN u. Mitarb. (1956)].

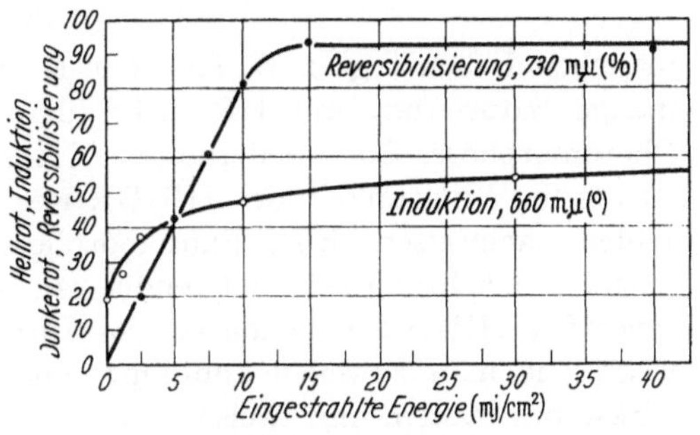

Abb. 5. Energie-Effektkurven für die HR-Induktion und die DR-Reversibilisierung. Objekt: Isolierte „Plumula-Haken" von Dunkelkeimlingen von *Phaseolus vulgaris*. [Nach WITHROW u. Mitarb. (1957)]

Genaue Untersuchungen zum HR-DR-System können nur vorgenommen werden, wenn die Bestrahlungszeiten relativ kurz sind (und somit in den Bereich der Gültigkeit des „Reizmengengesetzes" fallen), es sei denn, man übersieht genau die komplizierten Verhältnisse, die dadurch entstehen, daß nach einiger Zeit die „Empfindlichkeit" für HR sowohl im Licht wie auch im Dunkeln wieder „regeneriert", nachdem mit einer praktisch saturierenden HR-Dosis bestrahlt worden ist. In vielen Fällen hat sich gezeigt, daß die HR-Reaktion bei entsprechender Intensität innerhalb kurzer Zeit (Sekunden bis Minuten) praktisch saturiert werden kann. Nach relativ kurzer Zeit läßt sich jedoch wieder eine „Empfindlichkeit" für HR nachweisen [MOHR (1959a)]. Eine erneute HR-Bestrahlung steigert den durch die erste saturierende Bestrahlung erzielten Effekt entsprechend. Der in einem gewissen Zeitraum im HR-Dauerlicht über das HR-DR-System erzielbare Effekt entspricht etwa dem, der durch eine adäquate Zahl kurzer, saturierender „Lichtstöße" während dieses Zeitraums erzielt werden kann [MOHR (1959a)]. In manchen Fällen [vgl. z. B. TOOLE u. Mitarb. (1958a)] genügt eine einmalige Saturierung mit HR nicht für die Erzeugung einer meßbaren Photomorphose. Mit wiederholten HR-„Lichtstößen" läßt sich dann leicht die Wirksamkeit des HR-DR-Reaktionssystems auch in diesen Fällen nachweisen. Eine Photomorphose, zu deren Erzeugung eine längere Belichtungszeit notwendig ist, kann also durchaus auf das reversible HR-DR-Reaktionssystem zurückgehen und braucht keineswegs, wie man zuweilen annimmt, von vornherein auf ein anderes wirksames System hinzuweisen.

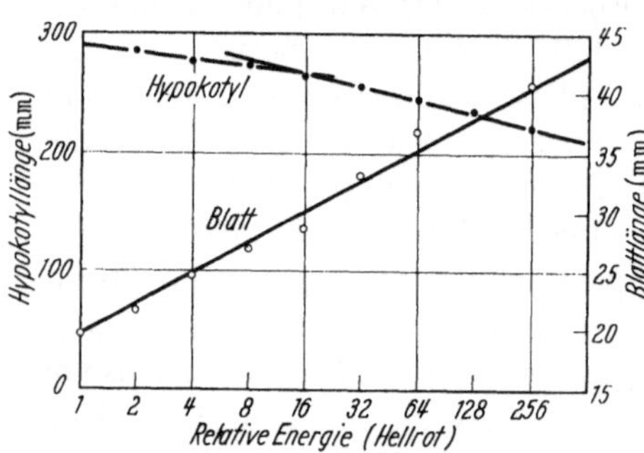

Abb. 6. Blattwachstum und Hypokotylwachstum in Abhängigkeit von der eingestrahlten Energie (Hellrot). Besonders zu beachten ist der Bruch in der log-Energie-Effekt-Kurve für das Hypokotylwachstum. Objekt: Dunkelkeimlinge von *Phaseolus*. [Nach DOWNS (1955)]

Wir haben gesehen, daß die HR-Induktion durch nachfolgendes DR annulliert (reversibilisiert) werden kann. Die eine Folge HR-DR kann nun noch weiter fortgesetzt werden: HR—DR—HR—DR—HR....Es wurde für verschiedene Objekte ausführlich gezeigt, daß dieser Wechsel mehrere Male wiederholt werden kann. Die Reaktion des Objekts hängt dann stets von der zuletzt gegebenen Strahlungsart ab (z. B. Tab. 1). — Wird die Keimung von *Lactuca*-Achänen mit HR induziert und wird anschließend an die HR-Induktion mit einer saturierenden Dosis DR bestrahlt, so hängt die Keimung von dem Zeitraum zwischen dem Ende der HR- und dem Beginn der DR-Bestrahlung ab. Bei den *Lactuca*-Achänen

Tabelle 1. *Die Induktion der Keimung durch Hellrot und die Reversibilisierung der Keiminduktion durch Dunkelrot, mehrmals wiederholt* (*Lactuca*-Achänen, Keimung bei 20° C, Bestrahlungen bei 26° bzw. 6°—8° C). [Nach BORTHWICK u. Mitarb. (1954)]

Bestrahlungsprogramm HR = Hellrot DR = Dunkelrot	Keimung in % bei 20° C nach Bestrahlung bei einer Temperatur von	
	26° C	6°—8° C
HR	70	72
HR—DR	6	13
HR—DR—HR	74	74
HR—DR—HR—DR	6	8
HR—DR—HR—DR—HR	76	75
HR—DR—HR—DR—HR—DR	7	11
HR—DR—HR—DR—HR—DR—HR	81	77
HR—DR—HR—DR—HR—DR—HR—DR	7	12

wurde z. B. gefunden [BORTHWICK u. Mitarb. (1954)], daß etwa 8 Std. nach dem HR 50% der Früchte nicht mehr durch DR gehemmt werden können (20° C). In anderen Fällen ergaben sich andere Werte für den zeitlichen Verlauf des Verlustes an Reversibilität. So kann bei der Beeinflussung der Blütenbildung bei *Xanthium* der HR-Effekt nach 30 min nicht mehr durch DR beeinflußt werden (Abb. 7), und bei Farnsporen kann erst nach etwa 16 Std. bei 50% der induzierten Sporen die HR-Wirkung nicht mehr durch DR annulliert werden (Abb. 8). In den eben zitierten Fällen war die Wirkung des DR unmittelbar nach dem HR

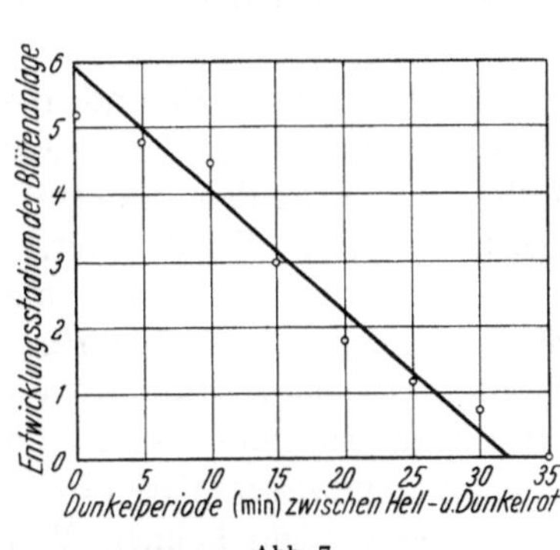

Abb. 7

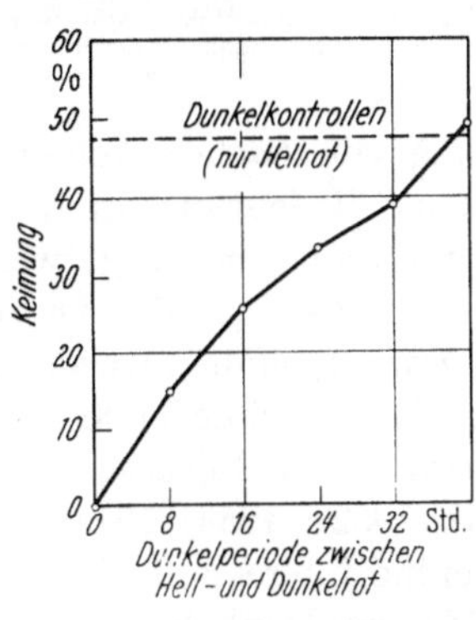

Abb. 8

Abb. 7. Die Reversibilisierung des Hemmeffekts von Hellrot (Zusatzlicht in der Mitte der Dunkelphase) auf die Blütenbildung von *Xanthium pensylvannicum* (Kurztagpflanze) durch nachfolgende Dunkelrotbestrahlung. Der Effekt der Dunkelrot-Dosis hängt von dem zeitlichen Intervall zwischen Hell- und Dunkelrot ab. [Nach DOWNS (1956)]

Abb. 8. Die Abhängigkeit der Hemmwirkung einer bestimmten Dosis Dunkelrot auf die hellrotinduzierte Keimung von Farnsporen (*Dryopteris filix-mas*, 20°C) von dem zeitlichen Intervall zwischen Ende der Hellrot- und Beginn der Dunkelrotbestrahlung. [Nach MOHR (1956)]

am stärksten. KLEIN u. Mitarb. (1957b) geben jedoch für die Öffnungsbewegung der isolierten Plumula-Haken von *Phaseolus* an, daß in diesem Fall der Reversibilisierungseffekt einer bestimmten Dosis DR zuerst ansteigt, nach 1—2 Std. ein Maximum erreicht und dann erst abfällt (Abb. 9). Auch bei der Samenkeimung soll ein entsprechender Effekt beobachtbar sein. Nach KADMAN-ZAHAVI (1957) ist auch hier eine Dosis DR, unmittelbar nach dem HR gegeben, weniger wirksam als wenn eine *kurze* Dunkelperiode (1—2 min) zwischen HR und DR eingelegt wird. Wichtig ist noch der Befund von DOWNS (1956) bei *Xanthium*: Wird während der Dunkelperiode zwischen HR und DR die Temperatur gesenkt (5°), so geht die Reversibilität sehr viel langsamer verloren.

Die Temperatur während der (stets relativ kurzen) Belichtung soll, wenigstens zwischen 6° und 26° C, keinen meßbaren Einfluß auf die

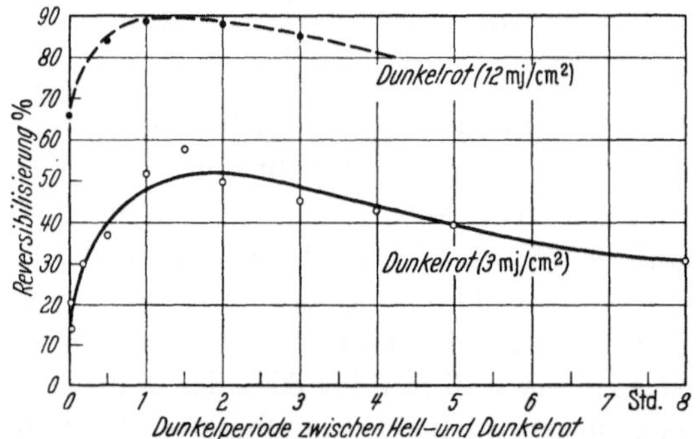

Abb. 9. Die Abhängigkeit des Reversibilisierungseffekts einer bestimmten Dosis Dunkelrot (3 bzw. 12 mj/cm², eingestrahlt während 3 bzw. 12 sec) von dem zeitlichen Intervall zwischen Hellrot- und Dunkelrotbestrahlung. Objekt: Isolierte „Plumula-Haken" von *Phaseolus*. Gemessen wird die Dunkelrot-Reversibilisierung der Hellrot-Induktion der Hakenöffnung. [Nach KLEIN u. Mitarb. (1957b)]

Keimung von *Lactuca*-Achänen haben (Tab. 1). Dieser Befund, dessen allgemeine Bedeutung an weiteren Objekten und für andere Bestrahlungsprogramme bestätigt werden müßte [vgl. z. B. den Einwand von KADMAN-ZAHAVI (1957)], ist von großer theoretischer Wichtigkeit.

In gewissen Fällen wurde nachgewiesen, daß die HR-Induktion auch durch eine anschließende zeitweilige Erhöhung der Temperatur (im Dunkeln) reversibilisiert (annulliert) werden kann [BORTHWICK u. Mitarb. (1952a, 1954), MOHR (1956)]. In diesem Zusammenhang ist auch der Befund interessant, daß *Lactuca*-Achänen gewisser Varietäten, die bei 20° C im Dunkeln keimen können, bei 30° C nicht keimen. Werden diese Früchte einige Zeit bei 30° C gehalten, so keimen sie auch nicht mehr, wenn man sie wieder nach 20° C bringt. Werden sie jedoch nun mit HR bestrahlt, so keimen sie aus [BORTHWICK u. Mitarb. (1952a)].

Einen genauen Vergleich der Reaktionsweise von Dunkelkeimlingen und grünen „Lichtpflanzen" derselben Art auf HR- bzw. DR-Bestrahlung lassen die Untersuchungen von DOWNS (1955) und DOWNS u. Mitarb. (1957) zu. Der Dunkelkeimling reagiert (z. B. hinsichtlich seines Hypokotylwachstums) nicht meßbar auf eine (kurze!) DR-Bestrahlung, wohl aber auf HR. Hat der Dunkelkeimling aber einmal HR erhalten, so kann er auch auf nachfolgendes DR reagieren: DR reversibilisiert die HR-Induktion der Etiolementverhinderung des Hypokotyls. Die im weißen Fluorescenzlicht hoher Intensität (Hauptlichtperiode z. B. 8 Std. täglich bei etwa 14000 Lux) gewachsenen Bohnenpflanzen reagieren (z. B. hinsichtlich des Längenwachstums ihrer Internodien) nicht auf nach der Hauptlichtperiode gebotenes HR, wohl aber auf DR. Nach der Bestrahlung mit DR reagiert die Lichtpflanze nun auch wieder auf HR: Der DR-Effekt wird durch nachfolgendes HR reversibilisiert. Daraus läßt sich allgemein schließen: Der Dunkelkeimling (ebenso z. B. unbelichtete, gequollene *Lactuca*-Früchte oder *Dryopteris*-Sporen) befindet sich (hinsichtlich des reversiblen HR-DR-Reaktionssystems) in einem DR-saturierten Zustand. Die Lichtpflanze dagegen befindet sich am Ende der Hauptlichtperiode in einem HR-saturierten Zustand. In beiden Fällen ist jedoch dasselbe reversible HR-DR-Reaktionssystem wirksam. Der Nachweis einer Wirksamkeit des HR-DR-Systems in grünen „Lichtpflanzen" (vgl. z. B. auch WASSINK und STOLWIJK 1956) zeigt an, daß die zugrunde liegenden Pigmente photostabil sind.

Es ist öfters beobachtet worden, daß die „Empfindlichkeit" von Samen für HR sich in Abhängigkeit von der Quellungszeit ändert [z. B. BORTHWICK u. Mitarb. (1954)]. Die Empfindlichkeit gegenüber HR scheint sich dabei entgegengesetzt zu verhalten wie die gegenüber DR. Neuerdings hat KÖNITZ (1958) gezeigt, daß die Empfindlichkeitsschwankungen der Kurztagpflanze *Chenopodium amaranticolor* gegenüber HR und DR offenbar mit der endogenen Tagesrhythmik zusammenhängen. Die Hemmwirkung einer bestimmten Dosis HR auf die Inflorescenzbildung während der skotophilen Phase ist in der Mitte dieser Phase am stärksten. (Eine entsprechende Dosis DR wirkt nicht meßbar.) DR hemmt die Inflorescenzbildung während der photophilen Phase (Kurztag-Hauptlichtperiode mit Fluorescenz-Weißlicht). Auch hier ist der Effekt einer bestimmten Dosis DR in der Mitte der Phase am größten. Der Effekt von HR kann durch nachfolgendes DR, der von DR durch nachfolgendes HR aufgehoben werden. Um einen bestimmten Hemmeffekt zu reversibilisieren, ist stets dieselbe Dosis HR (in der photophilen Phase) bzw. DR (in der skotophilen Phase) notwendig. Offenbar ist also das reversible HR-DR-Reaktionssystem sowohl in der skotophilen wie auch in der photophilen Phase in der Pflanze wirksam. Der Unterschied der beiden Phasen hinsichtlich dieses Reaktionssystems und hinsichtlich der Blütenbildung läßt sich folgendermaßen charakterisieren: Während der skotophilen Phase ist es für die optimale Blütenbildung notwendig, daß das System DR-saturiert vorliegt. Umgekehrt ist es für die optimale Blütenbildung notwendig, daß während der photophilen Phase das System im HR-saturierten Zustand vorliegt. Man kann sich die Sachlage mit der Annahme leicht verständlich machen, daß die durch HR induzierte Stoffwechselbeeinflussung während der photophilen Phase für die Blütenbildung fördernd, während der skotophilen Phase jedoch nachteilig ist. Eine Erklärung für die relativen Empfindlichkeitsschwankungen während der photophilen bzw. skotophilen Phase wird im nächsten Kapitel kurz diskutiert.

Die dem reversiblen HR-DR-Reaktionssystem zugrunde liegenden Photoreceptoren sind offenbar nicht an bestimmten Stellen der Pflanzen lokalisiert, sondern kommen wohl in Zellen mindestens aller oberirdischen Organe der höheren Pflanzen vor. Dies geht z. B. daraus hervor, daß isolierte Teile der Pflanze wie etwa Segmente der *Avena*-Koleoptile [LIVERMAN und BONNER (1953)], Scheiben aus Bohnenblättern (LIVERMAN u. Mitarb. 1955) oder der isolierte Plumula-Haken von Bohnenkeimlingen [WITHROW u. Mitarb. (1957)] reaktionsfähig sind. Auch in Albino-Dunkelkeimlingen liegt das Reaktionssystem vor [BORTHWICK u. Mitarb. (1951)].

d) Die theoretische Deutung des reversiblen HR-DR-Reaktionssystems

α) Die chemische Natur der dem HR- und dem DR-Effekt zugrunde liegenden Pigmente (Photoreceptoren) ist unbekannt. Für das HR-Pigment wurden verschiedene, z. T. rein spekulative, Ansätze zur Identifizierung gemacht, die jedoch alle nicht überzeugend sind. Man hat an Chlorophyll gedacht [PARKER u. Mitarb. (1946)] oder an Phycocyan-ähnliche Verbindungen [PARKER u. Mitarb. (1950), HENRICKS u. Mitarb. (1956)]. TODD und GALSTON (1954) vermuteten in dem von ihnen aus Samen isolierten Methyl-pyrophaeophorbid a das fragliche HR-Pigment. Jedoch spricht, unter anderem, die sehr starke Absorption dieser Verbindung im Blauviolett, der kein Gipfel im Wirkungsspektrum entspricht, gegen diese Vermutung. Nach den Befunden von DRUMM (1955) an *Tradescantia* (chlorophylldefekt bzw. grün) könnte man annehmen, daß Methyl-Pyrophaeophorbid a nur auftritt, wenn die Chlorophyllbildung aus irgendwelchen Gründen gehemmt ist. Auch Protochlorophyll kommt als HR-Pigment nicht in Frage [WITHROW u. Mitarb. (1957)]. Für die Identifizierung des fraglichen Pigments ist sicherlich die Feinstruktur des HR-Wirkungsspektrums (Abb. 3) bedeutsam. Die beiden Schultern könnten schwache Gipfel des Absorptionsspektrums anzeigen. Cyclische Tetrapyrrole kommen praktisch als HR-Pigment nicht in Frage, denn das fragliche Pigment darf im Blaubereich keine starke Absorption haben, falls man die Annahme aufrechterhält, daß die Quantenausbeute der Reaktion im Bereich des Sichtbaren nicht stark schwankt. Wenn man annimmt, daß das HR-Pigment einen Extinktionskoeffizienten in der üblichen Größenordnung hat, so ist sicher, daß die Konzentration des fraglichen Pigments in den auf HR reagierenden pflanzlichen Organen äußerst gering ist, denn bei spektralphotometrischen Untersuchungen an Gewebe von Albino-Pflanzen (*Hordeum vulgare*) ließen sich keine Anzeichen einer spezifischen Absorption im Rotbereich nachweisen, die man dem HR-Pigment hätte zuweisen können [BORTHWICK u. Mitarb. (1951)]. Auch die Anwendung anderer physikalischer Methoden (Fluorescenz-Untersuchung) zur Erfassung des Pigments hatte bisher keinen Erfolg (HENDRICKS, persönliche Mitteilung). Die offenbar sehr geringe Strahlungsabsorption durch das vorhandene Pigment hat den Vorzug, daß die Problematik einer Selbstbeschattung weitgehend wegfällt.

β) Obwohl also über die chemische Natur der fraglichen Pigmente nichts bekannt ist, kann man sich doch hinsichtlich der photochemischen Reaktionen, die auf die Quantenabsorption in diesen Pigmenten folgen, gewisse, gut begründete Vorstellungen machen. Diese Vorstellungen mögen spekulativen Charakter haben, doch sind sie als Arbeitshypothese unentbehrlich. Bisher haben sie sich sehr bewährt. Aus den experimentellen Befunden geht unmittelbar hervor, daß ein reversibles HR-DR-

Reaktionssystem vorliegen muß. Von HENDRICKS, BORTHWICK u. Mitarb. (seit 1952) ist die Vorstellung entwickelt worden, daß sich das reversible HR-DR-Reaktionssystem durch die Funktion eines reversiblen HR-DR-Pigmentsystems erklären lasse. Der Grundgedanke ist in Kürze folgender (vgl. Abb. 10): Ein HR-Quant wird von dem Pigmentmolekül P_{HR} absorbiert. Das angeregte Molekül kann nun (entweder durch Reaktion mit einem Reaktanten A oder durch Isomerisation) in ein anderes stabiles Molekül übergehen (P_{DR}), das ebenfalls ein Pigment ist und dessen Absorptionsmaximum, verglichen mit P_{HR}, etwas gegen das Langwellige hin verschoben ist. Absorbiert P_{DR} ein Quant, so ist es in der

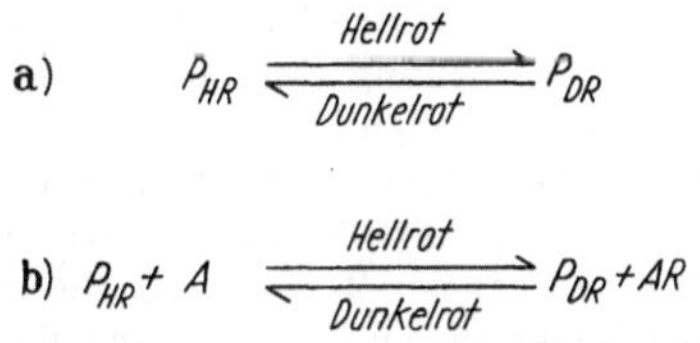

a) $$P_{HR} \underset{\text{Dunkelrot}}{\overset{\text{Hellrot}}{\rightleftharpoons}} P_{DR}$$

b) $$P_{HR} + A \underset{\text{Dunkelrot}}{\overset{\text{Hellrot}}{\rightleftharpoons}} P_{DR} + AR$$

Abb. 10. Zwei mögliche Reaktionsgleichungen des reversiblen Hellrot-Dunkelrot-Pigmentsystems. Die Reaktion nach links geht langsam auch im Dunkeln vonstatten, beschleunigt bei höherer Temperatur. In Gleichung (b) ist $P_{HR} = P_{DR} \cdot R$. (P_{HR}= Pigment mit Absorptionsmaximum bei 660 mμ, P_{DR}=Pigment mit Absorptionsmaximum bei 730 mμ). *AR* könnte ein Wasserstoffdonator sein, *A* ein Wasserstoffacceptor

Lage, sich wieder (entweder durch Reaktion mit dem Reaktanten AR oder durch Isomerisation) in das ursprüngliche Molekül P_{HR} zurückzuverwandeln. Kurz: es liegt ein reversibles Pigmentsystem vor. Zwei mögliche Reaktionsgleichungen zeigt Abb. 10. — P_{DR} ist die hinsichtlich einer Umsteuerung des Stoffwechsels aktive Form. Der praktisch ausschließlichen Anwesenheit von P_{HR} in der Zelle müssen wir den „Dunkelhabitus" des pflanzlichen Organismus (z. B. ruhender Same, Etiolement) zuordnen. Die biochemischen Reaktionen, die durch die Anwesenheit von P_{DR} möglich werden, äußern sich letztlich in den beobachtbaren Photomorphosen. Daß P_{DR} die physiologisch aktive Form ist, geht z. B. daraus hervor, daß die HR-Induktion stets nach einiger Zeit nicht mehr reversibilisiert werden kann. P_{DR} hat die von ihm ausgehenden Stoffwechseländerungen in Gang gebracht[1].

Da man noch nicht in der Lage ist, die Pigmente biochemisch oder biophysikalisch zu fassen, ist man einstweilen darauf angewiesen, die physiologische Reaktion (das quantitative Ausmaß der Photomorphose also) als Maß für die zugrunde liegende, photochemische Reaktion zu verwenden. Die große Schwierigkeit dabei ist, daß die quantitative Abhängigkeit der physiologischen Reaktion (z. B. Mittelwert einer Population) von der Menge an P_{DR} (ebenfalls Mittelwert einer Population) bekannt sein sollte. HENDRICKS u. Mitarb. (1956) haben versucht, für einige physiologische Reaktionen (z. B. Internodienwachstum von Lichtpflanzen, *Phaseolus*

[1] Einen direkten Beweis, daß nur das P_{DR} (und nicht das P_{HR}) die „aktive" Form sein kann, liefern die später zu besprechenden Befunde über die Wiederherstellung der HR-„Empfindlichkeit" nach einem saturierenden HR-„Lichtstoß" [MOHR (1959a)].

vulgaris) einen entsprechenden theoretisch begründeten Zusammenhang zwischen der relativen Menge der Pigmente und der beobachteten Reaktionsgröße aufzustellen. Dieser Analyse liegt aber bereits die Annahme zugrunde, daß die photochemische Reaktion im Sinn der Abb. 10 reversibel und in beiden Richtungen eine Reaktion 1. Ordnung (hinsichtlich der eingestrahlten Energie) ist.

Es wird also angenommen, daß das photochemische Reaktionsgeschehen in beiden Richtungen dem Gesetz einer Reaktion 1. Ordnung (hinsichtlich der eingestrahlten Energie) folgt [Hendricks u. Mitarb. (1956), Downs u. Mitarb. (1957)]. Die bei *Lactuca*-Achänen gefundene Temperaturunabhängigkeit der Reaktionen zwischen 6° und 26° C kann man als Hinweis dafür ansehen, daß wahrscheinlich die Reaktionsrate nicht durch Molekülzusammenstöße limitiert wird. Andererseits halten neuerdings Hendricks, Borthwick u. Mitarb. [z. B. Toole u. Mitarb. (1955a), Hendricks u. Mitarb. (1956), Downs und Mitarb. (1957)] die Gleichung (b) in Abb. 10 für die passendere. Die Frage ist nun, wie sich die Vorstellung einer Reaktion 1. Ordnung mit der Vorstellung einer bimolekularen Reaktion verträgt. Ein Beispiel für das Reaktionsgeschehen im Sinn der Gleichung (b) ist die, freilich nicht reversible, photochemische Bildung von Chlorophyll a aus Protochlorophyll. Diese Reaktion wird heute als eine bimolekulare angesehen. Sie gehorcht jedoch als solche dem Gesetz für eine Reaktion 2. Ordnung. Die Beteiligung von Molekülzusammenstößen an dieser Reaktion wird auch durch die (allerdings zwischen 0° und 20° relativ geringe) Temperaturabhängigkeit angezeigt [Smith und Benitez (1954), Virgin (1955); vgl. auch Rabinowitch (1956)]. Toole u. Mitarb. (1955a) und Hendricks u. Mitarb. (1956) ziehen die Annahme eines bimolekularen Reaktionsgeschehens insbesondere deshalb vor, weil es möglich ist, durch bestimmte Substanzen, z. B. KNO_3, die Lichtempfindlichkeit von Samen zu ändern. Im allgemeinen steigt die Empfindlichkeit für HR an, wenn die Empfindlichkeit für DR erniedrigt wird und umgekehrt. Ähnliche Änderungen der Empfindlichkeit für HR und DR lassen sich auch beobachten, wenn die Vorquellungszeit variiert wird. Es liegen Gründe für die Annahme vor, daß bei der Empfindlichkeitsänderung der für eine Keimungsauslösung notwendige Schwellenwert an P_{DR} nicht verändert wird, sondern daß sich die Quantenausbeute der Pigmentumwandlung ändert bei konstant bleibendem Pigmentgehalt. Diesen möglichen Veränderungen der Quantenausbeute suchen Toole u. Mitarb. (1955a) und Hendricks u. Mitarb. (1956) durch die Einführung der Reaktanten [Abb. 10 (b)] gerecht zu werden. Die Reaktanten sollen in großem Überschuß vorhanden sein, verglichen mit der Konzentration der Pigmente. Da das Reaktionsgeschehen temperaturunabhängig und von 1. Ordnung sein soll, so wird die zusätzliche Annahme gemacht, daß Reaktanten und Pigmentmoleküle sich in beständiger, engster Nachbarschaft befinden. Die Assoziationskonstanten für die 4 möglichen Kombinationen von Pigmentmolekülen und Reaktanten seien gleich. P_{HR} kann aber nur mit A, P_{DR} nur mit AR reagieren. Daraus folgt, daß die Quantenausbeute in jeder Richtung von dem Verhältnis der Konzentrationen der beiden Reaktanten abhängt. Wir können in Kürze sagen (ohne weiter auf die Schwierigkeiten dieses postulierten Reaktionsschemas einzugehen): Ein Reaktionsgeschehen 1. Ordnung (im Hinblick auf die eingestrahlte Energie) schließt ein, daß *ein* Quant in der Lage ist, die Photoreaktion ablaufen zu lassen. Freilich ist dies nicht so zu verstehen, daß *jedes* absorbierte Quant die photochemische Reaktion auslöst, sondern es ist damit gesagt, daß eine bestimmte Wahrscheinlichkeit besteht, daß ein absorbiertes Quant wirksam ist. Die Reaktanten A bzw. AR beeinflussen, so kann man sich vorstellen, durch ihre relative Konzentration diese Wahrscheinlichkeit, variieren also die Quantenausbeute der Reaktionen. Bestrahlt man mit monochromatischer Strahlung einer bestimmten Wellenlänge, die von beiden Pigmenten absorbiert werden kann (Überschneidung der Absorptionsspektren), so bestimmt das Verhältnis der Konzentrationen der Reaktanten die Lage des sich zwischen den beiden Pigmenten einstellenden dynamischen Gleichgewichts.

Die Quantenausbeute in beiden Richtungen könnte aber auch durch andere Einflüsse, z. B. durch die Konzentration fluorescenzlöschender Moleküle (quencher) beeinflußt werden. Die beobachteten Änderungen der Quantenausbeute könnten wohl auch dadurch erklärt werden. Die Photoreaktion als solche kann dabei monomolekular sein [Abb. 10 (a)].

Mit einer Beeinflussung der Quantenausbeute bei konstant bleibendem Pigmentgehalt lassen sich übrigens auch die im letzten Abschnitt kurz erwähnten Befunde über Empfindlichkeitsschwankungen gegenüber HR bzw. DR während der photo- bzw. skotophilen Phase erklären. Auch die Befunde von KLEIN u. Mitarb. (1957b) und KADMAN-ZAHAVI (1957), wonach die „Empfindlichkeit" für DR unmittelbar nach dem HR geringer ist als einige Zeit später (Abb. 9), können so gedeutet werden, daß die Quantenausbeute der Rückverwandlung zunächst geringer ist und dann ansteigt. Sie widersprechen also nicht unmittelbar der Vorstellung des reversiblen Pigmentsystems. Die der Änderung der Quantenausbeute zugrunde liegenden Reaktionen wären thermische Dunkelreaktionen. So würde sich auch die von KADMAN-ZAHAVI (1957) und KLEIN u. Mitarb. (1957c) gefundene Beeinflußbarkeit der DR-Reaktion durch die Temperatur erklären lassen.

Offenbar sind sowohl Gleichung (a) wie Gleichung (b) in Abb. 10 auf Grund der bisherigen experimentellen Resultate als Arbeitshypothesen geeignet. Um der einfacheren Formulierung willen ziehen wir die Gleichung (a) vor. Wir nehmen also (wenigstens formal) an, die photochemische Reaktion bestehe in einer phototropen Isomerisierung[1].

Der Ablauf der Reaktionen bei der Induktion der Keimung eines lichtbedürftigen Samens wäre nun folgender: In dem unbelichteten gequollenen Samen liegt das Gleichgewicht des reversiblen HR-DR-Pigmentsystems stark auf der linken Seite, so daß fast nur P_{HR} vorliegt. Damit in einem Samen die zur Keimung führenden Prozesse in Gang kommen können, ist ein bestimmter Schwellenwert an P_{DR} notwendig (Die Samenkeimung ist eine Alles-oder -Nichts-Reaktion). Wird der Same mit HR bestrahlt, so wird P_{HR} in P_{DR} umgewandelt. Wird dabei der Schwellenwert an P_{DR} erreicht, so können die zur Keimung führenden Prozesse einsetzen. Wird mit DR bestrahlt, ehe die zur Keimung führenden Prozesse angelaufen sind, so wird wegen der Rückwandlung von P_{DR} (Unterschreitung des Schwellenwertes an P_{DR}) die Keimung verhindert (Annullierung der HR-Induktion).

Die Umwandlung von P_{DR} in P_{HR} wird offenbar durch länger andauernde Temperaturerhöhung gefördert. [Wenigstens ist dies eine mögliche Erklärung der diesbezüglichen Befunde von BORTHWICK u. Mitarb. (1952a, 1954) und MOHR (1956)]. Die Einstellung des entsprechenden Dunkelgleichgewichts ist beschleunigt, verglichen mit normaler Temperatur, wo nur eine sehr langsame Umwandlung nach links vonstatten geht

[1] Bei reversiblen phototropen Umwandlungen braucht es sich nicht immer um Isomerisationen zu handeln [vgl. z. B. die Diskussion bei LINDEMANN (1955)].

[vgl. Borthwick u. Mitarb. (1954)]. — Wir haben gesehen, daß zahlreiche Photomorphosen des pflanzlichen Organismus auf die Funktion desselben reversiblen HR-DR-Reaktionssystems zurückgehen. Wir müssen uns vorstellen, daß die Anwesenheit von P_{DR} in der Zelle zu fundamentalen Umsteuerungen des Stoffwechsels führt. Als Resultat dieser Umsteuerung entstehen letztlich die von uns meßbaren Photomorphosen. Wie P_{DR} die Umschaltung des Stoffwechsels bewirkt, wissen wir nicht. Vielleicht sind Befunde, wonach zuweilen die HR-Wirkung durch gewisse Verbindungen, z. B. Kinetin, wenigstens teilweise ersetzt werden kann, für ein kausales Verständnis der Lichtwirkung von Bedeutung [Miller (1956), Hillman (1957); vgl. jedoch Miller (1958)]. Auch die Wirkung von Gibberellinsäure im Zusammenhang mit HR-DR-Wirkungen ist zu beachten [z. B. Kahn u. Mitarb. (1956), Downs u. Mitarb. (1957), Scott und Liverman (1957), Brian (1958). Es ist auch möglich, daß das reversible Pigmentsystem Teil eines Enzymsystems ist, das aktiviert wird, wenn P_{DR} entsteht

Die durch P_{DR} ermöglichte Umsteuerung des Stoffwechsels kann offenbar für dieselbe Photomorphose (z. B. Blütenbildung von Kurz- bzw. Langtagpflanzen) ganz verschiedene Konsequenzen haben, je nachdem, in welchem physiologischen Zustand sich der Organismus zum Zeitpunkt der Umsteuerung befindet [vgl. z. B. Parker u. Mitarb. (1946), Borthwick u. Mitarb. (1948), Leopold und Guernsey (1954), Könitz (1958)]. Bemerkenswert ist ferner, daß dieselbe pflanzliche Reaktion (z. B. Samenkeimung) in einem Fall von der HR-Induktion abhängt, im anderen Fall nicht. So sind z. B. die Samen von *Lepidium virginicum* obligate Lichtkeimer, die von *Sinapis alba* dagegen keimen im Dunkeln ebenso gut wie im weißen Fluorescenzlicht oder im DR. Das reversible HR-DR- Pigmentsystem ist jedoch in *Sinapis alba*-Keimlingen durchaus vorhanden und wirksam, wie die Untersuchung zahlreicher anderer Photomorphosen gezeigt hat. Bei *Sinapis alba* kann die Keimung offenbar ablaufen, ohne daß es zu einer erheblichen Bildung von P_{DR} kommt, denn im jungen Keimling liegt wenigstens das Gleichgewicht des reversiblen Pigmentsystems ganz auf der linken Seite [vgl. hierzu auch Toole u. Mitarb. (1955b)]. In unter den üblichen Keimbedingungen lichtunempfindlichen *Lactuca*-Achänen können durch mancherlei Agentien (z.B. hohe oder sehr tiefe Temperatur, Röntgenstrahlen, Chemikalien) Bedingungen geschaffen werden, unter denen die Wirksamkeit des HR-DR-Systems wieder nachgewiesen werden kann [z. B. Borthwick u. Mitarb. (1954), Haber (1958)]. Andererseits braucht natürlich P_{DR} nicht der einzige limitierende Faktor für die Auslösung der Keimung zu sein. Bei *Lepidium virginicum* z. B. ist bei gleicher HR-Behandlung der Prozentsatz der Keimung größer, wenn die Temperatur nicht konstant gehalten, sondern wenn ein bestimmter Temperaturwechsel vorgenommen wird [Toole u. Mitarb. (1955a, b)].

Die angenommene reversible Pigmentumwandlung ist molekülphysikalisch durchaus möglich. Welche angeregten Zustände und welche Übergänge dabei beteiligt sein könnten, ist unbekannt, da wir das Pigment nicht kennen und da bisher weder Fluorescenz noch eine Radikalbildung nachgewiesen wurden. Diskussionen darüber sind rein spekulativ [vgl. BORTHWICK u. Mitarb. (1954) und HENDRICKS u. Mitarb. (1956)]. Das Wissen um die Existenz reversibler Pigmentsysteme ist keineswegs neu. Derartige Erscheinungen werden von den Physikochemikern seit langem studiert. Um zu zeigen, wie berechtigt man ist, das von uns auf Grund unserer experimentellen physiologischen Befunde formulierte reversible HR-DR-Pigmentsystem mit einem phototropen Pigmentsystem zu vergleichen, sei nur auf die klassischen Untersuchungen von STOBBE (1908) über die Phototropieerscheinungen bei Fulgiden (im festen Aggregatzustand) hingewiesen: Wird z. B. das im Dunkeln orange-gelbe (also den grün-blau-violetten Teil des sichtbaren Spektrums absorbierende) Triphenylfulgid mit Blaulicht bestrahlt, so geht es in eine chemisch identische Form über, die blau aussieht (also den längerwelligen Teil des sichtbaren Spektrums absorbiert). Wird dieses blaue Fulgid mit Gelb-Rot bestrahlt, so geht es wieder in die orange-gelbe Form über. Im Grün-Gelb überschneiden sich die Absorptionsspektren der beiden Formen. Wird in diesem Bereich bestrahlt, so stellt sich ein dynamisches Gleichgewicht ein. Ebenso stellt sich im „Weißlicht" ein dynamisches Gleichgewicht ein, das von den Eigenschaften der beiden Formen und von der Zusammensetzung des „Weißlichts" abhängt. Allgemein fand STOBBE, daß zur Umwandlung der „Dunkelform" kürzere Wellenlängen nötig waren als zur Rückverwandlung der unter Lichteinwirkung entstandenen Form. STOBBE hat schon die Vorstellung geäußert, daß bei einer Verbindung, bei der das Hellrot die phototrope Umwandlung auslöst, die Rückwandlung durch Dunkelrot-nahes Infrarot erfolgen muß. Temperaturerhöhung begünstigt bei dem Triphenylfulgid den Übergang von der blauen zur gelborangen Form. Ähnliche Befunde über phototrope Pigmente sind seit STOBBE öfters gemacht worden. [Neuere Anschauungen und Literaturzusammenfassung z. B. bei LINDEMANN (1955)]. So fanden z. B. WYMAIN und BRODE (1951) bei Thioindigofarbstoffen eine reversible phototrope cis-trans- Isomerisierung in Lösung. Es sei noch erwähnt, daß auch für das bimolekulare Reaktionsgeschehen, [Abb. 10 (b)], sehr gut passende Beispiele aus der physikalischen Chemie bekannt sind [z. B. OSTER und WOTHERSPOON (1954)].

Die Deutung des beobachteten HR-DR-Reaktionssystems durch das reversible HR-DR-Pigmentsystem befriedigt nach meiner Auffassung die gegenwärtig vorliegenden Versuchsdaten am besten. Ob neue, überraschende experimentelle Resultate einmal eine neue Deutung erzwingen werden, ist eine Frage der Zukunft.

Freilich gibt es gegenwärtig auch noch andere theoretische Möglichkeiten; so kann z. B. die Vorstellung geäußert werden, es lägen zwei voneinander unabhängige Pigmente vor, die als Sensibilisatoren fungieren. Durch die Energieabsorption in dem einen Pigment würde ein bestimmter Stoffwechselvorgang ausgelöst. Dieser würde von dem durch die Energieabsorption in dem anderen Pigment induzierten Stoffwechselprozeß annulliert. Eine korrekte Deutung aller bisherigen Versuchsresultate auf der Basis einer solchen Grundvorstellung dürfte jedoch schwierig sein, z. B. die Deutung des Befundes, daß man den Induktionseffekt eines nur wenige Sekunden langen „HR-Stoßes" sofort wieder reversibilisieren kann. Alle mit der Theorie des reversiblen Pigmentsystems zunächst nicht ohne weiteres zu vereinbarenden Befunde (etwa KLEIN und PREISS 1958) müssen jedenfalls sorgfältig beachtet werden. Versuche über die Wirkung intermittierenden Lichts (Lichtblitze) liegen bisher nicht vor. Sie würden vielleicht wichtige Beiträge zu der Grundfrage liefern, ob das angenommene Reaktionsschema richtig sein kann.

γ) Wir wissen, daß Dunkelkeimlinge oder im Dunkeln gequollene Samen oder Sporen usw. HR-, nicht aber DR-empfindlich sind. Dies bedeutet, daß in diesen Fällen im Dunkeln das Gleichgewicht der Reaktion ganz auf der linken Seite liegt. Wenn man nun mit HR saturierend bestrahlt, so wird ein gewisser Prozentsatz des P_{HR} in das P_{DR} umgewandelt. Das sich einstellende dynamische Gleichgewicht der photochemischen Reaktion hängt von der spektralen Zusammensetzung des HR ab. Das P_{HR} stand vor der Belichtung in einem Gleichgewicht mit seiner Vorstufe, die wir $(P_{HR})_V$ nennen wollen. Dieses Gleichgewicht ist nun durch die (wenigstens partielle) Umwandlung von P_{HR} in P_{DR} gestört. Die Folge ist, daß P_{HR} aus seinen Vorstufen nachgebildet wird. Dadurch wird die „Empfindlichkeit" für HR wieder hergestellt. Bei erneuter HR-Bestrahlung findet eine erneute, dem im HR sich einstellenden Gleichgewicht entsprechende Umwandlung statt. Die Menge des entstandenen P_{DR} wird dadurch weiter erhöht [Mohr (1959a)]. Bei verschiedenen Organismen kann offenbar die Geschwindigkeit der Nachbildung etwas verschieden sein. Ein wesentlicher, direkter Einfluß des Lichts auf die Nachbildung von P_{HR} aus $(P_{HR})_V$ scheint in den ersten Stunden nach Einsatz der Belichtung nicht vorhanden zu sein [Mohr (1959a)]; nimmt man jedoch an, daß bei längerer Dauer des Lichts die Nachbildung gehemmt sein kann, so lassen sich vielleicht die Ergebnisse von Isikawa (1954) erklären, wonach die von ihm untersuchten Samen im Langtag bzw. Kurztag besser keimen als im Dauerlicht. Auch andere Fälle, wo durch Dunkelperioden getrennte Lichtstöße stärker wirken als Dauerlicht [z. B. Binet (1956), Isikawa und Oohusa (1956)], lassen sich so erklären.

Diese Vorstellungen über die Nachbildung von P_{HR}, die Mohr (1959a) zur Erklärung seiner entsprechenden, experimentellen Resultate entwickelt, deuten also die Befunde [z. B. auch Borthwick (1957), Toole u. Mitarb. (1958a, b)] über die Ersetzbarkeit längerer Bestrahlungszeiten durch kurze, saturierende Lichtstöße, die durch längere Dunkelperioden getrennt sind. Auf was es offenbar ankommt, ist die Menge an gebildetem P_{DR}. Man kann sich z. B. leicht vorstellen, daß für die Auslösung der Samen- bzw. Sporenkeimung oder der Haarbildung aus Epidermiszellen [Mohr (1959a)] ein bestimmter Schwellenwert an P_{DR} erreicht sein muß (Alles-oder-Nichts-Reaktion). Ist die Menge an P_{HR} in einer unbelichteten Epidermiszelle z. B. nicht entsprechend groß, so kann dieser Schwellenwert nicht mit einer einzigen Umwandlung des vorhandenen P_{HR} erreicht werden, sondern es bedarf mehrerer, saturierender HR-Belichtungen, zwischen denen die Zelle Gelegenheit haben muß, P_{HR} aus seinen Vorstufen nachzubilden.

Ein glänzendes Beispiel für die Richtigkeit der hier vorgetragenen Anschauungen über das reversible HR-DR-Pigmentsystem und über die

Nachbildung von P_{HR} ist auch die Samenkeimung bei *Paulownia tomentosa* [Toole u. Mitarb. (1958a)]. Eine Belichtung dieser obligaten Lichtkeimer mit einer HR-Dosis, die für 100% Keimung bei *Lepidium*-Samen ausreicht, bewirkt hier praktisch nichts. Wird jedoch für 48 Std. Dauerlicht gegeben, so ist die Keimung induziert. Das Dauerlicht kann durch mehrere saturierende Lichtstöße, über den enstprechenden Zeitraum verteilt, ersetzt werden. Werden nach 48 Std. Dauerlicht ein paar Minuten DR gegeben, so tritt keine Keimung ein. Gibt man nach dem DR wieder einige Minuten HR, so tritt volle Keimung ein. Die Deutung: Im Verlauf von 48 Std. wird bei der dauernden Umwandlung soviel P_{HR} nachgebildet, daß der Schwellenwert an P_{DR} bei den meisten Samen erreicht wird. Die zu Beginn der Belichtung vorliegende Menge von P_{HR} reicht dagegen nicht zur Erzielung des Schwellenwerts an P_{DR} aus. Nach 48 Std. Licht ist also der Schwellenwert an P_{DR} überschritten. Wird nun mit DR belichtet, so wird ein Großteil dieses P_{DR} wieder in P_{HR} verwandelt und der Schwellenwert unterschritten. Belichtung mit HR wandelt das P_{HR} wieder in P_{DR} um, wodurch der Schwellenwert wieder überschritten wird.

Eine gewisse Rolle könnte in diesem Zusammenhang auch die im Dunkeln, besonders bei höherer Temperatur, mögliche langsame Umwandlung von P_{DR} in P_{HR} spielen. Ob bei 20°C diese Dunkelreaktion bei Einstrahlung von „reinem" HR (damit ist solches HR gemeint, das vom P_{DR} nicht absorbiert werden kann!) als Gegenreaktion ein meßbares Ausmaß hat und damit die Einstellung eines Gleichgewichts erzwingt, ist nicht klar. Jedenfalls dürfte, falls überhaupt bei saturierender Einstrahlung von reinem HR infolge der Dunkelreaktion ein Gleichgewicht vorhanden ist, das Gleichgewicht ganz auf der rechten Seite der Gleichung liegen.

Wird ein breites Band HR bzw. DR eingestrahlt, so muß man erwarten, daß beide Pigmente die Strahlung absorbieren und daß sich deshalb ein der spektralen Verteilung des eingestrahlten Lichts entsprechendes Gleichgewicht des HR-DR-Pigmentsystems einstellt. Dies kann dazu führen, daß auch DR leicht im Sinn von HR wirkt, obgleich dasselbe DR eine starke HR-Induktion reversibilisiert [vgl. Mohr (1959a)]. Die einfache Erklärung: In den Epidermiszellen des Hypokotyls von *Sinapis alba*-Dunkelkeimlingen liegt, wie auch sonst, das Gleichgewicht des Pigmentsystems ganz auf der linken Seite. Der für eine Auslösung der Haarbildung aus diesen Epidermiszellen notwendige Schwellenwert an P_{DR} (d.h. die für die Haarentstehung aus bestimmten Zellen notwendige Menge an gebildetem und nicht wieder zurückverwandeltem P_{DR}) ist relativ niedrig. Strahlt man nun Standard-DR ein (Glühlampe konstanter Emission, Wasserfilter, je 2 Lagen roten und blauen Dupont-Cellophan), so stellt sich ein entsprechendes Gleichgewicht der Pigmente ein, das zwar auch noch stark auf der linken Seite liegt (vgl. immer Abb. 10), aber doch zur Bildung von so viel P_{DR} führt, daß bei einigen Zellen der Schwellenwert überschritten wird. Strahlt man statt DR

Standard-HR ein, so stellt sich ein Gleichgewicht ein, das in sehr vielen Zellen zur Überschreitung des Schwellenwerts führt. Wird unmittelbar nach dem HR mit DR saturierend bestrahlt, so stellt sich wieder das dem DR entsprechende Gleichgewicht ein, d. h. es kommt zu einer Reversibilisierung der HR-Induktion. Diese geht natürlich nicht bis zu dem (sehr kleinen) Dunkelwert, sondern bis zu dem Wert, der durch eine einmalige Bestrahlung mit DR allein erzielbar ist (vgl. Abb. 11 und Abb. 12). Wird einmal mit einem saturierenden DR-Lichtstoß das Gleichgewicht eingestellt, so wird, wegen der (in diesem Fall nur leichten)

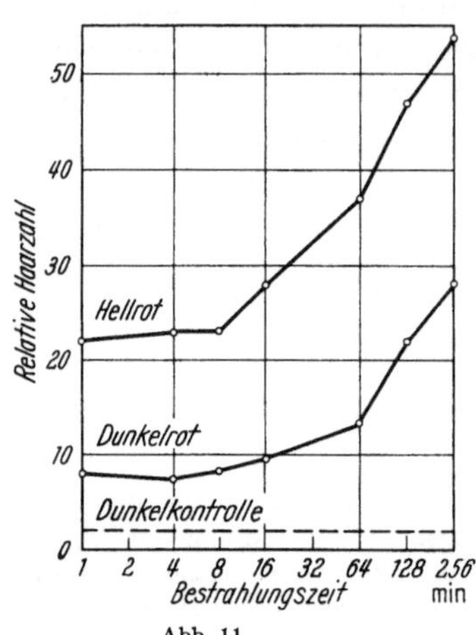

Abb. 11

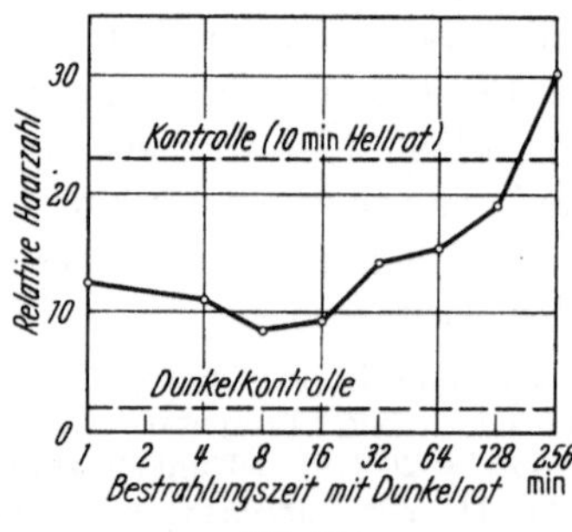

Abb. 12

Abb. 11. Die „Relative Haarzahl" (lichtabhängige Haarbildung aus Epidermiszellen des Hypokotyls der Dunkelkeimlinge von *Sinapis alba*) in Abhängigkeit von der Bestrahlungszeit im Standard-Hellrot- bzw. Standard-Dunkelrot-Feld. Das dem eingestrahlten Hellrot bzw. Dunkelrot entsprechende Gleichgewicht des reversiblen Hellrot-Dunkelrot-Pigmentsystems ist nach 1 min schon praktisch eingestellt. [Nach MOHR (1959a)]

Abb. 12. Die Reversibilisierung der Hellrot-Induktion (10 min Hellrot-Standard-Feld) durch unmittelbar folgende Dunkelrotbestrahlung (Objekt wie in Abb. 11). [Nach MOHR (1959a)]

Störung des Gleichgewichts zwischen P_{HR} und $(P_{HR})_V$, wieder P_{HR} nachgebildet. Nach einiger Zeit kann man also auch mit DR das Gleichgewicht erneut einstellen usw. Die (relativ schwachen) „HR-Effekte" des verwendeten DR lassen sich also in genau der Weise akkumulieren, wie man auf Grund der Theorie erwarten muß [vgl. MOHR (1959a)]. Wenn die Vorstellungen über die Wiederherstellung der HR-„Empfindlichkeit" infolge einer Nachbildung des P_{HR} aus seinen Vorstufen richtig sind, so muß eine Periode genügend tiefer Temperatur (+ 1° C), die unmittelbar auf die erste saturierende HR-Bestrahlung folgt, die Restaurierung der HR-Empfindlichkeit weitgehend verhindern. Dies ist tatsächlich der Fall (MOHR, 1959a).

e) Die Niederenergiereaktion und die Hochenergiereaktion

Es möchte scheinen, als sei mit dem offenbar universell verbreiteten HR-DR-Pigmentsystem dasjenige System erfaßt, auf das das gesamte photomorphogenetische Geschehen unter dem Einfluß längerwelligen

Lichts zurückgeführt werden könnte. Indessen gibt es in der modernen Literatur zum HR-DR-System zahlreiche, meist versteckte Angaben, aus denen man erschließen kann, daß noch andere Reaktionssysteme an der durch längerwelliges Licht ausgelösten Photomorphogenese beteiligt sein müssen.

Zweifellos dominiert bei Einstrahlung geringer Energiemengen das reversible HR-DR-Reaktionssystem. Wir nennen diese Reaktion deshalb „Niederenergiereaktion" = NER. Bei Einstrahlung großer Energiemengen (besonders ausgeprägt, wenn diese Energiemengen in kurzer Zeit eingestrahlt werden) tritt jedoch neben dem reversiblen HR-DR-Reaktionssystem mindestens noch ein anderes Reaktionssystem in Aktion („Hochenergiereaktion" = HER). In neuester Zeit ist es gelungen, durch komplizierte kinetische Studien ein derartiges Reaktionssystem, das erst bei hoher Energieeinstrahlung sich bemerkbar macht und das z. B. neben dem HR-DR-System die Photomorphogenese der Keimlinge von *Sinapis alba* steuert, zu isolieren und durch sein Wirkungsspektrum, das sich vom Wirkungsspektrum des HR-DR-Reaktionssystems grundsätzlich unterscheidet, zu charakterisieren. Wir dürfen mit einigem Recht vermuten, daß dieses System weit verbreitet ist. Darauf weisen zahlreiche Angaben in der Literatur hin, die bisher nicht interpretiert werden konnten und die sich, wenn man eine Funktion des „Hochenergie-Reaktionssystems" in Betracht zieht, leicht deuten lassen [z. B. WITHROW u. Mitarb. (1953), HENDRICKS und BORTHWICK (1954), DOWNS (1955, 1956), CATHEY und BORTHWICK (1957,) KÖNITZ (1958)]. Im nächsten Abschnitt werden einige der bisherigen experimentellen Befunde hinsichtlich dieses „Hochenergie-Reaktionssystems" zusammengefaßt.

4. Das Blau-Dunkelrot-Reaktionssystem

a) Einführung

Es wurde im letzten Abschnitt erwähnt, daß die Haarbildung aus gewissen Epidermiszellen des Hypokotyls der Dunkelkeimlinge von *Sinapis alba* durch das reversible HR-DR-Reaktionssystem beeinflußt wird [MOHR (1959a)]. Werden für HR und DR (streng standardisierte Lichtfelder) Energie-Effekt-Kurven aufgestellt, so ergeben sich bei einer Bestrahlungszeit größer als 8—10 min komplizierte Verhältnisse (Abb. 11 und 12). Bei kurzen Bestrahlungszeiten (bis etwa 10 min) lassen sich die Verhältnisse einfach mit dem reversiblen HR-DR-System erklären. Die Frage ist, ob sich der sekundäre Anstieg, der sowohl im HR wie auch im DR eintritt, ebenfalls auf die Wirkung des HR-DR-Systems zurückführen läßt (Nachbildung von P_{HR} und laufende Umwandlung zu P_{DR} im Rahmen des durch die beiden Strahlungstypen bedingten Gleichgewichts). Es muß also festgestellt werden, welcher Effekt im Verlauf von 4 Std. mit saturierenden Lichtstößen maximal zu erzielen ist (Tab. 2 und 3). Aus den Tabellen geht hervor, daß im HR fast der gesamte Effekt von 4 Std. Bestrahlung auf das Konto des reversiblen HR-DR-Pigmentsystems geht, daß jedoch im DR ein beträchtlicher Effekt übrig bleibt, der sich nicht auf dieses System zurückführen läßt. Das Problem war, diesen Effekt zu erklären.

Es hat sich bei der genauen Untersuchung verschiedener, leicht meßbarer Photomorphosen (Etiolementverhinderung des Hypokotyls, Haarbildung, Anthocyanbildung, Kotyledonenwachstum) der Keimlinge von *Sinapis alba* gezeigt [MOHR (1957), MOHR (1959a), MOHR (1959b)], daß sehr einheitliche Verhältnisse vorliegen. In allen Fällen konnte der beobachtete Gesamteffekt der Strahlung nicht ausschließlich auf die Wirkung des reversiblen HR-DR-Reaktionssystems zurückgeführt werden, sondern es wurde die Annahme nötig, daß ein weiteres Reaktionssystem vorliegt, über das sich zwar dieselben photomorphogenetischen Erscheinungen induzieren lassen wie über das reversible HR-DR-System, das jedoch erst bei Einstrahlung relativ hoher Energie registrierbar wird. Es sind also (mindestens!) zwei Reaktionssysteme an der Photomorphogenese der jungen Dunkelkeimlinge von *Sinapis alba* beteiligt: Einmal das reversible HR-DR-Reaktionssystem, das bei zeitlich adäquat verteilter Einstrahlung geringer Energiemengen die Photomorphogenese weitgehend allein durchführen kann (MOHR, noch unveröffentlicht). Andererseits aber ist ein Reaktionssystem vorhanden, über das sich Strahlung höherer Intensität und längerer Dauer auf die Photomorphogenese auswirken kann. Bei den Bestrahlungsverhältnissen, wie sie in der Natur normalerweise vorliegen, liefern beide Systeme gleichzeitig einen Beitrag zu der beobachteten Photomorphogenese. Wir nennen das der „Hochenergiereaktion" zugrunde liegende Reaktionssystem das „Blau-Dunkelrot-Reaktionssystem", weil, wie wir gleich sehen werden, das Wirkungsspektrum dieser Reaktion starke Gipfel im Blau und im dunkelroten Spektralbereich zeigt.

b) Das Wirkungsspektrum des Blau-Dunkelrot-Reaktionssystems

Bei *Sinapis alba* läßt sich die Hochenergiereaktion (HER) nicht unabhängig von der Niederenergiereaktion (NER) untersuchen. Immer wenn man (mit hoher Energieeinstrahlung) die HER untersucht, beeinflußt man auch die NER. Will man also die HER näher charakterisieren, z. B. das Wirkungsspektrum aufstellen, so muß man den Einfluß der NER weitgehend eliminieren. Dies kann dadurch geschehen, daß man, am besten für jede verwendete Wellenlänge, feststellt, wie weit bei einer gewissen Energieeinstrahlung und bei einem gewissen Bestrahlungsprogramm das erzielte Gesamtresultat auf die NER zurückgeht. Durch eine Subtraktion dieses „Korrekturfaktors" vom Gesamtresultat läßt sich der Anteil ermitteln, der auf das Konto der HER geht. (Das einfache Prinzip einer solchen „Korrektur" kann aus den Abb. 11 und 12 und den Tab. 2 und 3 entnommen werden.) Für die Herstellung des bisher brauchbarsten Wirkungsspektrums der HER bei *Sinapis alba* wurde der Lichteinfluß auf das Wachstum der Kotyledonen benutzt [MOHR (1959b)]. Folgendes Bestrahlungsprogramm wurde dabei angewandt: Die jungen Keimlinge wurden an drei aufeinanderfolgenden Tagen bestrahlt, pro Tag 6 Std. Das Bestrahlungsprogramm der einzelnen Tage war gleich. Zunächst wurden alle Keimlinge im HR-Feld bestrahlt (5 min). Damit ist die NER saturiert. Dann folgen 6 Std. monochromatische Bestrahlung (konstante Zahl von Quanten/cm^2 · sec). Anschließend werden nochmal alle Keimlinge für 5 min im HR-Feld bestrahlt. Damit ist, ehe die Keimlinge ins Dunkle kommen, wiederum für alle Keimlinge die NER mit HR (HR-Standard-Feld) saturiert. Der dadurch erzielte Totaleffekt ist in Abb. 13 angegeben. Die Frage ist, mit

Tabelle 2. *Aufteilung von 6 min Hellrot über einen Zeitraum von 4 Std.* (Untersuchter Effekt: Lichtabhängige Haarbildung aus Epidermiszellen des Hypokotyls von *Sinapis alba*-Dunkelkeimlingen). Aus Abb. 11 geht hervor, daß schon nach 1 min Hellrot (Standard-Lichtfeld) das dem hier verwendeten Hellrot entsprechende Gleichgewicht des Hellrot-Dunkelrot-Pigmentsystems praktisch eingestellt sein dürfte (HR = Hellrot, D = Dunkel; die Zahlen im Bestrahlungsprogramm bedeuten Minuten) [Nach MOHR (1959a)]

Bestrahlungsprogramm	Relative Haarzahl
6HR—240D	25,3
2HR—120D—2HR—120D—2HR	46,8
2HR—60D—1HR—60D—1HR—60D—1HR—60D—1HR	45,8
5HR—240D—5HR	42,0
240HR	51,5

Tabelle 3. *Aufteilung von 12 min Dunkelrot über einen Zeitraum von 4 Std.* (Objekt wie bei Tab. 2). Aus Abb. 11 geht hervor, daß schon nach 2 min Dunkelrot das dem hier verwendeten Dunkelrot (Standard-Lichtfeld) entsprechende Gleichgewicht des Hellrot-Dunkelrot-Pigmentsystems praktisch eingestellt sein dürfte. (DR = Dunkelrot, D = Dunkel; die Zahlen im Bestrahlungsprogramm bedeuten Minuten.) [Nach MOHR (1959a)]

Bestrahlungsprogramm	Relative Haarzahl
12DR—240D	9,5
4DR—120D—4DR—120D—4DR	18,2
4DR—60D—2DR—60D—2DR—60D—2DR—60D—2DR	18,2
5DR—240D—5DR	17,8
240DR	28,0

welchem Effekt die NER an dem Totaleffekt beteiligt ist („Korrekturfaktor"). Dieser „Korrekturfaktor" müßte für jede Wellenlänge gesondert festgestellt werden. Es war bisher wegen des großen Arbeitsaufwandes nur möglich, näherungsweise Korrekturfaktoren für den „Hellrotbereich" (560—700 mμ) und für den Dunkelrotbereich (700—800 mμ) unter Benutzung der Standard-HR- und Standard-DR-Felder festzulegen (vgl. Abb. 13). Es wurde, analog zu den erwähnten Versuchen über die Haarbildung, festgestellt, wieviel während der benützten

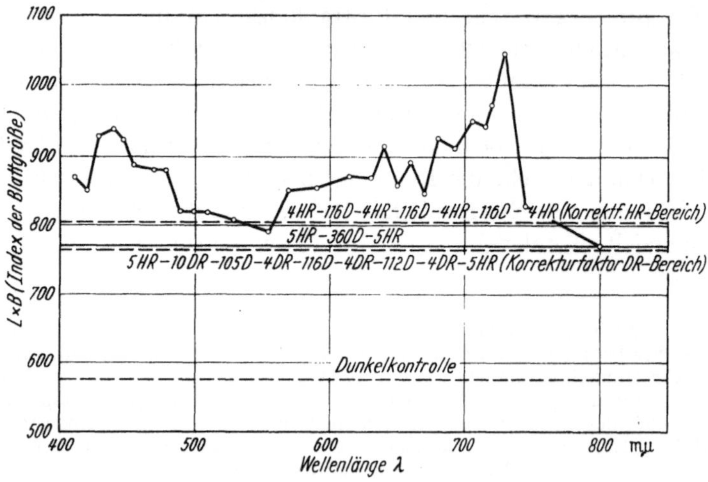

Abb. 13. $L \times B$ = „Index der mittleren Blattgröße" in Abhängigkeit von der Wellenlänge. Gemessen wurden die maximale Länge = L und die maximale Breite = B des deckenden, länger gestielten Kotyledon der Dunkelkeimlinge von *Sinapis alba* im relativen Maß. Folgendes Bestrahlungsprogramm wurde angewandt (24, 48 und 72 Std. nach Aussaat der Samen): 5 min Hellrot-Feld — 6 Std. monochromatische Strahlung gleicher Intensität in Quanten pro cm² · sec — 5 min Hellrot-Feld. Auswertung: 96 Std. nach Aussaat (25° C). Die Intensität der Strahlung bei λ = 412 mμ betrug 1070 erg/cm² · sec. — In die Abbildungen sind auch die Resultate anderer Bestrahlungsprogramme eingetragen, die als „Korrekturfaktoren" (vgl. Text) verwendet werden (*HR* Hellrot, *DR* Dunkelrot, *D* Dunkel, die Zahlen in den Bestrahlungsprogrammen bedeuten Minuten) [Nach MOHR (1959b)]

Bestrahlungszeit die NER im HR- bzw. DR-Feld maximal leisten kann. Der so gewonnene „Korrekturfaktor" für den HR-Bereich dürfte, für die kürzeren Wellenlängen zumindest, etwas zu niedrig sein, der für den DR-Bereich ist wohl für die längeren Wellenlängen etwas zu hoch, keinesfalls zu niedrig. Als Korrekturfaktor für den „Blaubereich" (400 bis 560 mμ) wurde, unter Berücksichtigung der sehr geringen Blaulichteinflüsse auf das HR-DR-Reaktionssystem und eingedenk der Möglichkeit, daß geringe Spuren von längerwelligem Streulicht durch die Blaulicht-Interferenzfilter gehen könnten, der Mittelwert zwischen dem Resultat des Bestrahlungsprogramms 5 min HR — 6 Std. Dunkel — 5 min HR und dem Saturierungswert für die HR-DR-Reaktion (HR-Feld) für den Zeitraum von 6 Std. (= Korrekturfaktor HR-Bereich) gewählt (vgl. Abb. 13).

Das derart korrigierte Wirkungsspektrum, welches das Wirkungsspektrum der HER repräsentiert, zeigt Abb. 14. Es ist zweigipfelig. Ein starker, steiler Gipfel liegt im Dunkelrot (zwischen 720 und 730 mμ), ein

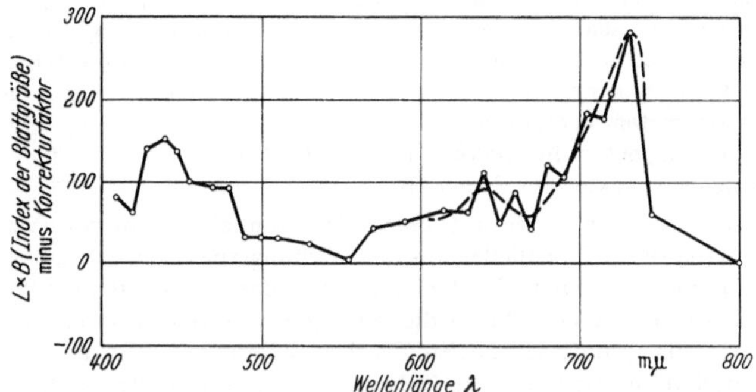

Abb. 14. Das Wirkungsspektrum des Blau-Dunkelrot-Reaktionssystems, näherungsweise korrigiert für den Einfluß des reversiblen *HR-DR*-Reaktionssystems. Verwendet sind die Angaben der Abb. 13. Zwischen 600 und 650 mμ befindet sich vielleicht ein schwacher Gipfel (gestrichelte Kurve). [Berechnet nach Mohr (1959 b).]

anderer Gipfel liegt im Blaubereich. Ein schwacher Gipfel könnte im Bereich zwischen 600 und 650 mμ vorhanden sein (gestrichelte Kurve). Dieses Wirkungsspektrum stimmt gut mit den (nur ungenügend hinsichtlich der NER korrigierten!) Hochenergie-Wirkungsspektren der anderen, bisher untersuchten Photomorphosen der Keimlinge von *Sinapis alba* überein [Mohr (1957, 1959a)].

Wir haben schon erwähnt, daß die HER sich nur nachweisen läßt, wenn mit relativ hoher Intensität bestrahlt wird. Dies zeigt deutlich eine Untersuchung der Intensitätsabhängigkeit der Reaktion (Abb. 15). Wir können daraus entnehmen, daß dieses Reaktionssystem praktisch nicht erfaßt werden kann, wenn nur monochromatische Strahlung niedriger Intensität zur Verfügung steht. (Man muß allerdings beachten, daß dem hier verwendeten „Weißlicht" das bei der HER besonders wirksame DR praktisch fehlt.)

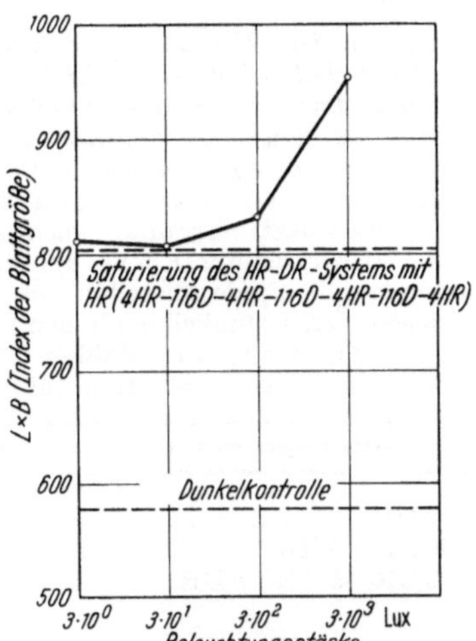

Abb. 15. Die Intensitätsabhängigkeit der „Hochenergiereaktion". Untersuchungsobjekt wie in Abb. 13. Bestrahlungsprogramm: 24, 48 und 72 Std. nach Aussaat der Samen je 6 Std. „Weißlicht" (Leuchtstoffröhren). Um die Niederenergiereaktion für alle Keimlinge gleichmäßig und ausreichend abzusättigen, erhalten alle Keimlinge während dieser 6 Std. 4 × 5 min die volle Intensität (3000 Lux). Der „Korrekturfaktor" für den Hellrot-Bereich (Abb. 13) ist zum Vergleich eingetragen. [Nach Mohr (1959b).]

c) Zur Theorie des Blau-Dunkelrot-Reaktionssystems

Wir haben bei der Besprechung des reversiblen HR-DR-Pigmentsystems gesehen, daß es offensichtlich, von P_{DR} ausgehend, zu einer fundamentalen Änderung des Stoffwechsels kommt, was sich, unmittelbar meßbar, in den Photomorphosen äußert. Alle die Photomorphosen, die mit dem HR-DR-System zusammenhängen, sind also nur Manifestationen einer grundsätzlichen Änderung des Stoffwechsels, die durch die Bildung von P_{DR} induziert wurde. Das Ausmaß der Stoffwechselumsteuerung ist dabei eine Funktion der Menge des P_{DR}. Manche Photomorphosen sind schon beobachtbar, wenn nur wenig P_{DR} gebildet wurde, andere verlangen eine mehrmalige, durch Dunkelperioden getrennte HR-Bestrahlung. Jedoch kann man sich leicht vorstellen, daß die gleichen, fundamentalen, quantitativ abgestuften Veränderungen im Stoffwechsel der Organe bzw. Zellen alle diese Photomorphosen im Gefolge haben. Die Art der Photomorphose hängt von der Organzugehörigkeit der Zellen ab.

Ähnlich kann es sich auch mit dem Blau-Dunkelrot-Reaktionssystem bei *Sinapis alba* verhalten. Wir schließen aus den Wirkungsspektren, daß die wirksame Absorption von einem Pigment durchgeführt wird, das bevorzugt im Blaubereich und im Dunkelrot absorbiert. Die in diesem Pigmentsystem absorbierte Energie verursacht ebenfalls grundsätzliche Stoffwechseländerungen, was zu zahlreichen beobachtbaren Photomorphosen führt. Dabei dürfte es so sein, daß zwar die primären Schritte verschieden sind, letztlich jedoch dieselben Stoffwechseländerungen durch das HR-DR-System erzeugt werden wie auch durch das Blau-Dunkelrot-System, denn beim Kotyledonenwachstum von *Sinapis alba* z. B. ließ sich zeigen, daß ein bestimmter Effekt sich sowohl durch die kombinierte Wirkung beider Systeme wie auch (bei Anwendung entsprechender Bestrahlungsprogramme) durch das reversible HR-DR-System allein erzielen läßt. Eine einfache Deutung wäre also die, daß das Blau-Dunkelrot-Pigmentsystem seine Anregungsenergie auf andere Moleküle überträgt (Sensibilisatorwirkung) und damit eine sensibilisierte photochemische Reaktion ermöglicht. Der Ablauf dieser photochemischen Reaktion würde zu Stoffwechselumsteuerungen führen, die letztlich die Ausbildung derselben Photomorphosen im Gefolge haben, wie sie auch als Resultat der Bildung von P_{DR} entstehen. Obwohl beide Systeme dieselben Photomorphosen erzeugen und diese Photomorphosen offenbar dieselbe fundamentale Veränderung des Stoffwechsels repräsentieren, muß zumindest der primäre Schritt, der auf die photochemische Reaktion folgt, bei den beiden Systemen verschieden sein, denn beide Systeme arbeiten, wenigstens relativ, unabhängig voneinander (vgl. z. B. Tab. 4).

Tabelle 4. *Die Wirkung einer kurzen Zusatzbestrahlung mit Hellrot bzw. mit Hellrot, gefolgt von Dunkelrot, nach einer langen Bestrahlung mit Dunkelrot.* Der durch die lange Dunkelrotbestrahlung induzierte Effekt, der zu einem gewissen Prozentsatz auf die Hochenergiereaktion zurückgeht (vgl. Tab. 3), wird durch die Zusatzbestrahlung nicht signifikant beeinflußt. (Untersuchter Effekt: Lichtabhängige Haarbildung aus Epidermiszellen des Hypokotyls von *Sinapis alba*-Dunkelkeimlingen.) HR = Hellrot, DR = Dunkelrot; die Zahlen im Bestrahlungsprogramm bedeuten Minuten. [Mohr (1959a)]

Bestrahlungsprogramm	Relative Haarzahl
240 DR	28
240 DR—5 HR	62
240 DR—5 HR—8 DR . . .	32

Wir haben soeben angenommen, das Blau-Dunkelrot-Pigment fungiere als Sensibilisator. Das Blau-Dunkelrot-Pigment könnte jedoch auch die prosthetische Gruppe (bzw. ein Teil der prosthetischen Gruppe) eines Enzyms sein. Die Aktivierung des Enzyms könnte das primäre Resultat der Energieabsorption sein. Enzymaktivierungen durch Licht sind öfters beobachtet worden [z. B. Bernheim und Dixon (1928), Tolbert und Burris (1950), Galston und Baker (1951),

APPLEMAN und PYFROM (1955)]. Die starke Wirkung des Blaulichts spricht dafür, daß die Strahlungsabsorption durch ein Flavin-haltiges Molekül erfolgt, der langwellige Gipfel des Wirkungsspektrums könnte auf die Absorption eines zu diesem Molekülverband gehörigen Schwermetalls, z. B. Cu, zurückgehen [vgl. SIEGELMAN und HENDRICKS (1957)]. Man kann natürlich dann nicht voraussetzen, daß die Quantenausbeute im kurzwelligen Spektralbereich dieselbe ist wie im langwelligen. Flavoproteide, die ähnliche Absorptionseigenschaften haben wie das aus dem Wirkungsspektrum der HER erschlossene Pigment, sind in jüngster Zeit bekannt geworden, z. B. die im oxydierten Zustand grüne Butyryl CoA-Dehydrase aus Lebermitochondrien [MAHLER (1954, 1956)]. Neuerdings wird jedoch bezweifelt, daß dieses gefärbte Enzym Cu enthält [STEYN-PARVE und BEINERT (1957)]. Die grüne Farbe des Enzyms soll auf der etwas veränderten Absorption des Flavin-Anteils beruhen. Unter den üblichen Pigmenten der höheren Pflanze dürfen wir, des langwelligen Gipfels wegen, das der Hochenergiereaktion zugrunde liegende Pigmentsystem nicht suchen.

d) Anthocyanbildung

Für die Keimlinge von *Sinapis alba* wurde gezeigt, daß auch die lichtabhängige Bildung von Anthocyan von beiden Reaktionssystemen beherrscht wird. Einmal ist ein Einfluß des reversiblen HR-DR-Reaktionssystems vorhanden, andererseits jedoch läßt sich auch die Hochenergiereaktion nachweisen [MOHR (1957)]. In jüngster Zeit haben SIEGELMAN und HENDRICKS (1957, 1958) bei Keimlingen von *Brassica rubra* und *Brassica rapa* sowie bei mehreren Apfel-Sorten den Lichteinfluß auf die Anthocyanbildung untersucht. Während sie nur bei *Brassica rubra* einen Einfluß des reversiblen HR-DR-Systems (NER) auf die Anthocyanbildung nachweisen konnten, fanden sie in allen Fällen Hochenergie-Wirkungsspektren, die Ähnlichkeit mit dem Wirkungsspektrum der Hochenergiereaktion bei *Sinapis alba* haben (starker Gipfel im langwelligen Teil des sichtbaren Spektrums, kleinerer Gipfel im Blau). Da bei *Brassica rubra* kein Versuch gemacht wurde, den Anteil der NER am Gesamtlichteffekt abzuschätzen und dementsprechend das gemessene Wirkungsspektrum zu korrigieren, so kann das Wirkungsspektrum für *Brassica rubra* nicht genauer mit dem Wirkungsspektrum der Abb. 14 verglichen werden. Dagegen scheint bei den untersuchten Keimlingen von *Brassica rapa* der Fall vorzuliegen, daß die Förderung der Anthocyanbildung durch Licht ausschließlich über das Blau-Dunkelrotpigmentsystem geht. Ein Effekt des reversiblen HR-DR-Systems konnte (unter den gewählten experimentellen Bedingungen) nicht nachgewiesen werden. Das längerwellige Wirkungsspektrum der Induktion der Anthocyanbildung bei diesen Keimlingen (Abb. 16) ist dem Wirkungsspektrum der Abb. 14 sehr ähnlich. Auch hier liegt ein weiterer, schwächerer Gipfel im Blaubereich ($\lambda < 500$ mμ). Auch die Anthocyanbildung bei Äpfeln [SIEGELMAN und HENDRICKS (1958)] wird offenbar nicht von der NER beeinflußt, sondern ausschließlich von einer Hochenergiereaktion. Unterhalb des Intensitätsschwellenwerts von etwa 1000 Lux wird die Anthocyanbildung, auch bei sehr langer Belichtung, nicht gefördert (vgl. den Intensitätsschwellenwert der Abb. 15). Das Wirkungsspektrum hat hier allerdings seinen langwelligen Hauptgipfel um 650 mμ. Sicher scheint zu sein, daß bei allen in neuerer Zeit genauer untersuchten Keimpflanzen bei Einstrahlung größerer Energiemengen die Anthocyanbildung über Reaktionssysteme beeinflußt wird, die dem an der Photomorphogenese von *Sinapis alba* beteiligten Blau-Dunkelrot-Reaktionssystem zumindest sehr ähnlich sind. Dafür sprechen auch die Befunde von KANDELER (1958).

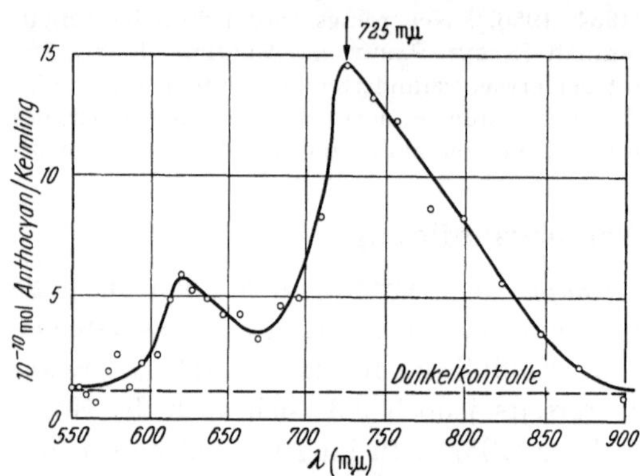

Abb. 16. Das Wirkungsspektrum der Induktion der Anthocyansynthese (Keimlinge von *Brassica rapa*) zwischen 550 und 900 mμ. Die Keimlinge wurden zuerst 2 Std. mit weißem Fluorescenzlicht bestrahlt (Induktionsperiode), anschließend für 8 Std. mit monochromatischer Strahlung. (Deren Intensität betrug etwa 2000 erg pro $cm^2 \cdot sec$.) [Nach SIEGELMAN und HENDRICKS (1957)]

Auf Grund der Ergebnisse bei *Sinapis alba*, wo gezeigt wurde, daß die durch das Blau-Dunkelrot-Pigmentsystem absorbierte Energie sich auf *alle* beobachtbaren Photomorphosen auswirkt, dürfen wir annehmen, daß auch der Einfluß der HER auf die Anthocyanbildung *indirekter* Art ist. Wir können uns vorstellen, es werde durch die vom Blau-Dunkelrot-System absorbierte Energie die photochemische Reaktion A → B ermöglicht (sensibilisierte Reaktion). B löst eine fundamentale Umsteuerung des Stoffwechsels aus. Das Ausmaß dieser Umsteuerung hängt von der Menge des Stoffes B ab. Die Anthocyanbildung ist *nur eine* Folge der Bildung von B mit Hilfe von Strahlungsenergie (vgl. den Abschnitt 4/c).

Eine weitere mögliche Auffassung sei kurz skizziert: Das Blau-Dunkelrot-Pigment arbeitet als Sensibilisator für ganz verschiedene photochemische Reaktionen. Die Anregungsenergie dieses Sensibilisators dient also als Energiequelle für ganz verschiedene, sensibilisierte, photochemische Reaktionen. Die verschiedenen beobachteten Photomorphosen des *Sinapis alba*-Keimlings hingen dann nicht über eine fundamentale, allen Photomorphosen zugrunde liegende Stoffwechselumsteuerung zusammen, sondern lediglich über den gemeinsamen Sensibilisator. Wäre diese Auffassung richtig, so könnte der Lichteinfluß auf die Anthocyan-Synthese (Hochenergiereaktion) auf einer sensibilisierten, photochemischen Reaktion im direkten Syntheseweg des Anthocyans beruhen.

e) Über die Verbreitung des Blau-Dunkelrot-Reaktionssystems

Ob das Blau-Dunkelrot-Reaktionssystem ähnlich weit verbreitet ist wie das reversible HR-DR-Reaktionssystem, kann heute noch nicht sicher beurteilt werden. Es spricht aber vieles dafür (vgl. auch die schon besprochenen Beispiele und die Hinweise in Abschnitt 3/d). So scheint ein Blau-Dunkelrot-System an der Lichthemmung der Samenkeimung von *Nemophila insignis* beteiligt zu sein [WAREING und BLACK (1958)], auch die Lichthemmung der Keimung von *Nigella*-Samen könnte auf dieses System zurückgehen [ISIKAWA (1957)]. Bei *Lactuca*-Keimlingen geht die Etiolementverhinderung des Hypokotyls wahrscheinlich über das Blau-Dunkelrot-System [FLINT und MORELAND (1939)]. Ebenso dürfte es bei der lichtbedingten Etiolementverhinderung von Kartoffelschößlingen sein [WASSINK u. Mitarb. (1950a)].

Gewisse Cruciferen sollen, hinsichtlich ihrer Blütenbildung, besonders auf dunkelrotes und blauviolettes Zusatzlicht (zusätzlich zu einer „Weißlicht"-Hauptlichtperiode hoher Intensität) reagieren. Diesbezügliche ältere Befunde von FUNKE (1936, 1937, 1938) wurden von WASSINK u. Mitarb. (1950b, 1951); STOLWIJK (1952, 1954) bestätigt und erweitert. Auch bei *Lemna gibba* hat KANDELER (1956) ein ähnliches Wirkungsspektrum für schwache Zusatzbestrahlung ausgearbeitet (starker Gipfel im DR). Nach STOLWIJK und ZEEVAART (1955) wirkten bei *Hyoscyamus niger*, im Langtag bei ausschließlich monochromatischer Bestrahlung hoher Intensität gezogen, nur Blauviolett und Dunkelrot blühfördernd. Nach den umfassenden Versuchen von MEIJER (1957), MEIJER und VAN DER VEEN (1957) und VAN DER VEEN und MEIJER (1958) über die Blütenbildung von Lang- und Kurztagpflanzen läßt sich ein „Langtag-Effekt" auf folgende 2 Arten zustande bringen: a) Die Pflanze erhält Kurztag-Hauptlichtperioden mit Licht, das ein genügend hohes Quantum an Blaulicht (oder Dunkelrot?) enthält. Als „Störlicht" erhält die Pflanze Hellrot. b) Die Pflanze erhält Langtag-Hauptlichtperioden mit Licht, das Blau oder Dunkelrot genügender Quantität enthält. In diesem Fall ist Hellrot für die Blütenbildung nicht notwendig. — Es ist sehr wohl möglich, daß in diesen Fällen das Blau-Dunkelrot-Reaktionssystem die entscheidende Rolle spielt.

Literatur

APPLEMAN, D., and H. T. PYFROM: Changes in catalase activity and other responses induced in plants by red and blue light. Plant Physiol. **30**, 543—549 (1955).

AVERY, G. S., P. R. BURKHOLDER and H. B. CREIGHTON: Polarized growth and cell studies in the first internode and coleoptile of *Avena* in relation to light and darkness. Bot. Gaz. **99**, 125—143 (1937).

BERNHEIM, F., and M. DIXON: Studies on xanthine oxidase X. The action of light. Biochem. J. **22**, 113—124 (1928).

BIEBEL, J. P.: Some effects of radiant energy in relation to etiolation. Plant Physiol. **17**, 377—396 (1942).

BINET, P.: Lumière et germination de *Zygophyllum album* L. C. R. Acad. Sci. (Paris) **243**, 397—399 (1956).

Borthwick, H. A.: Light effects on tree growth and seed germination. Ohio J. Sci. **57**, 357—364 (1957).

—, S. B. Hendricks and M. W. Parker: Action spectrum for photoperiodic control of floral initiation of a long day plant, Wintex barley (*Hordeum vulgare*) Bot. Gaz. **110**, 103—118 (1948).

— — — Action spectrum for inhibition of stem growth in darkgrown seedlings of albino and nonalbino barley (*Hordeum vulgare*). Bot. Gaz. **113**, 95—105 (1951).

— — — The reaction controlling floral initiation. Proc. nat. Acad. Sci. (Wash.) **38**, 929—934 (1952b).

— — — E. H. Toole, and V. K. Toole: A reversible photoreaction controlling seed germination. Proc. nat. Acad. Sci. (Wash.) **38**, 662—666 (1952a).

— — E. H. Toole and V. K. Toole: Action of light on lettuce-seed germination. Bot. Gaz. **115**, 205—225 (1954).

Boysen-Jensen, P.: Die Elemente der Pflanzenphysiologie. Jena: G. Fischer1939.

Brian, P. W.: Role of gibberellin-like hormones in regulation of plant growth and flowering. Nature (Lond.) **181**, 1122—1123 (1958).

Burkholder, P. R.: The role of light in the life of plants. Bot. Rev. **2**, 1—52, 97 to 172 (1936).

Caspary, R.: *Bullardia aquatica* D. C. Schr. königl. Physik.-ökonom. Ges. Königsberg **1**, 66—91 (1860).

Cathey, H. M., and H. A. Borthwick: Photoreversibility of floral initiation in *Chrysanthemum*. Bot. Gaz. **119**, 71—76 (1957).

Crocker, W.: Effect of the visible spectrum upon the germination of seeds and fruits. In: Biological Effects of Radiation, vol. II. B. M. Duggar ed. New York: McGraw-Hill Book Co. 1936.

— Growth of Plants. New York: Reinhold Publ. Corp. 1948.

Downs, R. J.: Photoreversibility of leaf and hypocotyl elongation of dark grown red kidney bean seddlings. Plant Physiol. **30**, 468—473 (1955).

— Photoreversibility of flower initiation. Plant Physiol. **31**, 279—284 (1956).

—, S. B. Hendricks, and H. A. Borthwick: Photoreversible control of elongation of Pinto beans and other plants under normal conditions of growth. Bot. Gaz. **118**, 199—208 (1957).

Drumm, K.: Vergleichende Untersuchung über Etiolement und Pigmentgehalt an chlorophylldefekten und normalen Sprossen von *Tradescantia albiflora* var. *albivittata*. Planta **46**, 92—112 (1955).

Duysens, L. N. M.: Energy transformations in photosynthesis. Ann. Rev. Plant Physiol. **7**, 25—50 (1956).

Evenari, M.: Seed Germination. In: Radiation Biology, Vol. III. A. Hollaender ed. New York: McGraw-Hill Book Co. 1956.

— G. Neumann and S. Klein: The influence of red and infrared light on the respiration of photoblastic seed. Physiol. Plantarum **8**, 33—47 (1955).

Flint, L. H., and E. D. McAlister: Wave length of radiation in the visible spectrum inhibiting the germination of light-sensitive lettuce seed. Smithsonian Misc. Coll. **94**, No. 5, 1—11 (1935).

— — Wave length of radiation in the visible spectrum promoting the germination of light-sensitive lettuce seed. Smithsonian Misc. Coll. **96**, No. 2,1—8 (1937).

— and C. F. Moreland: Response of lettuce seedlings to 7600 Å radiation. Amer. J. Bot. **26**, 231—233 (1939).

Funke, G. L.: Proeven over photoperiodiciteit by verschillend gekleurd licht. Biol. Jb. nat. Genoot. Dodonaea **3**, 225—261 (1936); **4**, 345—359 (1937); **5**, 404—424 (1938).

Galston, A. W., and R. S. Baker: Studies on the physiology of light action III. Light activation of a flavoprotein enzyme by reversal of a naturally occuring inhibition. Amer. J. Bot. **38**, 190—195 (1951).

Goodwin, R. H.: On the inhibition of the first internode of *Avena* by light. Amer. J. Bot. **28**, 325—332 (1941).

— and O. v. H. Owens: An action spectrum for inhibition of the first internode of *Avena* by light. Bull. Torrey bot. Club **75**, 18—21 (1948).

— — The effectiveness of the spectrum in *Avena* internode inhibition. Bull. Torrey bot. Club **78**, 11—21 (1951).

Gordon, S. A., and K. Surrey: Red, far-red interaction in oxidative phosphorylation. Plant Physiol. (Suppl.) **33**, 24 (1958).

Haber, A.: Rendering light-insensitive lettuce seed sensitive to light. Plant Physiol. (Suppl.) **33**, 23 (1958).

Haupt, W.: Hellrot-Dunkelrot-Antagonismus bei der Auslösung der Chloroplastenbewegung. Naturwissenschaften **45**, 273—274 (1958).

Hendricks, S. B., and H. A. Borthwick: Time dependencies in photoperiodism. 8. Congrès int. Bot., Sect. 11, p. 323—324. Paris **1954**.

— — and R. J. Downs: Pigment conversion in the formative responses of plants to radiation. Proc. nat. Acad. Sci. (Wash.) **42**, 19—26 (1956).

Hillman, W. S.: Nonphotosynthetic light requirement in *Lemna minor* and its partial satisfaction by kinetin. Science **126**, 165—166 (1957).

— and A. W. Galston: Inductive control of indole acetic acid oxidase by red and near infrared light. Plant Physiol. **32**, 129—135 (1957).

Ikuma, H., and K. V. Thimann: The mechanism of germination in light sensitive lettuce seed. Plant Physiol. **33** (Suppl.), **25** (1958).

Isikawa, S.: Light-sensitivity against the germination I. „Photoperiodism" of seeds. Bot. Mag. (Tokyo) **67**, 51—56 (1954).

— Interaction of temperature and light in the germination of *Nigella* seeds I, II. Bot. Mag. (Tokyo) **70**, 264—275 (1957).

— and T. Oohusa: Effects of light upon the germination of spores of ferns II. Two lightperiods of *Dryopteris crassirhizoma* Nakai. Bot. Mag. (Tokyo) **69**, 132—137 (1956).

Johnston, E. S.: Growth of *Avena* coleoptile and first internode in different wavelength bands of the visible spectrum. Smithsonian Misc. Coll. **96**, No. 6, 1—19 (1937).

Jones, M. B., and L. F. Bailey: Light effects on the germination of seeds of henbit (*Lamium amplexicaule* L.). Plant Physiol. **31**, 347—349 (1956).

Kadman-Zahavi, A.: Effects of red and far-red radiation on seed germination. Nature (Lond.) **180**, 996—997 (1957).

Kahn, A., J. A. Goss and D. E. Smith: Light and chemical effects on lettuce seed germination. Plant Physiol. (Suppl.) **31**, 37 (1956).

Kandeler, R.: Über die Blütenbildung bei *Lemna gibba* L. II. Das Wirkungsspectrum blühförderndem Schwachlicht. Z. Bot. **44**, 153—174 (1956).

— Die Wirkung von farbigem und weißem Licht auf die Anthocyanbildung bei Cruciferen-Keimlingen. Ber. dtsch. bot. Ges. **71**, 34—44 (1958).

Kinzel, W.: Frost und Licht als beeinflussende Kräfte bei der Samenkeimung. Ludwigsburg: E. Ulmer 1913—1926.

Klein, S., and J. V. Preiss: Reversibility of the red-far-red reaction by irradiation at different sites. Nature (Lond.) **181**, 200—201 (1958).

Klein, W. H., R. B. Withrow and V. B. Elstad: Response of the hypocotyl hook of bean seedlings to radiant energy and other factors. Plant Physiol. **31**, 289—294 (1956).

KLEIN, W. H., R. B. WITHROW, V. B. ELSTAD and L. PRICE: Photocontrol of growth and pigment synthesis in the bean seedling as related to irradiance and wavelength. Amer. J. Bot. **44**, 15—19 (1957a).

— — A. P. WITHROW, and V. ELSTAD: Time course of far-red inactivation of photomorphogenesis. Science **125**, 1146—1147 (1957b).

— — and V. ELSTAD: Kinetics of the far-red inactivation of photomorphogenesis in the beanhook. Plant Physi ol. (Suppl.) **32**, 9 (1957c).

KÖNITZ, W.: Blühhemmung bei Kurztagpflanzen durch Hellrot- und Dunkelrotlicht in der photo- und skotophilen Phase. Planta **51**, 1—29 (1958).

KOSKI, V. M., C. S. FRENCH and J. H. C. SMITH: The action spectrum for the transformation of protochlorophyll to chlorophyll a in normal and albino corn seedlings. Arch. Biochem. Biophys. **31**, 1—17 (1951).

LEHMANN, E., u. F. AICHELE: Keimungsphysiologie der Gräser. Stuttgart: F. Enke 1931.

LEOPOLD, A. C., and F. S. GUERNSEY: Respiratory responses to red and infrared light. Physiol. Plantarum **7**, 30—40 (1954).

LINDEMANN, G.: Die Phototropie der Anile, untersucht am Salicylal-m-toluidin. Z. wiss. Photographie, Photophysik und Photochemie **50 II**, 347—386 (1955).

LIVERMAN, J. L., and J. BONNER: The interaction of auxin and light in the growth response of plants. Proc. nat. Acad. Sci. (Wash.) **39**, 905—916 (1953).

— N. P. JOHNSON and L. STARR: Reversible photoreaction controlling expansion of etiolated bean leaf disks. Science **121**, 440—441 (1955).

LOCKHART, J. A.: The influence of red and far-red radiation on the response of *Phaseolus vulgaris* to gibberellic acid. Physiol. Plantarum **11**, 487—492 (1958).

LÖRCHER, L.: Die Wirkung verschiedener Lichtqualitäten auf die endogene Tagesrhythmik von *Phaseolus*. Z. Bot. **46**, 209—241 (1958).

LOOFBOUROW, J. R.: Effects of ultraviolet radiation on cells. Growth **12** (Suppl.), 75—149 (1948).

MAHLER, H. R.: Studies on the fatty acid oxidizing system of animal tissues IV. The prosthetic group of butyryl coenzyme A dehydrogenase. J. biol. Chem. **206**, 13—26 (1954).

— Nature and function of metalloflavoproteins. Advanc. Enzymol. **17**, 233—291 (1956).

MEIJER, G.: The influence of light quality on the flowering response of *Salvia occidentalis*. Acta bot. neerl. **6**, 395—406 (1957).

— and R. VAN DER VEEN: Wavelength dependence on photoperiodic responses. Acta bot. neerl. **6**, 429—433 (1957).

MEISCHKE, D.: Über den Einfluß der Strahlung auf Licht- und Dunkelkeimer. Jb. Bot. **83**, 359—405 (1936).

MILLER, C. O.: Similarity of some kinetin and red light effects. Plant Physiol. **31**, 318—319 (1956).

— The relationship of the kinetin and red-light promotions of lettuce seed germination. Plant Physiol. **33**, 115—117 (1958).

MOH, C. C., and R. B. WITHROW: Interaction of red and far-red radiant energy in modifying x-ray induced chromatid aberrations in Broad bean. Plant Physiol. (Suppl.) **32**, 11 (1957).

MOHR, H.: Die Beeinflussung der Keimung von Farnsporen durch Licht und andere Faktoren. Planta **46**, 534—551 (1956).

— Der Einfluß monochromatischer Strahlung auf das Längenwachstum des Hypokotyls und auf die Anthocyanbildung bei Keimlingen von *Sinapis alba* L. (= *Brassica alba* Boiss.) Planta **49**, 389—405 (1957).

MOHR, H.: Der Lichteinfluß auf die Haarbildung am Hypokotyl von *Sinapis alba* L. Planta **53**, 109—124 (1959a).
— Der Lichteinfluß auf das Wachstum der Keimblätter bei *Sinapis alba* L. Planta **53**, 219—245 (1959b).
— and G. SCHOSER: Eine Interferenzfilter-Monochromatoranlage für photobiologische Zwecke. Planta **53**, 1—17 (1959).
MONK, G. S., and C. F. EHRET: Design and performance of a biological spectrograph. Radiation Res. **5**, 88—106 (1956).
OSTER, G., and N. WOTHERSPOON: Photobleaching and photorecovery of dyes. J. chem. Physics **22**, 157—158 (1954).
PARKER, M. W., S. B. HENDRICKS and H. A. BORTHWICK: Action spectrum for the photoperiodic control of floral initiation of the long day plant *Hyoscyamus niger*. Bot. Gaz. **111**, 242—252 (1950).
— — — and N. J. SCULLY: Action spectrum for the photoperiodic control of floral initiation in Biloxi soybean. Science **102**, 152—155 (1945).
— — — — Action spectra for photoperiodic control of floral initiation in shortday plants. Bot. Gaz. **108**, 1—26 (1946).
— — — and F. W. WENT: Spectral sensitivities for leaf and stem growth of etiolated pea seedlings and their similarity to action spectra for photoperiodism. Amer. J. Bot. **36**, 194—204 (1949).
PFEFFER, W.: Pflanzenphysiologie, Bd. II. 2. Aufl. Leipzig: W. Engelmann 1904.
PIRINGER, A. A., and P. H. HEINZE: Effect of light on formation of a pigment in the tomato cuticle. Plant Physiol. **29**, 467—472 (1954).
POPP, H. W., and F. BROWN: Effects of different regions of the visible spectrum upon seed plants. In: Biological Effects of Radiation, Vol. II. B. M. DUGGAR ed. New York: McGraw-Hill Book Co. 1936.
RABINOWITCH, E. I.: Photosynthesis and related Processes. Vol. II, part 2. New York: Interscience Publ. 1956.
REIMERS, J.: Vergleichende zellphysiologische Studien an einigen etiolierten und grünen mono- und dikotylen Pflanzen. Planta **49**, 455—488 (1957).
RESÜHR, B.: Beiträge zur Lichtkeimung von *Amaranthus caudatus* L. und *Phacelia tanacetifolia* Benth. Planta **30**, 471—506 (1939).
SACHS, J.: Lehrbuch der Botanik 4. Aufl. Leipzig: W. ENGELMANN 1874.
SCOTT, R. A., and J. L. LIVERMAN: Control of etiolated bean leaf disk expansion by gibberellins and adenine. Science **126**, 122—124 (1957).
SIEGELMAN, H. W., and S. B. HENDRICKS: Photocontrol of anthocyanin formation in turnip and red cabbage seedlings. Plant Physiol. **32**, 393—398 (1957).
— — Photocontrol of anthocyanin synthesis in apple skin. Plant Physiol. **33**, 185—190 (1958).
SMITH, J. H. C., and A. BENITEZ: The effect of temperature on the conversion of protochlorophyll to chlorophyll a in etiolated barley leaves. Plant Physiol. **29**, 135—143 (1954).
STEYN-PARVE, E., and H. BEINERT: Enzyme substrate interaction in a flavoprotein system. Fed. Proc. **16**, 256 (1957).
STOBBE, H.: Phototropieerscheinungen bei Fulgiden und anderen Stoffen. Liebigs Ann. Chem. **359**, 1—48 (1908).
STOLWIJK, J. A. J.: Photoperiodic and formative effects of various wavelength regions in *Cosmos bipinnatus*, *Spinacea oleracea*, *Sinapis alba* and *Pisum sativum*. Proc. kon. Ned. Akad. Wetensch. **55 C**, 489—502 (1952).
— Wave length dependence of photomorphogenesis in plants. Meded. Landb. Hogeschool Wageningen **54**, 181—244 (1954).

STOLWIJK, J. A. J.; and J. A. D. ZEEVAART: Wave length dependence of different light reactions governing flowering in *Hyoscyamus niger*. Proc. kon. Ned. Akad. Wetensch. **58 C**, 386—396 (1955).

THOMSON, B. F.: The relation between age at time of exposure and response of parts of the *Avena* seedling to light. Amer. J. Bot. **38**, 635—638 (1951).

TODD, G. W., and A. W. GALSTON: A porphyrin pigment from photosensitive non-chlorophyllous plant tissues. Plant Physiol. **29**, 311—318 (1954).

TOLBERT, N. E., and R. H. BURRIS: Light motivation of the plant enzyme which oxidises glycolic acid. J. biol. Chem. **186**, 791—804 (1950).

TOOLE, E. H., S. B. HENDRICKS, H. A. BORTHWICK, and V. K. TOOLE: Physiology of seed germination. Ann. Rev. Plant Physiol. **7**, 299—324 (1956).

— V. K. TOOLE, H. A. BORTHWICK, and S. B. HENDRICKS: Photocontrol of *Lepidium* seed germination. Plant Physiol. **30**, 15—21 (1955a).

— — — Interaction of temperature and light in germination of seeds. Plant Physiol. **30**, 473—478(1955b).

— — — — and R. J. DOWNS: Action of light on germination of seeds of *Paulownia tomentosa*. Plant Physiol. (Suppl.) **33**, 23 (1958a).

TOOLE, V. K., H. A. BORTHWICK, E. H. TOOLE, and A. G. SNOW: The germination response of seeds of *Pinus taeda* to light. Plant Physiol. (Suppl.) **33**, 23 (1958b).

TORREY, J. G.: Effects of light on elongation and branching in pea roots. Plant Physiol. **27**, 591—602 (1952).

VEEN, R. VAN DER, u. G. MEIJER: Licht und Pflanzen. Philips' techn. Bibl. Eindhoven 1958.

VIRGIN, H. I.: The conversion of protochlorophyll to chlorophyll a in continuous and intermittent light. Physiol. Plantarum **8**, 389—403 (1955).

— Studies on the formation of protochlorophyll and chlorophyll a under varying light treatments. Physiol. Plantarum **11**, 347—362 (1958).

WAREING, P. F., and M. BLACK: Similar effects of blue and infrared radiation on light sensitive seeds. Nature (Lond.) **181**, 1420—1421 (1958).

WASSINK, E. C.: On plants grown exclusively in light of restricted spectral regions. 8. Congrès int. de Bot., Sect. 11, p. 316—317. Paris 1954.

— N. KRIJTHE and C. VAN DER SCHEER: On the effect of light of various spectral regions on the sprouting of potato tubers. Proc. kon. Ned. Akad. Wetensch. **53**, 1228—1239 (1950a).

— C. M. J. SLUIJSMANS, and J. A. J. STOLWIJK: On some photoperiodic and formative effects of coloured light in *Brassica rapa*, f. *oleifera*, subf. *annua*. Proc. kon. Ned. Akad. Wetensch. **53**, 1466—1475 (1950b).

— and J. A. J. STOLWIJK: Effects of light quality on plant growth. Ann. Rev. Plant Physiol. **7**, 373—400 (1956).

— — and A. B. R. BEEMSTER: Dependence of formative and photoperiodic reactions in *Brassica rapa* var., *Cosmos* and *Lactuca* on wavelength and time of irradiation. Proc. kon. Ned. Akad. Wetensch. **54 C**, 421—432 (1951).

WEINTRAUB, R. L., and E. D. MCALISTER: Developmental physiology of the grass seedling I. Inhibition of the mesocotyl of *Avena sativa* by continuous exposure to light of low intensities. Smithsonian Misc. Coll. **101**, 1—10 (1942).

— and L. PRICE: Developmental physiology of the grass seedling II. Inhibition of mesocotyl elongation in various grasses by red and by violet light. Smithsonian Misc. Coll. **106**, 1—15 (1947).

WENT, F. W.: Effects of light on stem and leaf growth. Amer. J. Bot. **28**, 83—95 (1941).

WETMORE, R. H.: Growth and development in the shoot system of plants. In: Cellular Mechanisms in Differentiation and Growth. B. RUDNICK ed. Princeton: Princeton Univ. Press. 1956.

WILLIAMS, W. T.: Etiolation phenomena and leaf expansion. In: The Growth of Leaves. F. L. MILTHORPE ed. London: Butterworth 1956.

— F. A. BARRETT, and L. G. MOCKETT: The effect of light on the osmotic behaviour of the plumular hook. J. exp. Bot. **8**, 368—372 (1957).

— R. D. PRESTON, and G. W. RIPLEY: A biophysical study of etiolated Broad Bean internodes. J. exp. Bot. **6**, 451—457 (1955).

WITHROW, R. B.: An interference - filter monochromator system for the irradiation of biological material. Plant Physiol. **32**, 355—360 (1957).

— W. H. KLEIN, and V. ELSTAD: Action spectra of photomorphogenic induction and its photoinactivation. Plant Physiol. **32**, 453—462 (1957).

— — L. Price, and V. ELSTAD: Influence of visible and near infrared radiant energy on organ development and pigment synthesis in bean and corn. Plant Physiol. **28**, 1—14 (1953).

— and A. P. WITHROW: Generation, control, and measurement of visible and near visible radiant energy. In: Radiation Biology, Vol. III. A. HOLLAENDER ed. New York: McGraw-Hill Book Co. 1956.

— J. B. WOLFF, and L. PRICE: Elimination of the lag phase of chlorophyll synthesis in dark-grown bean leaves by a pretreatment with low irradiances of monochromatic energy. Plant Physiol. (Suppl.) **31**, 13 (1956).

WYMAN, G. M., and W. R. BRODE: The relation between the absorption spectra and the chemical constitution of dyes. XXII. Cis-trans-isomerism in thioindigo dyes. J. Amer. chem. Soc. **73**, 1487—1493 (1951).

Die Bitterstoffe der Cucurbitaceen

Von SIGMUND REHM

(Pretoria Horticultural Research Station, Division of Horticulture, Department of Agriculture)

Inhaltsübersicht

I. Einleitung

Die Familie der Cucurbitaceen ist in erster Linie bekannt, weil sie zahlreiche Kulturpflanzen mit eßbaren Früchten liefert (Gurken, Melonen, Kürbisse; in wärmeren Ländern auch Arten der Gattungen *Lagenaria, Luffa, Sechium, Trichosanthes* u. a.). Daneben kommen in der Familie viele Arten vor, deren Samen, Früchte, Blätter oder Wurzeln giftige Bitterstoffe enthalten [HEGNAUER, REHM et al. (2)]. Einige dieser bitteren Arten sind als Ursache von Viehvergiftungen bekannt

(Watt und Breyer-Brandwijk), daneben werden viele in der Volksmedizin gebraucht. In den meisten Pharmakopöen ist jedoch nur noch eine Art, die Koloquinte, offizinell. In den letzten Jahren ist das Interesse an den Bitterstoffen der Cucurbitaceen wegen ihrer nekrotisierenden Wirkung auf Tumorgewebe neu erwacht [Lavie et al. (2)].

Dieselben Bitterstoffe wie in den giftigen Wildpflanzen kommen gelegentlich in kultivierten Formen vor. Diese Tatsache kann ein ernsthaftes wirtschaftliches Problem darstellen, wie bei der Produktion von Treibhausgurken (Andeweg und de Bruyn), sie kann aber auch eine schwere Gefährdung für Gesundheit und Leben sein [Steyn (3)]. Da in Südafrika jedes Jahr Fälle von Vergiftung durch Kürbisse, Wassermelonen und andere kultivierte Arten vorkommen, werden seit einigen Jahren bei der Pretoria Horticultural Research Station, in Zusammenarbeit mit dem National Chemical Research Laboratory des South African Council for Scientific and Industrial Research, die Chemie der Bitterstoffe und die Vererbung der Bitterkeit untersucht. Der folgende Bericht beruht im wesentlichen auf den Ergebnissen der südafrikanischen Arbeitsgruppe.

Da zu Beginn unserer Arbeiten so gut wie nichts über die Chemie der Cucurbitaceen-Bitterstoffe bekannt war und gar nichts über ihre Biogenese und Vererbung, mußte die Untersuchung weit über den Rahmen des praktischen Problems, das den Ausgangspunkt bildete, hinausgehen. Die bisherigen Ergebnisse haben die Richtigkeit unseres Vorgehens bewiesen. Bei praktisch-züchterischen Problemen stehen wir heute auf festem Boden, und darüber hinaus ergaben sich Resultate von allgemeinem chemischem und biologischem Interesse.

Dieser Bericht kann über die chemischen Erkenntnisse nur eine kurze Übersicht geben, soweit sie für das Verständnis der biologischen Arbeiten nötig sind. Die Wiedergabe der biologischen Untersuchungen ist so vollständig wie möglich, weist aber natürlich noch viele Lücken auf, da die Arbeiten auf den meisten Gebieten noch nicht abgeschlossen sind.

II. Die Chemie der Bitterstoffe

Eine Übersicht über alle bekannten Inhaltsstoffe der Cucurbitaceen hat kürzlich Hegnauer gegeben. Von allen bisher untersuchten Stoffen (Triterpen-Saponine, vielleicht auch Alkaloide) sind nur die Bitterstoffe typisch für die Familie.

Die 14 bis jetzt isolierten Bitterstoffe scheinen alle chemisch nahe verwandt zu sein. Sie werden vorläufig als Cucurbitacine A—N bezeichnet. Die Mehrzahl der Cucurbitacine sind C_{30}-Verbindungen. Vier Cucurbitacine [A, B, C und E (α-Elaterin)] enthalten auch eine Acetylgruppe und sind deshalb C_{32} -Verbindungen [Enslin et al. (5)].

Die Struktur der Cucurbitacine ist sehr kompliziert und noch in keinem Fall völlig geklärt. Dehydrogenierungs- und Oxydationsversuche ergaben, daß diese Verbindungen sehr wahrscheinlich zu der Klasse der tetracyclischen Triterpene gehören.

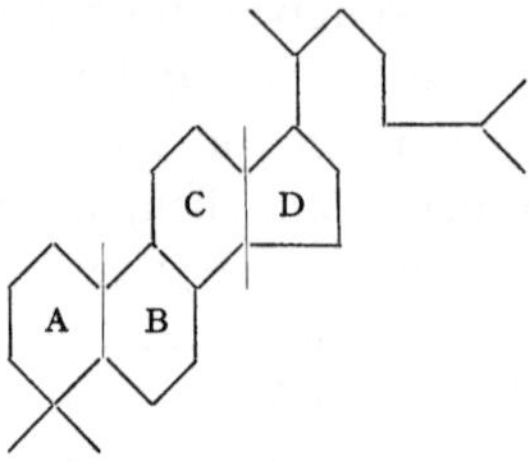

Grundskelet der tetracyclischen Triterpene

Im Vergleich mit anderen Verbindungen dieser Klasse werden die Cucurbitacine durch die große Zahl von Sauerstoffunktionen (7 bis 9) gekennzeichnet. Diese Funktionen bestehen aus einer Acetylgruppe, 3 bis 5 Hydroxyl- und 2 bis 3 Keto-Gruppen.

Vor kurzem wurde die Seitenkette verschiedener Cucurbitacine völlig aufgeklärt [ENSLIN et al. (3)].

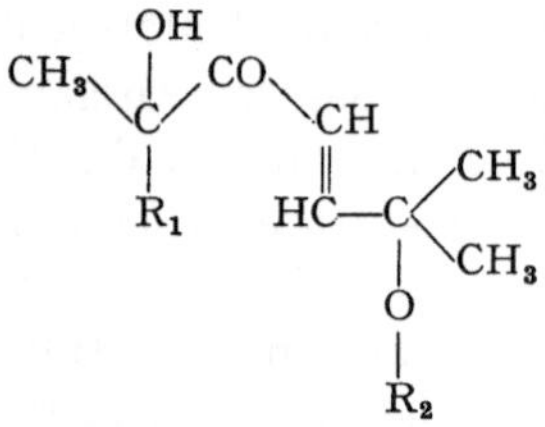

(R_2 = H in Cucurbitacin D und I)
(R_2 = $CO \cdot CH_3$ in Cucurbitacin A, B, C und E)

Seitenkette einiger Cucurbitacine

Das folgende Schema faßt einige der wichtigsten strukturellen Unterschiede der Cucurbitacine und ihre biogenetischen Beziehungen zusammen. Die durch ausgezogene Pfeile verbundenen Stoffe sind rein chemisch und/oder enzymatisch ineinander übergeführt worden. Die biogenetischen Ableitungen, die durch unterbrochene Pfeile angedeutet sind, werden später besprochen (S. 120). Eine Übersicht über die neuesten chemischen Arbeiten ist kürzlich erschienen [LAVIE et al. (1)].

Cucurbitacin M wurde nicht in das Schema aufgenommen, da über seine Struktur noch zu wenig bekannt ist; es gehört zweifellos zu der B-Serie. Eine ganze Reihe neuer Stoffe wurde vor kurzem in einer Form von *Lagenaria siceraria* („Maranka") gefunden. Da sie größtenteils noch nicht isoliert wurden, und über ihre Struktur noch nichts bekannt ist, sprechen wir vorläufig nur von „Maranka-Stoffen". Sie sind aber sehr wahrscheinlich mit den Cucurbitacinen verwandt (s. u.).

Die Cucurbitacine, mit Ausnahme von A, C, F, M und N, kommen häufig als Glykoside vor. Im Falle von Cucurbitacin E handelt es sich um ein mono-Enol-β-Glucosid [ENSLIN et. al. (1)]. Trotz weitgehender Reinigung ist es noch nicht gelungen, ein Cucurbitacin-Glykosid kristallin zu erhalten.

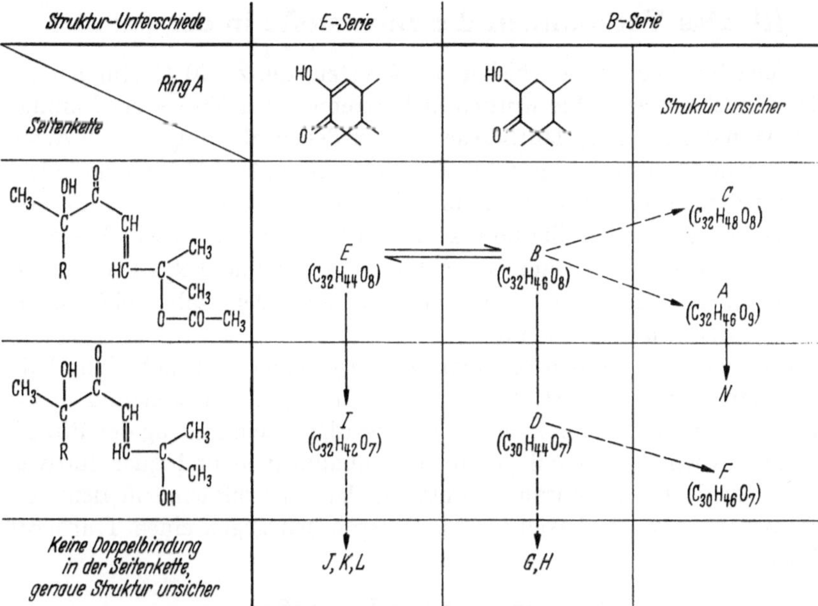

Schema. Die biogenetischen Beziehungen zwischen den Cucurbitacinen

Die Arten der Cucurbitaceen können in zwei Gruppen eingeteilt werden. In der einen Gruppe werden im Preßsaft der Früchte hauptsächlich Aglykone gefunden, in der andern ausschließlich Glykoside. Die Arten der ersten Gruppe enthalten eine sehr spezifische β-Glucosidase, die von BERG den Namen „Elaterase" erhielt. Elaterase wurde in den letzten Jahren isoliert und näher charakterisiert [ENSLIN et al. (1)].

Die Untersuchungen über das Vorkommen und die Biogenese der Cucurbitacine und ganz besonders die genetischen Arbeiten hätten ohne eine schnelle und zuverlässige papierchromatographische Methode, welche die Identifizierung und grob-quantitative Bestimmung der Bitterstoffe gestattet, nicht durchgeführt werden können. R_f-Werte mit verschiedenen Lösungsmitteln und Farbreaktionen mit einer Reihe von Reagenzien erlauben die Bestimmung aller 14 Cucurbitacine und der „Maranka-Stoffe" [REHM et al. (2)].

Wir wissen, daß die Papierchromatographie kein Ersatz für die chemische Charakterisierung einer Verbindung ist. Aus 16 Arten wurden aber verschiedene Cucurbitacine auch in kristalliner Form isoliert und chemisch charakterisiert. Die Ergebnisse stimmten ohne Ausnahme völlig mit der Interpretation der Papierchromatogramme überein. Wir sind deshalb überzeugt, daß die Papierchromatographie der Cucurbitacine durchaus zuverlässige Resultate für biologische Arbeiten liefert.

III. Das Vorkommen der Bitterstoffe in der Familie

Cucurbitacine wurden bisher in 64 Arten, die zu 20 Gattungen gehören, nachgewiesen. Sie wurden in Vertretern aller Tribus der Familie, mit Ausnahme der Cyclanthereae — von denen allerdings nur 2 Arten der Gattung *Cyclanthera* geprüft wurden — gefunden. Allgemeine Folgerungen können einstweilen nur mit Vorbehalt gezogen werden, da erst 7% der Arten der Familie untersucht sind. Listen, die unsere Kenntnis bis 1957 wiedergeben, sind bei Rehm et al. (2) und Enslin et al. (4) veröffentlicht. Im folgenden Bericht sind außerdem zahlreiche unveröffentlichte Befunde verwendet.

Cucurbitacine werden offenbar *in loco* gebildet und nicht innerhalb der Pflanze transportiert. Pfropfungsversuche mit bitteren und nichtbitteren Gurken ergaben keinerlei gegenseitige Beeinflussung der Pfropfpartner. Auch die überall gefundenen qualitativen und quantitativen Unterschiede in der Zusammensetzung der Cucurbitacine in den verschiedenen Organen derselben Pflanze sprechen gegen einen Transport dieser Stoffe.

A. Cucurbitacine in Keimpflanzen

Für das Verständnis der Biogenese und Vererbung der Cucurbitacine war die Entdeckung, daß die Keimpflanzen vieler Arten und kultivierter Varietäten Bitterstoffe enthalten, auch wenn die erwachsenen Pflanzen keine Spur davon zeigen, von großer Bedeutung [Rehm et al. (3)]. Soweit untersucht, sind die Embryonen der reifen Samen immer völlig frei von Bitterstoffen. Sobald die Keimung beginnt, werden oft große Mengen von Cucurbitacinen gebildet. Dieser Prozeß kann bei schnell keimenden Samen in wenigen Stunden ablaufen. Das Maximum der Bitterstoffkonzentration wird in den Keimwurzeln bald (bei einer Länge von 2—5 cm) erreicht. Danach nimmt der Bitterstoffgehalt langsam ab. Bei den meisten kultivierten Varietäten sind die älteren Wurzeln überhaupt nicht mehr bitter. Ähnlich liegen die Verhältnisse bei den Kotyledonen, die meist die höchste Bitterstoffkonzentration kurz nach der Entfaltung aufweisen. Alte Keimblätter haben manchmal nur noch einen sehr geringen Bitterstoffgehalt. Man erhält den Eindruck, daß die Bitterstoffsynthese während der ersten Keimungsstadien sehr intensiv abläuft und in den meisten Fällen nach kurzer Zeit, oft schon nach wenigen Tagen, ganz aufhört.

Die Organe der Keimpflanzen (Keimwurzeln und Kotyledonen) enthalten meist nur *ein* Cucurbitacin, und zwar B oder E. Diese beiden Stoffe sind offenbar die primären Cucurbitacine. Ausnahmsweise kommen B und E gemeinsam vor (einige *Cucurbita*-Arten, *Luffa acutangula*), und manchmal werden kleine Mengen von D zusammen mit B, und von I zusammen mit E gefunden. Das einzige andere Cucurbitacin, das in Keimpflanzen *einer* Art gefunden wurde, ist C, und zwar in den Kotyledonen von *Cucumis sativus*. Cucurbitacin C ist aber offenbar auch hier nicht die zuerst gebildete Verbindung, denn sehr junge Kotyledonen enthalten hauptsächlich B, höchstens mit Spuren von C, und erst in den entfalteten Kotyledonen ist B verschwunden und C nun der einzige Bitterstoff. Die Keimwurzeln von *Cucumis sativus* enthalten immer nur Cucurbitacin B.

Der eben genannte Fall, daß Keimwurzeln einen andern Bitterstoff enthalten als die Kotyledonen, kommt in mehreren Arten vor. So findet sich z. B. in den Keimwurzeln von *Cucurbita palmata* nur E, in den Kotyledonen hauptsächlich B, zusammen mit kleinen Mengen von E. Auch sind nicht immer alle Teile der Keimpflanzen bitter. *Cucurbita ficifolia* hat völlig bitterstofffreie Keimwurzeln, in den Kotyledonen aber eine große Menge Cucurbitacin B; eine Form von *Luffa cylindrica* hatte sehr bittere Keimwurzeln (Cucurbitacin B) und völlig „süße“ Kotyledonen. Die Keimblätter von *Cucumis sativus* sind der einzige bisher bekannt gewordene Fall von qualitativer Änderung der Bitterstoffe während der Entwicklung der Keimpflanzen.

In den Keimpflanzen kommen die Cucurbitacine hauptsächlich als Aglykone vor, auch bei solchen Arten, die später nur noch glykosidische Bitterstoffe bilden.

B. Cucurbitacine in vegetativen Organen

1. Blätter und Sprosse

Über die Zusammensetzung der Cucurbitacine in Blättern und Sprossen liegen nur wenige Angaben vor. Das hat zwei Gründe: einmal sind diese Organe bei vielen Cucurbitaceen mit bitteren Wurzeln oder Früchten frei von Bitterstoffen, oder sie enthalten so geringe Spuren, daß deren Nachweis mit der üblichen Aufbereitungsmethode unsicher oder unmöglich ist; zum andern sind Blätter und Sprosse wirtschaftlich meist bedeutungslos, so daß unter diesem Gesichtspunkt kein Anlaß zu einer gründlicheren Untersuchung vorlag. Allerdings ist die leichte Bitterkeit der Blätter und jungen Sprosse in zwei Fällen für genetisch-züchterische Arbeiten verwendet worden [Enslin et al. (2), Andeweg und de Bruyn].

Hoher Bitterstoffgehalt in den Blättern findet sich selten, und daher sind die beiden Fälle mit sehr hoher Bitterstoffkonzentration in diesen Organen besonders auffällig, und das um so mehr, da in beiden Fällen ein ungewöhnliches Cucurbitacin auftritt. Die Blätter von *Cucumis dinteri* Cogn. enthalten 0,06%[1] Cucurbitacin F zusammen mit anderen Bitterstoffen (vgl. Tab. 8). Die Blätter von *Cucumis asper* Cogn. enthalten 0,1% Cucurbitacin M zusammen mit 0,005% B.

Junge, schnell wachsende Blätter von *Colocynthis vulgaris* Schrad. [= *Citrullus colocynthis* (L.) Schrad.] und *Colocynthis ecirrhosa* (Cogn.) Chakr. (= *Citrullus ecirrhosus* Cogn.) haben einen relativ niedrigen Bitterstoffgehalt (0,01%), alte Blätter und Sprosse am Ende der Vegetationszeit können sehr bitter sein (0,1—0,3% Bitterstoff). Verhältnismäßig hohen Bitterstoffgehalt (0,01—0,03%) haben auch die Blätter mehrerer *Cucurbita*-Arten (*C. cylindrata, foetidissima, lundelliana* und *okeechobeensis*) und die Blätter von *Ecballium elaterium.*

Abgesehen von dem Verhalten der „Spezialisten" *Cucumis asper* und *C. dinteri* sind die Blattbitterstoffe durch das Vorherrschen der Desacetyl-Verbindungen charakterisiert. So bestehen z. B. die Bitterstoffe in den Früchten von *Cucurbita palmata* hauptsächlich aus Cucurbitacin B mit nur 3% seines Desacetyl-Homologen D; die Blätter dieser Art enthalten keine Spur der primären Cucurbitacine, sondern ausschließlich D und etwas I. Die Bitterstoffe der Früchte von *Cucurbita andreana* setzen sich aus 88% B und 12% D, die der Blätter aus 30% B und 70% D zusammen. Zu dieser Beobachtung paßt, daß in den Blättern mehrerer nicht-bitterer *Cucurbita*-Arten eine besonders hohe Acetylesterase-Aktivität gefunden wurde. Doch genügen die untersuchten Fälle wohl noch nicht, eine allgemeine Gesetzmäßigkeit festzustellen. Es gibt auch Arten, bei denen in den Blättern nur die primären Bitterstoffe gefunden wurden *(Colocynthis vulgaris, C. ecirrhosa, Cucurbita cylindrata, C. foetidissima, Lagenaria siceraria).* In den vegetativen Teilen von *Cucumis sativus* kommt nur Cucurbitacin C vor, wie in den Früchten; in den Blättern einer bitterfrüchtigen Form der Gurke („Hanzil") wurde dieselbe Konzentration (0,001%) gefunden wie in denen einer „süßen" Kulturform ("Early Fortune").

2. Wurzeln

Dicke, verholzte Wurzeln oder auch fleischige Wurzelknollen finden sich bei den meisten ausdauernden Arten der Cucurbitaceen. Solche mehrjährige Wurzeln sind häufig bitter. Oft ist die Wurzel das einzige bittere Organ der Pflanzen; Rehm et al. (2) führen 14 Arten mit dieser Bitterstoffverteilung an. Aber auch das umgekehrte Verhalten kommt vor.

---

[1] Wenn nicht anders vermerkt, sind alle Angaben über Bitterstoffgehalt auf das Frischgewicht bezogen.

Cucumis kalahariensis A. MEEUSE ined.[1] und drei *Coccinia*-Arten *(C. adoensis, hirtella* und *quinqueloba)* sind Beispiele für Arten mit bitteren Früchten und bitterstofffreien Wurzelknollen.

In manchen Fällen ist der Bitterstoffgehalt solcher mehrjähriger Wurzeln außerordentlich hoch. *Colocynthis naudiniana* (SOND.) O. KTZE. [= *Citrullus naudinianus* (SOND.) HOOK. f.] hat 1,4% Bitterstoffe in den frischen Wurzeln (auch die Form mit nicht-bitteren Früchten!). *Acanthosicyos horrida,* die Naras, liefert 1,1%, und *Colocynthis ecirrhosa* 0,9% Bitterstoffe aus den lufttrockenen Wurzeln.

Qualitativ unterscheiden sich die Bitterstoffe in den mehrjährigen Wurzeln von denen der oberirdischen Organe oft durch einen besonders hohen Anteil der abgeleiteten Cucurbitacine. Tab. 1 gibt zwei Beispiele.

Tabelle 1. *Bitterstoffe in Früchten und Wurzeln derselben Art*

Colocynthis ecirrhosa			*Cucumis heptadactylus*		
Cucurbitacin	Frucht (frisch) %	Wurzel (trocken) %	Cucurbitacin	Frucht (frisch) %	Wurzel (frisch) %
E	0,3	0,1	B	0,07	0,02
I u. L	—	0,4	D	0,02	0,07
J	—	0,2	G	—	0,05
K	—	0,2	H	—	0,05

Die Listen in REHM et al. (2) und ENSLIN et al. (4) führen noch mehrere Fälle von hohem Gehalt an abgeleiteten Bitterstoffen in Wurzeln an. Mehrjährige Wurzeln waren auch die Quelle für die Isolation dieser Cucurbitacine [ENSLIN et al. (5)].

Wenn Wurzeln auf verschiedenen Altersstufen untersucht werden konnten, wurde immer gefunden, daß am Anfang ausschließlich die primären Cucurbitacine (B und E) auftreten. Nicht nur die Keimwurzeln, sondern auch zwei Monate alte Wurzeln von *Colocynthis ecirrhosa* enthalten ausschließlich Cucurbitacin E (vgl. Tab. 1 für mehrjährige Wurzeln). Keimwurzeln von *Bryonia dioica* enthalten nur E. Mit drei Monaten enthalten die Wurzeln dieser Art etwa gleiche Mengen B und E, und erst mit 18 Monaten etwa 20% D und 20% I neben B und E.

Das Auftauchen von B neben E, das bei *Bryonia* die älteren Wurzeln von den Keimwurzeln unterscheidet, findet sich in entsprechender Weise auch bei anderen Arten. So enthalten die Keimwurzeln von *Cucurbita foetidissima* nur Spuren von B zusammen mit E, die Wurzeln drei Monate alter Pflanzen aber 0,09% B und nur noch 0,006% E.

[1] Herr Dr. A. D. J. MEEUSE ist mit einer Revision der südafrikanischen Cucurbitaceen beschäftigt, die voraussichtlich noch in diesem Jahr erscheinen wird. Dort wird die Diagnose von *C. kalahariensis* veröffentlicht werden. Angaben über diese Art finden sich bei STORY (S. 50, „*Cucumis* spec.").

Die Wurzeln einjähriger Cucurbitaceen wurden nur gelegentlich untersucht. In den meisten Fällen waren die Wurzeln nicht oder nur sehr wenig bitter, auch bei solchen Formen, die bittere Früchte tragen. Ein Zusammenhang mit der Bitterkeit der Früchte wurde nur bei *Lagenaria siceraria* gefunden; bei dieser Art haben alte Pflanzen mit bitteren Früchten auch deutlich bittere Wurzeln (Bitterstoffgehalt 0,01%), Pflanzen mit „süßen" Früchten haben bitterstofffreie Wurzeln.

C. Cucurbitacine in Früchten

Früchte sind die am meisten untersuchten Organe. Bittere Früchte sind besonders auffallend und praktisch am wichtigsten, sowohl in Hinsicht auf Vergiftungen von Menschen und Tieren als auch für züchterische Arbeiten. Außerdem stellen Früchte die am leichtesten zugängliche Quelle von Bitterstoffen für chemische Arbeiten dar.

Es ist nicht die Absicht, in diesem Bericht Einzelheiten über alle untersuchten Arten zu geben. Es soll aber versucht werden, ein allgemeines Bild von der quantitativen und qualitativen Biogenese der Bitterstoffe zu geben. Das ist freilich nur beschränkt möglich, weil die einzelnen Arten eine große Mannigfaltigkeit und manchmal ein geradezu gegensätzliches Verhalten zeigen.

1. Quantität der Bitterstoffe

Uns ist nur ein Fall bekannt [*Cucumis anguria* L. var. *longipes* (HOOK. f.) A. MEEUSE (MEEUSE (1))], daß die Fruchtknoten einer Art zur Zeit der Blüte keine Spur von Bitterstoffen enthalten und erst nach der Befruchtung mit der Bildung von Cucurbitacin beginnen (vgl. Abb. 2, S. 119). Häufiger ist der Fall, daß die Fruchtknoten vor der Befruchtung einen gewissen, manchmal recht niedrigen Bitterstoffgehalt haben, der in den ersten Tagen nach der Befruchtung schnell ansteigt, so daß die jungen Früchte eine besonders hohe Bitterstoffkonzentration aufweisen; mit zunehmender Fruchtgröße fällt dann die Konzentration wieder, während der Gesamtbitterstoffgehalt der Frucht zunimmt, solange das Wachstum anhält. Besonders ausgeprägt ist dieses Verhalten bei der von uns untersuchten bitteren Form der Wassermelone „Hawkesbury" mit ihrem schnellen Fruchtwachstum (tägliche Gewichtszunahme in der Hauptwachstumszeit rund 500 g!) (Abb. 1).

Gewöhnlich ist der Gesamtbitterstoffgehalt am höchsten in den reifen Früchten. Es kommt aber auch vor, daß mit einsetzender Fruchtreife die Cucurbitacine in wenigen Tagen völlig verschwinden. Das auffallendste Beispiel ist die Naras, deren voll ausgewachsene, aber noch grüne Früchte eine recht hohe Bitterstoffkonzentration (0,12%) haben. Die gelblichen, reifen Früchte sind aber nicht mehr bitter und bilden ein wichtiges Nahrungsmittel der Hottentotten in der Namib. Ein anderes

Beispiel ist *Coccinia adoensis*, bei der die Farbänderung der reifenden Früchte von Grün nach Rot mit dem völligen Verschwinden des Bitterstoffes gepaart ist. Ebenso verlieren die Früchte von *Cucumis asper* jede Spur von Bitterkeit bei der Reife.

Bei Arten, von denen genetisch reine Linien zur Verfügung stehen (Wassermelone, *Lagenaria siceraria*, *Cucurbita pepo*) findet sich eine deutliche negative Korrelation zwischen Fruchtgröße und Bitterstoffkonzentration ($r = -0{,}45$ bei der bitteren Wassermelone „Hawkesbury"). Jedoch ist der Gesamtbitterstoffgehalt großer Früchte trotzdem meist höher als der kleiner Früchte ($r = +0{,}79$ für Bitterstoffgehalt/Fruchtgewicht bei der Wassermelone).

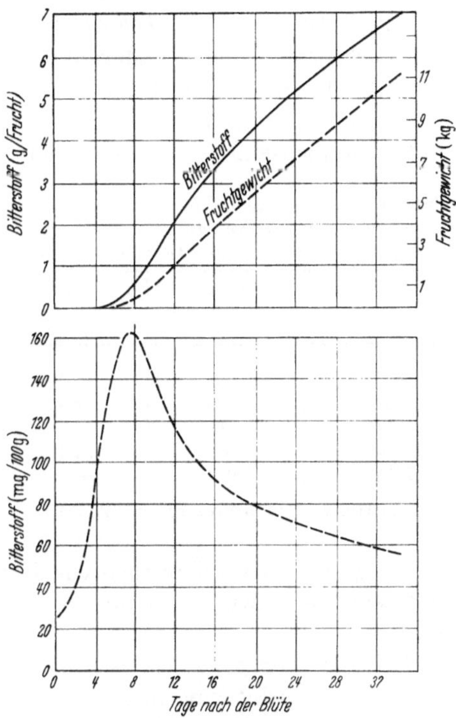

Abb. 1. Bittere Mutante der Wassermelone „Hawkesbury". Bitterstoffbildung in den Früchten von der Blüte bis zur Reife

Die Verteilung des Bitterstoffes innerhalb der Frucht kann auch stark variieren. Am häufigsten findet man, daß das Fruchtfleisch ziemlich gleichmäßig bitter ist, während die Schale gewöhnlich nicht oder nur schwach bitter ist. Bei andern Arten ist die Bitterstoffkonzentration in dem lockeren Gewebe, das die Samen umgibt, höher als in dem festen äußeren Fleisch der Frucht (z. B. Formen von *Cucurbita pepo* und *Lagenaria siceraria*). Bei einer Form von *Cucurbita mixta* ist dieser Teil der Frucht sogar völlig frei von Bitterstoffen, während das Placentargewebe recht bitter schmeckt. Die Verteilung kann aber auch umgekehrt sein: *Cucumis humifructus* hat bitteres äußeres Fruchtfleisch und nichtbitteres saftiges Gewebe um die Samen, eine Verteilung, der hier möglicherweise biologische Bedeutung zukommt, da das Erdferkel *(Orycteropus afer)* offenbar den inneren Teil der Früchte als Wasserquelle gebraucht [Meeuse (2)].

2. *Qualitative Zusammensetzung der Bitterstoffe*

Im Vergleich zu den vegetativen Organen fällt zunächst auf, daß in Früchten die primären Cucurbitacine B und E stark vorherrschen. Eine ganze Reihe von Arten haben ausschließlich B oder E in ihren Früchten.

Häufig kommen kleine Mengen der Desacetylverbindungen zusammen mit den primären Bitterstoffen vor. Manchmal finden sich auch Spuren von G und H oder J und K in Früchten, aber nie in ähnlichen Mengen wie in manchen Wurzeln. Die Fälle, in denen abgeleitete Bitterstoffe die Hauptmenge der Cucurbitacine in Früchten bilden, sind selten. Wir kennen nur: *Cucumis dinteri* (Cucurbitacin D), *Cucumis hookeri*, *C. leptodermis* und *C. myriocarpus* (Cucurbitacin A), *Cucumis sativus* (Cucurbitacin C), *Cucurbita okeechobeensis* (Cucurbitacin D und I), und die Form von *Lagenaria siceraria* („Maranka") mit den neuen, noch nicht identifizierten Stoffen (s. o. S. 110).

In einigen dieser Fälle wurde die qualitative Änderung in der Bitterstoffzusammensetzung während der Fruchtentwicklung verfolgt. Mit Ausnahme von *Cucumis sativus*, der vom Blühstadium an nur Cucurbitacin C enthält, ergab sich, daß die abgeleiteten Bitterstoffe erst im Laufe der Fruchtentwicklung entstehen, während die primären Bitterstoffe zurücktreten. Kleine, grüne Früchte von *Cucumis myriocarpus* enthalten 70% B und 30% A, reife, braun-gelbe Früchte 25% B und 75% A. Das Vorherrschen von B in jungen Früchten von Marankas und sein Verschwinden während der Fruchtentwicklung gibt Tab. 2 wieder.

Tabelle 2. *Änderung in der Bitterstoffzusammensetzung von bitteren Marankas (Lagenaria siceraria) während der Fruchtentwicklung*

Fruchtgewicht g	Gesamt-Bitterstoff mg	Anteil von Cucurbitacin B %
7	5	70
18	16	40
29	32	15
60	73	12
1500[1]	1800[1]	0—5

[1] Mittelwert mehrerer reifer Früchte.

Bei *Cucurbita pepo* gibt es Linien, deren überreife Früchte 50% und mehr der Cucurbitacine in der desacetylierten Form enthalten. In den jungen Früchten dieser Linien findet man nur B und E, trotz hoher Acetylesterase-Aktivität des Fruchtfleisches. Offenbar ist das Enzym in den Zellen der jungen Früchte von den Cucurbitacinen getrennt, und erst mit dem Einsetzen plasmatischer Veränderungen in den überreifen Früchten kann die Acetylesterase auf die Cucurbitacine einwirken.

Im Zusammenhang mit den biogenetischen Beziehungen, die zwischen den beiden primären Cucurbitacinen bestehen, ist die Änderung, die in Früchten von *Cucurbita pepo* im Verhältnis von B und E während der Fruchtentwicklung stattfindet, von Interesse. Fruchtknoten einer Linie, die im reifen Zustand fast nur E in den Früchten enthält, haben zur Blütezeit 15% B neben E. Ähnlich verhält sich eine Linie, bei der in den reifen Früchten 85% B und 15% E gefunden werden; ihre Fruchtknoten enthalten zur Blütezeit ausschließlich B; E wird erst nach der Befruchtung allmählich gebildet.

3. Glykoside und Aglykone

Bei der Besprechung der Chemie der Cucurbitacine wurde erwähnt, daß die Bitterstoffe in Früchten als Glykoside oder als Aglykone vorkommen können. Ausschließlich glykosidisch sind die Bitterstoffe in den meisten der untersuchten Arten von *Coccinia, Colocynthis, Echinocystis* und *Peponium.* Vorwiegend oder ausschließlich als Aglykone finden sich die Cucurbitacine in den Gattungen *Acanthosicyos, Cucumis* und *Lagenaria.* Allen Früchten mit glykosidischen Bitterstoffen fehlt Elaterase. Eine Ausnahme ist *Ecballium elaterium,* das glykosidische Bitterstoffe trotz hoher Elaterase-Aktivität in den Früchten enthält. Da bei Autolyse des Fruchtbreis die Glykoside schnell gespalten werden, muß in den Zellen der frischen Früchte die Elaterase von den Bitterstoffen räumlich getrennt sein.

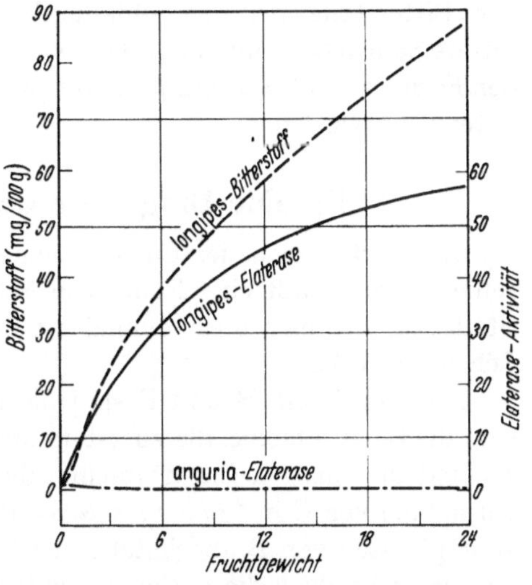

Abb. 2. Bitterstoffkonzentration und Elaterase-Aktivität in wachsenden Früchten der beiden Varietäten von *Cucumis anguria*

Mehrere Arten von *Cucumis* und *Lagenaria* kommen in bitteren und nicht-bitteren Formen vor. Ohne Ausnahme zeigen die nicht-bitteren Formen eine sehr niedrige Elaterase-Aktivität[1]. Die Gegenwart der Cucurbitacine scheint für die Bildung der Elaterase nötig zu sein. Sehr schön wird der Zusammenhang von Bitterstoff und Elaterase in der Entwicklung der Früchte von *Cucumis anguria* deutlich (Abb. 2). *C. anguria* var. *anguria* hat nicht-bittere Früchte. Die Elaterase-Aktivität bleibt von der Blüte bis zur Fruchtreife niedrig. Bei *C. anguria* var. *longipes* sind die Fruchtknoten zur Blütezeit ebenfalls nicht bitter und besitzen dieselbe niedrige Elaterase-Aktivität wie die Fruchtknoten von var. *anguria.* Nach der Befruchtung wird Bitterstoff gebildet, und gleichzeitig steigt die Elaterase-Aktivität. Wie weit echte biochemische Enzyminduktion durch das Substrat oder eine ungewöhnliche doppelte Genwirkung auf Enzym- und Bitterstoff-Bildung vorliegen, konnte noch nicht geklärt werden [REHM (1)].

[1] Die Elaterase-Aktivität bestimmen wir durch die Zeit (in Minuten), die nach Beifügung von 1 ml Fruchtsaft zu einer Standard-Elaterinidlösung verstreicht, bis eine Trübung eintritt. Die Trübung ist das Zeichen für den Beginn der Elaterin-Ausfällung. Zahlenmäßig wird die Elaterase-Aktivität ausgedrückt als $\frac{1}{\text{min}} \times 100$.

D. Cucurbitacine in Samen

Unter den von uns untersuchten Arten befinden sich nur drei mit bitteren Samen[1]. Auch bei diesen sind die Embryonen völlig frei von Bitterstoffen. Bei *Luffa acutangula* und *Sicyos angulata* ist der Bitterstoff im Perisperm, einem dünnen Häutchen, das zum Gewebe der Mutterpflanze gehört, lokalisiert. Bei *Telfairia pedata* ist nur das netzartige Fasergewebe, das den Samen umhüllt, bitter. Es wäre von großem Interesse, wenn die Lokalisation und chemische Natur der Bitterstoffe bei den südamerikanischen Gattungen, von denen bittere Samen bekannt sind (*Anisosperma, Cayaponia, Fevillea*), und bei den Formen von *Cucurbita pepo* mit bitteren Samen [GREBENŠČIKOV (1)] untersucht würden.

IV. Die Biogenese der Cucurbitacine

Das Vorkommen der Cucurbitacine in verschiedenen Organen derselben Pflanze läßt eine Reihe von Schlüssen auf die Biogenese dieser Stoffe zu, die hier noch einmal zusammengefaßt werden sollen (vgl. Schema S. 111).

1. Cucurbitacin B und E sind die beiden primären Bitterstoffe. Es sind die Cucurbitacine, die zuerst in keimenden Samen gebildet werden. Es sind die einzigen Cucurbitacine, die allein vorkommen können, mit Ausnahme von C in *Cucumis sativus*, das aber nach dem Verhalten der Keimpflanzen von B abgeleitet erscheint. Endlich sind B und E oft die einzigen oder doch die vorherrschenden Bitterstoffe in jungen Organen, in denen später auch andere Cucurbitacine auftreten.

2. Cucurbitacin B und E können ineinander übergeführt werden. Diese Reaktion verläuft *in vitro* mit Fruchtsaft verschiedener Arten (z. B. *Lagenaria siceraria, Cucumis melo*). Das Verhalten der Bitterstoffe in manchen Wurzeln oder in Früchten von *Cucurbita pepo*, das oben beschrieben wurde, deutet auf dieselbe Umsetzung in der Pflanze.

3. Von B leitet sich zunächst die Hauptreihe B→D→G und H ab. Die Abspaltung der Acetylgruppe (Bildung von D) ist sehr häufig; die meisten Arten, die B enthalten, weisen daneben kleine Mengen von D in den Früchten und manchmal erhebliche Mengen davon in Blättern oder Wurzeln auf. Von D seinerseits scheinen G und H abgeleitet zu sein, die niemals ohne die Gegenwart von D vorkommen.

Neben der Hauptreihe gibt es mehrere seltene Cucurbitacine, die auf anderen Wegen mit B verbunden zu sein scheinen. Cucurbitacin A kommt immer mit B zusammen vor. In jungen Früchten von *Cucumis myriocarpus* erscheint es zunächst in kleinen Mengen neben B und wird erst

[1] Die Angabe bei HEGNAUER, daß *Cucumis africanus, C. leptodermis* und *C. myriocarpus* bittere Samen besitzen, beruht auf einem Irrtum.

in den reifen Früchten der vorherrschende Stoff. Die Keimpflanzen dieser Art enthalten nur B.

Der Zusammenhang von Cucurbitacin C mit B wird durch die eben erwähnte Zusammensetzung der Bitterstoffe in den Keimpflanzen von *Cucumis sativus* deutlich. Cucurbitacin F, der charakteristische Bitterstoff in den Blättern von *Cucumis dinteri*, ist wahrscheinlich auch von B abzuleiten. B ist der Bitterstoff in den Keimpflanzen dieser Art und kommt in ihren Früchten, Blättern und Wurzeln zusammen mit Stoffen der B-Hauptreihe vor. Das gleiche Argument gilt für den Zusammenhang von Cucurbitacin M mit B. Die Früchte von *Cucumis asper* enthalten hauptsächlich B und etwas M, und auch in den Blättern ist immer etwas B neben M vorhanden. Außerdem wurde M in kleinen Mengen in mehreren Arten, die Bitterstoffe der B-Hauptserie enthalten, und nur in solchen, gefunden.

Eine letzte Gruppe von Stoffen, die mit B in Verbindung zu stehen scheinen, sind die „Maranka-Stoffe". In *Lagenaria siceraria* herrschen allgemein Cucurbitacine der B-Serie vor. In den jungen Früchten der „Marankas" erscheint zunächst hauptsächlich Cucurbitacin B (Tab. 2). Die Blätter und Wurzeln der Form mit den „Maranka-Stoffen" in den Früchten enthalten dieselben Cucurbitacine (hauptsächlich B) wie die Formen, die nur B in den Früchten enthalten.

4. Von E leitet sich eine der B-Hauptserie völlig parallele Serie (E→I→J und K) ab. Die Desacetylverbindung (Cucurbitacin I) kommt ebenso häufig neben E vor wie D neben B. J und K erscheinen nie ohne die Gegenwart von I. Die völlige Parallelität der B- und E-Serien wird sehr schön durch einige Fälle, in denen beide Serien zusammen auftreten, beleuchtet. *Colocynthis naudiniana* enthält nur Stoffe der B-Hauptserie in den Früchten, in den Wurzeln B, D, G, H und E, I, J, K. *Kedrostis* spec. (= *Toxanthera natalensis*) enthält B, D, G und H in den Wurzeln, die verwandte *Kedrostis africana* enthält sowohl B, D, G, H als auch E, I, J, K.

Ein weiteres Cucurbitacin, das von E abzuleiten ist, scheint L zu sein. Es wurde bisher nur in Wurzeln von *Colocynthis ecirrhosa* gefunden, die daneben fast ausschließlich Verbindungen der E-Serie enthält; über seine Beziehungen zu anderen Cucurbitacinen der E-Serie können wir noch nichts sagen.

V. Die Vererbung der Cucurbitacine

Die Erforschung der Vererbung der Bitterkeit steht naturgemäß noch in den ersten Anfängen, da die Kenntnis der Cucurbitacine und der Methoden für ihre schnelle Bestimmung (Papierchromatographie) erst wenige Jahre alt ist. Der folgende Bericht kann nur ein vorläufiges Bild geben, das in Zukunft noch manche Veränderungen erfahren wird. Die Darstellung beruht zum größten Teil auf unveröffentlichten Arbeiten.

Die verschiedenen Gene, die auf die Bildung der Cucurbitacine Einfluß haben, können in fünf Hauptgruppen gegliedert werden:

1. Das Hauptgen für Bitterstoffbildung, das gewöhnlich in Keimpflanzen zum Ausdruck kommt. Bisher ist nur ein Fall bekannt, daß Pflanzen mit nicht-bitteren Keimblättern später bittere vegetative Organe bildeten (*Cucumis sativus*, ANDEWEG und DE BRUYN); doch wurde in diesem Fall die Bitterkeit der Keimwurzeln nicht geprüft, so daß möglicherweise ein ähnliches Verhalten wie bei *Luffa cylindrica* (s. S. 113) vorliegt. Die meisten kultivierten Varietäten von Wassermelonen, Kürbissen und Gurken haben dieses Gen.
2. Gene, welche die Bildung von Bitterstoffen in späteren Entwicklungsstadien und in einzelnen Organen unterdrücken. Solche Gene sind nur bei Früchten untersucht, müssen aber auch für andere Organe (Wurzeln, Blätter) wirksam sein.
3. Gene, welche die Quantität der Cucurbitacine beeinflussen.
4. Gene, welche die Art der Cucurbitacine beeinflussen.
5. Gene für die Bildung von Elaterase, die auch darüber entscheiden können, ob die Cucurbitacine als Glykoside oder Aglykone vorkommen.

A. Die Vererbung der Bitterkeit von Keimpflanzen

Varietäten mit bitteren und nicht-bitteren Keimpflanzen der folgenden Arten wurden gekreuzt: *Cucumis melo* (zwei Kreuzungen; ANDEWEG und DE BRUYN und eigene Untersuchungen); *Cucumis sativus* (eine Kreuzung; ANDEWEG und DE BRUYN); *Cucurbita maxima* (eine Kreuzung); *Cucurbita moschata* (eine Kreuzung); *Cucurbita pepo* (vier Kreuzungen). In allen Fällen war die F_1 bitter und die F_2 spaltete im Verhältnis 3 bitter : 1 nicht-bitter. Reziproke Kreuzungen bei *Cucurbita pepo* ergaben identische Resultate. Die Dominanz des Gens für Bitterkeit scheint bei *Cucumis* vollkommen zu sein. Bei allen *Cucurbita*-Kreuzungen hatten die Heterozygoten einen wesentlich niedrigeren Bitterstoffgehalt als die homozygot bitteren Eltern (Tab. 3).

Bei den in Tab. 3 angeführten und bei allen andern von uns untersuchten Kreuzungen fanden sich keine nennenswerten Unterschiede in der qualitativen Zusammensetzung der Cucurbitacine in bitteren Eltern und F_1-Pflanzen.

B. Die Vererbung der Fruchtbitterkeit

Nach unserer Hypothese [REHM et al. (3)] beruht die Bitterkeit der Früchte nicht auf einem Gen, das selbständig Bitterstoffbildung in den Früchten bewirkt. Das Fehlen von Bitterstoffen in den Früchten von Pflanzen, deren Keimpflanzen oder andere vegetative Teile bitter sind, wird auf ein Suppressor-Gen zurückgeführt. Bittere Früchte entstehen, wenn das inaktive Allel des Suppressor-Gens anwesend ist. Der Erfolg von ANDEWEG und DE BRUYN, die auf dem Weg über die Auslese nichtbitterer Keimpflanzen Gurken erhielten, die keinen Bitterstoff in den Früchten bilden, bestätigt unsere Hypothese.

Tabelle 3. *Bitterstoffkonzentration in homozygot bitteren und heterozygoten Keimpflanzen von Cucurbita-Arten*

Cucurbita maxima

Varietät	Iron Bark				Iron Bark X Big Tom F_1			
Organ	Wurzeln		Kotyledonen		Wurzeln		Kotyledonen	
Cucurbitacin	B	E	B	E	B	E	B	E
Konzentration	0,09%	—	0,05%	—	0,001%	—	0,002%	—

Cucurbita moschata

Varietät	Ceylon				Ceylon X Butternut F_1			
Organ	Wurzeln		Kotyledonen		Wurzeln		Kotyledonen	
Cucurbitacin	B	E	B	E	B	E	B	E
Konzentration	0,1%	0,05%	0,02%	—	0,01%	Spur	0,01%	—

Cucurbita pepo

Varietät	Little Gem				Little Gem X White Custard F_1			
Organ	Wurzeln		Kotyledonen		Wurzeln		Kotyledonen	
Cucurbitacin	B	E	B	E	B	E	B	E
Konzentration	—	0,05%	0,05%	0,03%	—	0,005%	0,008%	0,003%

Eine weitere Klärung der Zusammenhänge erhoffen wir von einer Kreuzung der Kürbisse „Little Gem" (bitter-früchtige Form mit bitteren Keimpflanzen) × „Table Queen" (nicht-bittere Keimpflanzen und Früchte). Die F_1 dieser Kreuzung hatte erwartungsgemäß bittere Keimpflanzen und bittere Früchte. Die F_2 liegt noch nicht vor.

Bei allen Kreuzungen, über die in der Literatur berichtet wird, wurde gefunden, daß die Fruchtbitterkeit monofaktoriell-dominant vererbt wird. Bei früheren Autoren beruht diese Angabe nur auf dem Geschmack, durch den die Konzentration unsicher und die Art des Bitterstoffes gar nicht festgestellt werden kann [*Cucurbita maxima* × *andreana*: CONTARDI; *Lagenaria siceraria*: PATHAK und SINGH; *Cucumis sativus*: BARHAM; *Colocynthis citrullus, Cucurbita pepo*: GREBENŠČIKOV (2, 3); *Cucurbita mixta* × *pepo*: GREBENŠČIKOV (4)].

Auch in allen von uns untersuchten Fällen wurde monofaktoriell-dominante Vererbung der Fruchtbitterkeit gefunden. Die Spaltungszahlen für die F_2 folgender Kreuzungen liegen vor:

Colocynthis citrullus. Wassermelone „Hawkesbury", bittere Mutante × nicht-bittere normale Form.

Cucumis anguria var. *anguria* (nicht-bitter) × var. *longipes* (bitter).

Cucumis sativus. Gurke „Hanzil" (bitter) × „Early Fortune" (nicht-bitter).

Cucurbita pepo. Mehrere Kreuzungen von bitteren „Little Gem" und bitteren Zierkürbissen × nicht-bitteren Formen derselben Varietäten.

Lagenaria siceraria. Großfrüchtige und kleinfrüchtige bittere Kalebasse × großfrüchtige nicht-bittere Kalebasse; großfrüchtige bittere Kalebasse × nicht-bittere „Maranka".

Mit einer Ausnahme, bei der zusätzliche Gene zur Wirkung kamen (s. nächster Abschnitt), war der Bitterstoffgehalt der F_1 ebenso hoch wie der des homozygot-bitteren Elters. Unterschiede in der Fruchtgröße der Eltern machten sich in der oben (S. 117) angegebenen Weise bemerkbar. Zum Beispiel hat die wilde bittere Gurke „Hanzil" ein mittleres Fruchtgewicht von 65 g und eine Bitterstoffkonzentration von 112 mg/100 g Frischgewicht. Der F_1-Bastard mit „Early Fortune" hat ein durchschnittliches Fruchtgewicht von 165 g und eine Bitterstoffkonzentration von 36 mg/100 g. Hier ist zufällig der Bitterstoffgehalt pro Frucht bei bitterem Elter und F_1 ungefähr derselbe (rund 70 mg pro Frucht).

C. Quantitative Gene

Der Zusammenhang zwischen Fruchtgröße und Bitterstoffkonzentration, wobei erstere genetisch oder modifikatorisch beeinflußt sein kann, erschwert die Analyse quantitativer Gene. Auch andere Faktoren können den Bitterstoffgehalt modifizieren. Bekannt ist die Labilität der Bitterkeit europäischer Treibhausgurken (ANDEWEG und DE BRUYN), bei der die Hydratur der Pflanzen während der Fruchtentwicklung entscheidend zu sein scheint (VAN WINDEN). Nach eigenen Untersuchungen erniedrigt die Reduktion der Blattfläche (durch Wegschneiden der Blätter) den Bitterstoffgehalt der Früchte; Blattkrankheiten (Mehltau, Anthraknose) haben vermutlich denselben Effekt.

Daß Gene oder Allele, welche die Bitterstoffkonzentration verändern, neben dem Hauptgen für Bitterkeit der Früchte vorkommen, unterliegt in mehreren Fällen keinem Zweifel. Das Gen, welches die stabile Bitterkeit der wilden Gurke „Hanzil" verursacht, muß verschieden sein von dem genetischen Mechanismus, welcher der labilen Bitterkeit der Treibhausgurken zugrunde liegt. In unserem Versuchsmaterial finden sich bei drei Arten *(Cucurbita pepo, Colocynthis citrullus, Lagenaria siceraria)* Linien, deren unterschiedliche Bitterstoffkonzentration genetisch bedingt ist.

Bei *Cucurbita pepo* haben wir Linien mit niedrigem, mittlerem und hohem Bitterstoffgehalt. Tab. 4 gibt den mittleren Bitterstoffgehalt der drei Linien, zusammen mit der Bitterstoffkonzentration in den Fruchtknoten einen Tag nach der Blüte. Es ist bemerkenswert, daß die Bitterstoffkonzentration der niedrigen und mittleren Linie am Anfang der Fruchtentwicklung identisch ist. Bei der niedrigen Linie sinkt der Bitterstoffgehalt auf rund ein Fünftel des Ausgangswertes, während er bei der mittleren Linie ungefähr konstant bleibt.

Tabelle 4. *Bitterstoffkonzentration in jungen und reifen Früchten von drei Linien von Cucurbita pepo*

Zuchtnummer	Bitterstoff im Fruchtknoten 1 Tag nach der Blüte mg/100 g	Bitterstoff in reifen Früchten mg/100 g
5-2-11	70	15
12-22-8	70	60
9-6-15	290	170

In der Kreuzung niedrig × mittel ist die höhere Bitterstoffkonzentration dominant. Die Kreuzung niedrig × hoch ergibt eine F_1 mit der etwa intermediären Bitterstoffkonzentration von 50. In Kreuzungen mit nicht-bitteren Linien ist die niedrige und mittlere Bitterstoffkonzentration völlig dominant, die hohe intermediär. Nur von der Kreuzung hoch × niedrig und hoch × nicht-bitter liegt die F_2 vor. In beiden Fällen finden sich darin nur wenige Pflanzen mit der hohen Bitterstoffkonzentration des einen Elters (3 aus 60, und 3 aus 47); für die hohe Bitterstoffkonzentration scheinen demnach zwei recessive Gene verantwortlich zu sein.

Von *Colocynthis citrullus* haben wir die bittere Mutante der Wassermelone „Hawkesbury“ und bittere Wildformen („Tsamma“, „Karkoer“).

Tabelle 5. *Bitterstoffgehalt der bitteren Mutante von Colocynthis citrullus Var. „Hawkesbury“, einer wilden „Tsamma“ und der Kreuzung bittere „Hawkesbury“ × nicht bittere „Tsamma“*

Varietät	Fruchtgewicht g	Bitterstoffkonzentration mg/100 g	Bitterstoffgehalt einer Frucht mg
Hawkesbury	10500	63	6615
Tsamma	500	77	385
Hawkesbury × Tsamma	3000	36	1080

Alle analysierten Wildformen hatten einen wesentlich niedrigeren Bitterstoffgehalt als die Mutante von „Hawkesbury“. Tab. 5 gibt Daten für die bittere Mutante, eine Tsamma, und die Kreuzung bittere Mutante × nicht-bittere Tsamma. Die F_2 dieser Kreuzung liegt noch nicht vor. Wir haben aber zahlreiche Nachkommen natürlicher Hybriden von kultivierten Wassermelonen × Tsamma, die wir von Farmern erhielten.

untersucht und fanden darin immer einen hohen Anteil von Pflanzen, deren Früchte einen sehr niedrigen Bitterstoffgehalt hatten (z. B. 8 mg Bitterstoff/100 g bei einem Fruchtgewicht von 2100 g, Bitterstoffgehalt der ganzen Frucht also 168 mg). Gegenüber der bitteren Mutante der Wassermelone „Hawkesbury" muß in den Tsamma-Kreuzungen mindestens ein anderes dominantes Allel oder Gen vorliegen, das den Bitterstoffgehalt erniedrigt. Goldhausen behauptet, daß in Kreuzungen zwischen kultivierten und wilden Formen von *Colocynthis citrullus* die Bitterkeit durch polymere dominante Gene vererbt werde; freilich hat sie keine quantitativen Bestimmungen des Bitterstoffgehaltes ausführen können, und sie gibt auch keine Spaltungszahlen für die F_2.

Bei *Lagenaria siceraria* liegen noch keine Kreuzungsergebnisse vor, doch haben wir reine Linien, deren Bitterstoffkonzentration sich signifikant unterscheidet. Zwei Beispiele gibt Tab. 6.

Tabelle 6. *Zuchtlinien von Lagenaria siceraria mit verschiedenem Bitterstoffgehalt*

Zuchtnummer	Fruchtgewicht g	Bitterstoff-konzentration mg/100 g	Bitterstoffgehalt einer Frucht mg
D 6—8	1250	350	4375
K 2—1	1700	120	2040

D. Qualitative Gene

Über genetisch reine Linien mit qualitativen Unterschieden in den Bitterstoffen verfügen wir bei *Cucurbita pepo* und *Lagenaria siceraria.* Die Linien von *Cucurbita pepo* sind die folgenden:

1. 5% Cucurbitacin B und 95% E;
2. 80% Cucurbitacin B und 20% E;
3. Linien mit hohem oder niedrigem B-Gehalt und hohem Anteil von Desacetylverbindungen (D oder I) im überreifen Zustand.

In der F_1 der Kreuzung von (1) × (2) erweist sich der hohe E-Gehalt als völlig dominant. Die F_2 ergibt eine eindeutige Spaltung von 3 hoch E: 1 niedrig E. Das Gen für hohen E-Gehalt ist offenbar unabhängig von den quantitativen Genen.

Über die Gene, welche die Bildung der Desacetylverbindungen lenken, kann noch nicht viel gesagt werden. In der F_1 scheint die Fähigkeit zur Bildung von D und I zu dominieren. Die Werte zeigen aber starke Variabilität, da es unmöglich ist, alle Analysen an Früchten desselben Reifezustandes durchzuführen. Bei nicht-bitteren Formen der Varietät „Little Gem" konnten Linien mit hoher und mit niedriger Acetylesterase-Aktivität in den reifen Früchten isoliert werden. Kreuzungen dieser Linien untereinander und mit bitteren Linien liegen noch nicht vor.

Bei *Lagenaria siceraria* ergab die Kreuzung bittere Kalebasse (Hauptbitterstoff B) × nicht-bittere Maranka eine F_1, die nur die „Maranka-Stoffe" enthielt. Die F_2 spaltete in der in Tab. 7 angegebenen Weise auf. Die Spaltung zeigt, daß die Bildung der „Maranka-Stoffe" von einem einzigen dominanten Gen abhängig ist. Da gleich eine ganze Serie neuer

Tabelle 7. F_2 *der Kreuzung bittere Kalebasse × nicht-bittere „Maranka"*

	Bitter, Maranka-Stoffe	Bitter, Cucurbitacin B	nicht bitter
Gefunden	93	28	35
Erwartet für 9 : 3 : 4	88	29	39

Verbindungen gebildet wird, muß dieses Gen eine biochemische Umschaltung zu einem neuartigen Ausgangsstoff bewirken, aus dem in der Pflanze vorhandene chemische Systeme sofort die ganze Serie neuer Verbindungen entstehen lassen.

Welch komplizierte Verhältnisse in qualitativer und quantitativer Beziehung bei der genetischen Kontrolle der Bitterstoffbildung herrschen können, zeigt das Verhalten der Hybriden zwischen *Cucumis angolensis* × *C. dinteri*. Die Bitterstoffzusammensetzung in Blättern und Früchten

Tabelle 8. *Bitterstoffzusammensetzung in Cucumis angolensis, C. dinteri und im Bastard C. angolensis × C. dinteri*

Cucurbitacine	*angolensis* %	*dinteri* %	*angolensis* × *dinteri* %
		Blätter	
B	0,0005	0,01	0,008
D	—	0,08	0,0025
F	—	0,06	—
G	—	0,005	Spur
H	—	0,005	Spur
		Früchte	
B	0,015	0,005	0,064
D	Spur	0,06	0,003
G	—	0,005	Spur
H	—	0,005	Spur

der Eltern und des Bastards wird in Tab. 8 gegeben. Die hohe Bitterstoffkonzentration von *C. dinteri* ist in den Früchten dominant, in den Blättern intermediär. Der Faktor für Bildung von D ist in den Blättern und Früchten recessiv, und recessiv ist auch der Faktor für die Bildung von F in den Blättern.

E. Genetik der Elaterasebildung

Die Versuche, die Genetik der Elaterasebildung zu klären, haben noch zu keinen sicheren Resultaten geführt. In Kreuzungen von *Lagenaria siceraria* zwischen bitteren Formen mit hoher × nicht-bitteren mit niedriger Elaterase-Aktivität[1] zeigten die nicht-bitteren Pflanzen der F_2 und F_3 ohne Ausnahme niedere Enzymaktivität. Von den bitteren Pflanzen hatten die meisten hohe Enzymaktivität (70—100) und rund 10% mittlere (5—10). Die Zahlenverhältnisse wechselten erheblich in verschiedenen Kreuzungen, so daß wir keine Hypothese über die genetischen Grundlagen der Elaterasebildung formulieren können. Fruchtgröße und Bitterstoffkonzentration beeinflussen zweifellos die Elaterase-Aktivität. Die Bedeutung anderer Faktoren, z. B. des Fruchtalters, ist noch nicht untersucht worden.

Interessant wären Kreuzungen von *Cucurbita*-Arten, bei denen sehr viel größere Unterschiede in der Elaterase-Aktivität als bei den bisher von uns verwendeten Formen vorkommen. In reifen Früchten von *Cucurbita andreana, lundelliana* und *okeechobeensis* liegen die Cucurbitacine vorwiegend als Aglykone vor, und die Elaterase-Aktivität ist mittel bis hoch (2,5—100); *Cucurbita mixta, palmata* und *pepo* haben rein glykosidische Bitterstoffe und keine oder sehr niedrige Elaterase-Aktivität.

VI. Die Bedeutung der Cucurbitacine für die Züchtung

Drei züchterische Probleme bei den Cucurbitaceen stehen in unmittelbarem Zusammenhang mit den Untersuchungen über Biochemie und Genetik der Cucurbitacine:

1. Das Bitterwerden der Treibhausgurken;
2. Das Entstehen von bitteren Formen der Kulturvarietäten; und
3. Die Benützung bitterer Wildformen in der Züchtung.

1. Die große Variabilität der Bitterkeit von Treibhausgurken erschwerte von Anfang an die züchterische Bearbeitung dieses Problems, das besonders in Holland große wirtschaftliche Bedeutung hat. ANDEWEG und DE BRUYN haben nun die Auslese über die Bitterstofffreiheit der Keimpflanzen bzw. der vegetativen Teile der jungen Pflanzen durchgeführt und unter 15000 geprüften Pflanzen *eine* völlig bitterstofffreie gefunden. Diese Pflanze soll die Grundlage für die künftige Züchtung bilden.

2. Über die Entstehung der bitteren Formen kultivierter Varietäten haben wir die folgenden Feststellungen gemacht:

---

[1] Definition der in unseren genetischen Arbeiten benutzten Einheit für Elaterase-Aktivität in Anm. 1 S. 119.

a) Wassermelonen. In der Varietät „Hawkesbury" haben wir mehrere Fälle von Bitterkeit untersucht. Alle bitteren Früchte und die Nachkommen aus ihren Samen waren reine Hawkesbury, und die Bitterkeit wurde durch ein einziges Allel vererbt. Die Bitterkeit muß also als Mutation entstanden sein. Da alle Fälle, von denen wir Kenntnis erhielten, in einem Umkreis von etwa 60 km um Pretoria vorkamen, halten wir es für wahrscheinlich, daß sie alle auf eine einzige Mutation zurückzuführen sind, die sich durch süße Früchte, die z. T. mit „bitterem" Pollen befruchtet waren, verbreitet haben kann. Da die Mutation dominant ist, kann ihre Verbreitung durch Selbstung nicht-bitterer Pflanzen ausgeschaltet werden. Von einer entsprechenden Mutation bei der Wassermelone „Uzbeskij" berichtet ARASIMOVIČ.

Bitterkeit bei Wassermelonen und Futtermelonen („Kaffir Watermelons") kommt auch durch Kreuzung mit bitteren „Tsammas" zustande. Solche Fälle sind immer an der abweichenden Form und Zeichnung der Früchte und an der großen Variabilität der Nachkommen zu erkennen.

b) Kürbisse *(Cucurbita pepo)*. STEYN berichtet von Bitterkeit bei mehreren Varietäten [vollständige Literatur bei ENSLIN et al. (2)]. Wir selbst haben nur bittere Formen der Varietät „Little Gem" untersuchen können. In drei Fällen handelte es sich zweifellos um Mutationen, da die Nachkommenschaft völlig sortenecht war. In mehreren anderen Fällen war die Bitterkeit durch Kreuzung mit Zierkürbissen entstanden.

c) „Marankas", „Calebash Marrows" *(Lagenaria siceraria)*. Da auch die häufig bittere Kalebassen-Form von *Lagenaria siceraria* angebaut und verwildert in Südafrika vorkommt, liegt es nahe, Kreuzung mit dieser als Ursache der Bitterkeit bei der Gemüse-Form zu vermuten. Wir haben zwei Fälle von bitteren „Marankas" genauer untersuchen können. In dem einen sind nie irgendwelche morphologischen Charaktere der Kalebasse aufgetreten (nun schon in der 3. Generation); er muß als Mutation gedeutet werden. Auch der andere Fall ist zweifellos nicht eine F_1 (Maranka × Kalebasse) gewesen; doch traten unter den Nachkommen vereinzelt Formen mit „Flaschenhälsen" auf, aber keine mit der dicken Schale der Kalebassen. Wir halten in diesem Fall eine Kreuzung mit bitterer Kalebasse für möglich, die allerdings vor zwei oder mehr Generationen stattgefunden haben müßte.

Die Verhütung des Auftretens von bitteren Formen bei all diesen Arten ist einfach, wenn nur bei der Saatproduktion darauf geachtet wird, daß keine Bestäubung durch bittere Pollenspender vorkommen kann. Da die „bitteren" Mutationen in allen Fällen dominant sind, lassen sie sich in einer Generation ausschließen. Aber auch bei den hier besprochenen Arten sollte vielleicht die Züchtung von Varietäten mit nicht-bitteren Keimpflanzen angestrebt werden. Eine Mutation des Gens für Fruchtbitterkeit würde dann — wenn unsere oben (S. 122) besprochene Hypothese richtig ist — noch keinen Schaden tun, und die Wahrscheinlichkeit des gleichzeitigen Mutierens von zwei Genen ist so gering, daß diese Möglichkeit praktisch vernachlässigt werden kann. Ein weiterer Vorteil von Varietäten mit nicht-bitteren Keimpflanzen wäre, daß die Prüfung auf Abwesenheit von bitteren Pflanzen bei den Samenkontrollstellen durchgeführt werden könnte.

3. Die Verwendung bitterer Wildformen in der Züchtung setzt die Kenntnis des Erbganges der Bitterkeit und einer möglichen genetischen oder chemischen Koppelung der Bitterkeit mit erstrebenswerten Eigenschaften voraus. Barham fand Bitterkeit in der Gurke unabhängig von der Resistenz gegen *Peronoplasmopora cubensis*. Wir fanden Bitterkeit in der Gurke, der Wassermelone und in *Lagenaria siceraria* unabhängig von der Resistenz gegen *Colletotrichum lagenarium*. Reine Glykoside von Cucurbitacin B und E hatten keinerlei Wirkung auf das Wachstum dieses Pilzes in Plattenkulturen.

Zwischen Bitterkeit und dem Befall durch Melonenfliegen *(Dacus ciliatus* und *vertebratus)* besteht keinerlei Zusammenhang. Die bittere Wassermelone Hawkesbury wird ebenso gern zur Eiablage benützt wie die nicht-bittere, und die Maden der Fliegen gedeihen ausgezeichnet in ihr. *Dacus* zeigt auf unserer Versuchsstation sogar eine besondere Vorliebe für die sehr bitteren Koloquinten. Die Resistenz einiger Arten *(Lagenaria siceraria, Cucumis anguria, C. melo* var. *agrestis)* gegen *Dacus* beruht auf besonderen morphologischen und histologischen Eigenschaften der Früchte. Die Larven der Coccinellide *Epilachna argus*, die recht lästig werden können, fressen ebenso gern bittere wie nicht-bittere Blätter — sogar *Cucumis dinteri* schreckt sie nicht ab. Über eine ähnliche Feststellung berichtet Contardi. Im Zusammenhang mit diesen Beobachtungen ist es interessant, daß Extrakte aus Cucurbitaceen mit Lösungsmitteln, welche die Cucurbitacine lösen mußten, meist keine insecticiden Eigenschaften hatten; wo sich gewisse Wirkungen fanden, können sie durch andere Stoffe im Pflanzenextrakt hervorgerufen worden sein (Literatur bei Jacobson).

Von einigen anderen züchterisch wichtigen Eigenschaften ist ebenfalls bekannt, daß sie unabhängig von der Bitterkeit vererbt werden. Arasimovič fand in Kreuzungen von *Colocynthis citrullus* keinen Zusammenhang zwischen Bitterkeit und Zuckergehalt. Wir fanden in der Kreuzung *Cucumis anguria* var. *anguria* × var. *longipes* keine Koppelung zwischen der starken Wüchsigkeit von *longipes* und der Bitterkeit.

Wir kennen bisher keinen Fall, in dem Bitterkeit mit einer wünschenswerten Eigenschaft gekoppelt ist, so daß der Züchter nach Belieben unter den bitteren Wildformen wählen kann. Bitterkeit wird in allen ausreichend untersuchten Fällen einfach-dominant vererbt und kann deshalb schon in der F_2 völlig ausgeschaltet werden.

VII. Die Bedeutung der Cucurbitacine für die Systematik der Cucurbitaceen

Cucurbitacine oder Stoffe, die mit ihnen nahe chemische Verwandtschaft zeigen, sind von keiner anderen Familie der höheren Pflanzen bekannt. Sie können als Charakterstoffe der Cucurbitaceen bezeichnet werden. Sie kommen natürlich nicht in allen Arten der Familie vor (nicht-bittere Arten, Gattungen mit anderen Bitterstoffen wie *Momordica* und *Raphanocarpus*), sind aber doch in den meisten Tribus verbreitet.

Zunächst ist man geneigt anzunehmen, daß der Verbreitung der einzelnen Cucurbitacine kaum große systematische Bedeutung zukommen könne, da die chemischen Unterschiede zwischen ihnen klein sind, und da in den untersuchten Fällen nur wenige Gene die Bildung der verschiedenen Cucurbitacine steuern; ob Cucurbitacin B oder E der Hauptbitterstoff in *Cucurbita pepo* ist, hängt von einem Gen ab, und ebenso bestimmt ein Gen darüber, ob in *Lagenaria siceraria* Cucurbitacin B oder die „Maranka-Stoffe" gebildet werden.

Es gibt aber doch einige Fälle, in denen ein klarer Zusammenhang zwischen der systematischen Stellung einer Art und dem Vorkommen von bestimmten Cucurbitacinen besteht. So ist die Gattung *Cucumis* durch die Cucurbitacine der B-Serie charakterisiert, die Gattung *Colocynthis* durch die Cucurbitacine der E-Serie. Die „seltenen" Cucurbitacine können sehr wohl als Art-Charakteristika gebraucht werden. Cucurbitacin A kommt nur in den nah verwandten Arten *Cucumis hookeri*, *C. leptodermis* und *C. myriocarpus* vor, Cucurbitacin C nur in *Cucumis sativus*, Cucurbitacin F nur in *C. dinteri* und Cucurbitacin M in großen Mengen nur in *C. asper*. Im Falle von *Cucumis dinteri* hat die Kenntnis der Bitterstoffe wesentlich zur Klärung der Abgrenzung zweier Arten beigetragen. *Cucumis angolensis* Hook. f. ex Cogn. und *C. dinteri* Cogn. sind morphologisch ähnlich, so daß ihre Bestimmung nach Beschreibungen und Herbarmaterial recht schwierig ist. Wir selbst waren früher über die korrekte Namengebung unsicher [Rehm et al. (2), Enslin et al. (4)]. Inzwischen wurde neues Material aus Südwestafrika verfügbar, wir konnten die Arten an kultivierten Exemplaren studieren und fanden, daß sie neben anderen Merkmalen auch durch die Cucurbitacine in ihren Blättern und Früchten gut charakterisiert sind (vgl. Tab. 8).

In einem anderen Fall ist die Quantität des Bitterstoffes ein nützliches Hilfsmittel zur Unterscheidung zweier Arten. Über die Abgrenzung und Verbreitung von Koloquinte (*Colocynthis vulgaris* Schrad.) und Wassermelone [*Colocynthis citrullus* (L.) O. Ktze.] besteht keineswegs Klarheit.

Beide Arten kommen in einer großen Zahl von Formen vor, die morphologisch schwer zu klassifizieren sind. GREBENŠČIKOV (2) findet, „daß fast alles, was in europäischen botanischen Gärten unter dem Namen *Citrullus colocynthis* (= *Colocynthis vulgaris* SCHRAD.) läuft, meist mit der echten Koloquinte nichts zu tun hat. Es sind bittere oder nicht bittere Formen von *Citrullus colocynthoides* PANG. [=*Colocynthis citrullus* (L.) O. KTZE.]." Wir haben zahlreiche Herkünfte der echten Koloquinte aus Nordafrika und Israel, und viele Formen der wilden Wassermelone aus Südafrika untersucht. Alle Koloquinten hatten hohen Bitterstoffgehalt (0,1—0,3%), in keiner wilden Wassermelone wurde mehr als 0,08% Bitterstoff gefunden. Die einzige andere Art der Gattung, die einen ebenso hohen Bitterstoffgehalt wie die Koloquinte hat (0,3%), ist *Colocynthis ecirrhosa* (COGN.) CHAKR.; diese ist aber morphologisch gut von der Koloquinte zu unterscheiden (mehrjährig, keine Ranken). Wir können auf Grund unserer heutigen Kenntnis mit Sicherheit sagen, daß alle Angaben über das Vorkommen der echten Koloquinte in Südafrika auf irrtümlichen Bestimmungen beruhen.

In der Gattung *Cucurbita* scheint ein deutlicher Zusammenhang zwischen dem Auftreten von Cucurbitacin B und E in den Keimpflanzen und der systematischen Stellung der Arten zu bestehen. Die primitiveren Arten haben ausschließlich B in den Keimwurzeln und Kotyledonen, die abgeleiteten Arten B und E oder auch nur E (Tab. 9). Auch die kultivierten *Cucurbita*-Arten zeigen ähnliche Verschiedenheiten in der Zusammensetzung der Cucurbitacine in ihren Keimpflanzen (Tab. 10), so daß der Gedanke naheliegt, daraus Schlüsse auf ihre Abstammung zu ziehen. *Cucurbita ficifolia, mixta* und *maxima* erscheinen demnach primitiv, und mögen mit den Wildarten *andreana, sororia* oder *lundelliana* in Verbindung gebracht werden. *C. moschata* gehört schon zu den mehr abgeleiteten Formen, und *C. pepo* schließt sich deutlich an *C. texana*

Tabelle 9. *Cucurbitacine in den Keimpflanzen wilder Cucurbita-Arten*

Species	Wurzeln		Kotyledonen	
	B %	E %	B (+ D) %	E (+ I) %
andreana	0,06	—	0,09	—
sororia	0,03	—	0,03	—
lundelliana	0,01	0,007	0,27	—
texana	0,02	0,03	0,04	Spur
cylindrata	0,01	0,009	0,009	0,04
foetidissima	0,008	0,04	0,007	0,07
palmata	—	0,02	0,11	0,03
okeechobeensis	—	0,014	1,2	0,06

(= *C. pepo* L. convar. *microcarpina* GREB.) an, hat sogar einen höheren Anteil von E als seine Wildform.

Die Bitterstoffe in den Früchten der wilden *Cucurbita*-Arten zeigen ähnliche Verhältnisse. Früchte von *C. andreana* und *C. lundelliana* enthalten ausschließlich Cucurbitacine der B-Serie (*C. sororia* konnte noch nicht untersucht werden), und zeigen damit wieder Beziehungen zu *C. mixta*, die nur B in den Früchten enthält. *C. texana* enthält nur E in den Früchten, das Cucurbitacin, das auch in *C. pepo* vorherrscht. Von *C. ficifolia*, *C. maxima* und *C. moschata* sind uns keine Formen mit bitteren Früchten bekannt.

Tabelle 10. *Cucurbitacine in den Keimpflanzen kultivierter Cucurbita-Arten*

Species	Wurzeln		Kotyledonen	
	B %	E %	B (+ D) %	E (+ I) %
ficifolia	nicht bitter		0,14	—
mixta	nicht bitter		0,003	—
maxima	0,02—0,09	0,0—0,01	0,02—0,05	0,0—0,005
moschata	0,1	0,05	0,02	—
pepo	0,0—0,005	0,02—0,05	0,03—0,05	0,005—0,03

Da noch keineswegs alle *Cucurbita*-Arten untersucht werden konnten, tragen unsere Folgerungen vorläufigen Charakter. Es wird zweifellos lohnend sein, die Zusammensetzung der Cucurbitacine in weiteren Arten und in Material verschiedener Herkünfte zu untersuchen. [Zur Diskussion über die Ableitung der kultivierten *Cucurbita*-Arten vgl. GREBENŠČIKOV (3, 4) und WHITAKER (1, 2).]

VIII. Schlußbetrachtungen

Die Bitterstoffe der Cucurbitaceen sind zweifellos ein Gebiet, das besonders günstige Möglichkeiten für die Untersuchung der Biogenese sekundärer Pflanzenstoffe bietet. Schon die kultivierten Arten liefern ein sehr mannigfaltiges Material, und die rund 900 Wildarten der Familie mögen noch viel Interessantes bergen. Die große Zahl der Cucurbitacine, die Einfachheit ihrer Bestimmung und die Zugänglichkeit wenigstens eines Teiles der Enzyme, die bei den Umsetzungen der Cucurbitacine in der Pflanze eine Rolle spielen, lassen für die Zukunft noch manche Vertiefung unseres Wissens über die Verknüpfung von Genen mit den Endprodukten des Stoffwechsels erwarten.

Einige Punkte von allgemeiner Bedeutung, die sich jetzt schon abzeichnen, seien hier kurz zusammengefaßt.

1. Die Bildung aller Enzyme und der verschiedenen Cucurbitacine wird durch Kerngene gesteuert. Die meisten Kreuzungen wurden reziprok durchgeführt, und nie haben wir plasmatische Einflüsse auf die Art und Menge des gebildeten Bitterstoffes gefunden. Bei der Elaterase-Bildung spielt neben genetischen Faktoren vielleicht Substratinduktion eine Rolle.

2. Zu der üblichen Kette Gen → Enzym → Stoff tritt bei den Cucurbitacinen noch ein weiteres Glied, die strukturelle Verteilung von Enzym und Cucurbitacin in der Zelle. Sowohl die Elaterase wie die Acetylesterase zeigen in manchen Fällen hohe Aktivität, ohne auf das Subtrat in der lebenden Zelle einwirken zu können.

3. Von besonderem Interesse scheint uns die organspezifische Genwirkung zu sein, die sich in der verschiedenen qualitativen und quantitativen Zusammensetzung der Cucurbitacine in Wurzeln, Kotyledonen, Blättern und Früchten äußert. Hier muß eine Wechselwirkung zwischen organspezifischem Stoffwechsel und Genen oder primären Genprodukten stattfinden. Vergleichbare Beobachtungen liegen für viele andere sekundäre Pflanzenstoffe vor (z. B. Alkaloide: Šantavý et al.; Terpenoide: Stahl). Wenn erst mehr über die Chemie der abgeleiteten Cucurbitacine bekannt ist, kann eine vergleichende Studie über das Verhalten der Cucurbitacine, anderer Inhaltsstoffe und charakteristischer Stoffwechselprozesse und -bedingungen in den verschiedenen Organen vielleicht Licht auf diese Vorgänge werfen. Der Vergleich mit dem organspezifischen Verhalten von Alkaloiden, Flavonoiden und anderen Stoffgruppen mag zur Aufdeckung allgemeiner Gesetzmäßigkeiten führen.

In den einleitenden Worten wurde schon darauf hingewiesen, daß unsere Kenntnis der Cucurbitacine noch große Lücken aufweist. Mit der Aufklärung der chemischen Struktur der Haupt-Cucurbitacine sind zwar große Fortschritte gemacht, aber unter den abgeleiteten Cucurbitacinen findet sich noch ein weites Feld für künftige chemische Arbeiten. Ein weiteres chemisches und biochemisches Gebiet, das reizvolle Möglichkeiten verspricht, wäre die Untersuchung des Zusammenhanges zwischen den Cucurbitacinen und anderen Triterpenoiden, besonders den Saponinen, in der Familie.

Ein letztes Gebiet, auf das hier verwiesen sei, ist die Erforschung der Toxikologie und Pharmakologie der Cucurbitacine. Ein Beginn ist mit der toxikologischen (David und Vallance) und pharmakologischen [Lavie et al. (2)] Untersuchung der Wirkung reiner Cucurbitacine bereits gemacht. Besonders interessant wären Untersuchungen über die Wirkung der Cucurbitacine bei Zuführung *per os*, da hierbei große Unterschiede in der Giftwirkung derselben Pflanze auf verschiedene Tiere festgestellt sind [Steyn (1, 2), Quin]. Vielleicht lohnt es sich nun auch, manchen Anwendungen von Cucurbitaceen in der Volksmedizin von neuem nachzugehen (Hegnauer).

Der Verfasser dankt seinen Kollegen vom National Chemical Research Laboratory und von der Pretoria Horticultural Research Station für viele Anregungen und für die Möglichkeit, unpubliziertes Material für die Abfassung dieses Berichtes zu benützen. Herr Dr. P. R. Enslin und Herr Dr. A. D. J. Meeuse waren so freundlich, das Manuskript durchzusehen. Der Artikel wird veröffentlicht mit Genehmigung des Chief, Division of Horticulture.

Literatur

Andeweg, J. M., and J. W. de Bruyn: Breeding of non-bitter cucumbers. Euphytica **8**, 13—20 (1959).

Arasimovič, V. V.: Gesetzmäßigkeiten in der Vererbung chemischer Merkmale bei Kürbisgewächsen im Hinblick auf Selektion für chemische Beschaffenheit. Izv. Akad. Nauk, Ser. biol. **1937**, 1835—1851.

Barham, W. S.: The inheritance of a bitter principle in cucumbers. Proc. Amer. Soc. hort. Sci. **62**, 441—442 (1953).

Berg, A.: Sur le mode de formation de l'élatérine dans l'Ecballium élatérium. Bull. Soc. chim. France (III), **17**, 85—88 (1897).

Contardi, H. G.: Estudios genéticos en *Cucurbita* y consideraciones agronómicas. Physis **18**, 331—347 (1939).

David, A., and D. K. Vallance: Bitter principles of Cucurbitaceae. J. Pharm. Pharmacol. **7**, 295—296 (1955).

Enslin, P. R., F. J. Joubert and S. Rehm: (1) Bitter principles of the Cucurbitaceae. III. Elaterase, an active enzyme for the hydrolysis of bitter principle glycosides. J. Sci. Food Agric. **7**, 646—655 (1956).

— T. G. Joubert and — (2) Bitter principles of the Cucurbitaceae. II. Paper chromatography of bitter principles and some applications in horticultural research. J. S. Afr. chem. Inst., N. S. **7**, 131—138 (1954).

— and K. B. Norton: (3) Structure of the side chain of the cucurbitacins. Chem. and Ind. **1959**, 162—163.

— and S. Rehm: (4) The distribution and biogenesis of the cucurbitacins in relation to the taxonomy of the Cucurbitaceae. Proc. Linn. Soc. (London) **169**, 230—238 (1958).

— — and D. E. A. Rivett: (5) Bitter principles of the Cucurbitaceae. VI. The isolation and characterization of six new crystalline bitter principles. J. Sci. Food Agric. **8**, 673—678 (1957).

Goldhausen, M.: Interspecific hybrids in watermelons. C. R. (Doklady) Acad. Sci. URSS. **20**, 595—597 (1938).

Grebenščikov, I.: (1) Zur Kenntnis der Kürbisart *Cucurbita pepo* L. nebst einigen Angaben über Ölkürbis. Züchter **20**, 194—207 (1950).

— (2) Notulae cucurbitologicae I. Zur Vererbung der Bitterkeit und Kurztriebigkeit bei *Cucurbita pepo* L. Kulturpflanze **2**, 145—154 (1954).

— (3) Notulae cucurbitologicae II. Über *Cucurbita texana* A. Gr. und ihre Kreuzung mit einer hochgezüchteten *C. pepo*-Form. Kulturpflanze **3**, 50—59 (1955).

— (4) Über zwei *Cucurbita*-Artkreuzungen. Züzhter **28**, 233—237 (1958).

Hegnauer, R.: Chemotaxonomische Übersichten. 4. Cucurbitaceae. Pharm. Acta Helv. **32**, 334—350 (1957).

Jacobson, M.: Insecticides from Plants. U. S. Dep. Agric. Washington 1958.

Lavie, D., Y. Shvo, D. Willner, P. R. Enslin, J. M. Hugo and K. B. Norton: (1) Interrelationships in the cucurbitacin series. Chem. and Ind. **1959** (im Druck).

— D. Willner, M. Belkin and W. G. Hardy: (2) Symposium on the chemotherapy of cancer. Tokyo, October 1957. Abstracts S. 53 (1959) (im Druck).

Meeuse, A. D. J.: (1) The possible origin of *Cucumis anguria* L. Blumea, Suppl. IV, 196—205 (1958).
— (2) A possible case of interdependence between a mammal and a higher plant. Arch. Néerl. Zool. **13**, Suppl. 1, 314—318 (1958).
Pathak, G. N., and B. Singh: Genetical studies in *Lagenaria leucantha* (Duchs.) Rusby (*L. vulgaris* Ser.). Indian J. Genet. **10**, 28—35 (1950).
Quin, J. I.: The toxicity of *Cucumis myriocarpus* Naud. S. Afr. J. Sci. **25**, 242—245 (1928).
Rehm, S.: (1) Genetic control or biochemical induction of elaterase formation in cucurbits. Proc. 1. Congr. S. Afr. gen. Soc. Publ. Univ. Pretoria **7**, 75—76 (1958).
— P. R. Enslin, A. D. J. Meeuse and J. H. Wessels: (2) Bitter principles of the Cucurbitaceae. VII. The distribution of bitter principles in this plant family. J. Sci. Food Agric. **8**, 679—686 (1957).
— and J. H. Wessels: (3) Bitter principles of the Cucurbitaceae. VIII. Cucurbitacins in seedlings — occurrence, biochemistry and genetical aspects. J. Sci. Food Agric. **8**, 687—691 (1957).
Šantavý, F., F. A. Kincel u. A. R. Shinde: Isolierung der Substanzen aus verschiedenen Teilen der indischen *Gloriosa superba* L. Arch. Pharm. u. Ber. dtsch. pharm. Ges. **290**, 376—385 (1957).
Stahl, E.: Chemische Rassen bei Pflanzen mit terpenoiden Inhaltsstoffen. Pharm. Weekbl. **92**, 829—842 (1957).
Steyn, D. G.: (1) The Toxicology of Plants in South Africa. Central News Agency (S. A.) 1934.
— (2) Vergiftiging van Mens en Dier. Pretoria: Van Schaik 1949.
— (3) The toxicity of bitter-tasting cucurbitaceous vegetables (vegetable marrow, watermelons etc.) for man. S. Afr. med. J. **24**, 713—715 (1950).
Story, R.: Some Plants Used by the Bushmen in Obtaining Food and Water. Bot. Survey of S. Afr., Mem. **30**, Pretoria: Govt. Printer 1958.
Watt, J. M., and M. G. Breyer-Brandwijk: The Medicinal and Poisonous Plants of Southern Africa. Edinburgh: E. & S. Livingstone 1932.
Winden, W. P. van: Bitterhijd bij komkommers vraagt de aandacht. Groenten en Fruit **12**, 1331 (1957).
Whitaker, Th. W.: (1) The origin of the cultivated *Cucurbita*. Amer. Naturalist **90**, 171—176 (1956).
— and G. Bohn: (2) The taxonomy, genetics, production and uses of the cultivated species of *Cucurbita*. Econom. Bot. **4**, 52—81 (1950).

Die cytologischen und genetischen Konsequenzen von Inversionen

Von Reinhard Panitz, Gatersleben

Institut für Kulturpflanzenforschung der Deutschen Akademie der Wissenschaften zu Berlin in Gatersleben

Mit 27 Abbildungen

Inhaltsübersicht

I. Einleitung

Neben den mikroskopisch unsichtbaren Genmutationen haben die sichtbaren Aberrationen der linearen Chromosomenstruktur einen erheblichen Anteil an der Gesamtmutabilität des Genotyps. Diese als Chromosomenmutationen bekannten Erscheinungen stehen den Genmutationen wahrscheinlich zahlenmäßig nicht nach.

Eine besondere Stellung unter den Chromosomenmutationen nehmen die Inversionen ein, die durch umgekehrte Einlagerung eines Chromosomenbruchstückes innerhalb eines Chromosoms keine quantitative Veränderung seines Genmaterials zur Folge haben. Obwohl Inversionen unter den intrachromosomalen Aberrationen an erster Stelle stehen und bei den Dipteren den häufigsten Aberrationstyp darstellen, wurden sie im Vergleich zu anderen Aberrationen sehr spät entdeckt. Sturtevant (1926) gelang es im genetischen Experiment, den ersten Nachweis einer Inversion bei *Drosophila* zu bringen, indem er einen Crossing-over reduzierenden C-Faktor als Inversion eines Chromosomensegments identifizierte. Die cytologische Bestätigung dieses Befundes war erst nach der Entdeckung der Riesenchromosomen der Dipteren möglich (Heitz und Bauer 1933). Painter (1933, 1934) fand cytologisch die ersten Inversionen in den Speicheldrüsenchromosomen von *Drosophila*. Bei Pflanzen wurden Inversionen durch die cytologische Untersuchung der Meiose nachgewiesen. Ihren ersten Nachweis verdanken wir McClintock (1931). Sie wies auch zuerst auf den Zusammenhang zwischen Inversionsheterozygotie und Meiosestörungen hin (McClintock 1933).

Die zahlreichen cytologischen Untersuchungen haben ergeben, daß spontane Inversionen relativ selten auftreten. Erst die Möglichkeit, Chromosomenaberrationen in beliebiger Anzahl durch Röntgenstrahlen zu induzieren, schuf die Voraussetzung zur eingehenden Analyse der cytologischen und genetischen Konsequenzen von Inversionen.

II. Die cytologischen Phänomene

1. Die Klassifizierung der Inversionstypen

Unter einer Inversion versteht man die nach zwei Chromosomenbrüchen erfolgende umgekehrte Einordnung eines interkalaren Chromosomenbruchstückes. In der cytologischen Praxis wird dieser Terminus auch im weiteren Sinn für das Paarungsbild einer heterozygoten Inversion verwendet. Die Grenzpunkte einer Inversion werden im allgemeinen als ihre Inversionspunkte bezeichnet. Durch die Beziehung der Inversionspunkte zueinander sowie zum Centromer ist ein Einteilungsprinzip der mannigfaltigen Inversionstypen gegeben.

a) nach der Beziehung der Inversionspunkte zueinander

Den einfachsten und zugleich häufigsten Inversionstyp in natürlichen Populationen und im Strahlenexperiment stellt die *einfache Inversion* dar, die durch die umgekehrte Einlagerung eines interkalaren Chromosomensegments innerhalb eines Chromosoms entsteht. Aus der normalen Gensequenz ABCDEF ergibt sich die neue Folge A/EDCB/F (Abb. 1). Inversionshomozygote Chromosomen paaren normal. Während die Bestimmung ihrer Inversionspunkte nur unter den günstigen Bedingungen der Riesenchromosomen an der im Vergleich zum Standardtyp geänderten Reihenfolge der Querscheiben mit Sicherheit gelingt, lassen sich Inversionen im heterozygoten Zustand auf Grund ihrer typischen Paarungsbilder auch in der Meiose nachweisen. Der Paarungsmodus inversionsheterozygoter Chromosomen wird von der Größe der Inversion und dem Längenverhältnis des invertierten und nichtinvertierten Bereichs bestimmt (Abb. 2). Bei ausreichender Länge der Inversion entsteht bei der Paarung der für eine Inversion heterozygoten, homologen Chromosomen eine rückläufige Schlinge, die im Pachytän und in den Riesenschromosomen der Dipteren cytologisch erkennbar ist. Mit abnehmender Länge

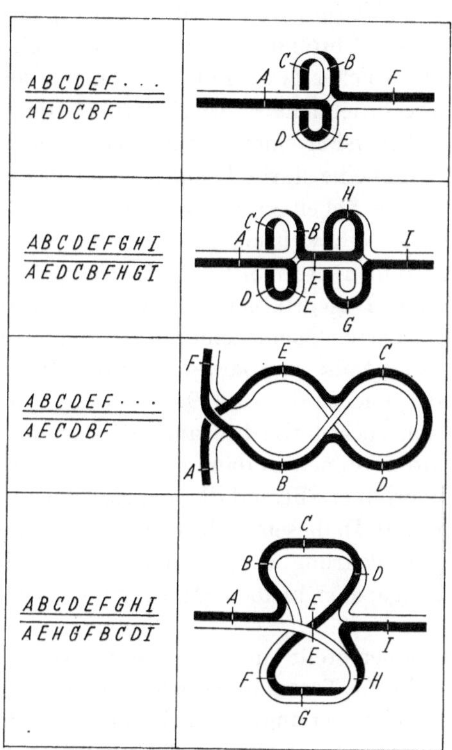

Abb. 1

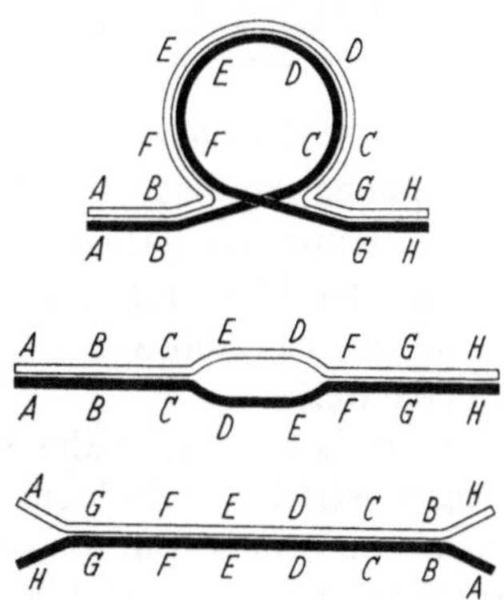

Abb. 2

Abb. 1. Paarungsbild heterozygoter Inversionen. Von oben nach unten: Einfache Inversion, unabhängige Inversionen, eingeschlossene und übergreifende Inversionen (nach DOBZHANSKY und SOKOLOV 1939)

Abb. 2. Paarungsmöglichkeiten heterozygoter Inversionen

des invertierten Segments wird die Bildung einer Inversionsschleife erschwert; im Inversionsbereich bleibt eine Paarungslücke bestehen. Jedoch kann die Paarungsaffinität der angrenzenden Segmente so groß sein, daß besonders bei sehr kleinen Inversionen aus mechanischen Gründen ein Aneinanderlegen erfolgt und die heterozygote Inversion eine Paarung vortäuscht. So fanden HORTON (1939) und PAINTER (1939) im Bastard *Drosophila simulans* × *D. melanogaster* eine vollständig gepaarte heterozygote Inversion am Ende des X-Chromosoms, die aus zwei Scheiben bestand. Während das Auffinden solcher Zweibandinversionen besonders bei Querscheiben gleicher Stärke erhebliche Schwierigkeiten bereitet, dürfte es kaum möglich sein, die Inversion einer einzigen Querscheibe nachzuweisen. Nur dem Umstand, daß eine starke Doppelscheibe in zwei schwächere Scheiben zerlegt wurde, ist der eindeutige Nachweis einer Zweibandinversion zuzuschreiben (Abb. 3) (HORTON 1939, PAINTER 1939). Dagegen bleibt die Interpretation einer tordierten Querscheibe im X-Chromosom von *Drosophila melanogaster* durch GOLDSCHMIDT und HANNAH (1944) als Einbandinversion spekulativ. Ein anderer Paarungsmodus inversionsheterozygoter Chromosomen ist dann gegeben, wenn die Inversion nahezu das ganze Chromosom erfaßt und die nichtinvertierten Segmente nur von geringer Länge sind. In diesem Fall erfolgt der Zusammenhalt der Homologen durch die Paarung der Inversion, wobei die heterologen Chromosomenenden ungepaart bleiben (MCCLINTOCK 1931, KOLLER 1935).

Abb. 3. Zweibandinversion am linken Ende des X-Chromosoms des Bastards *Drosophila melanogaster* × *D. simulans* (nach PAINTER 1939)

Die Existenz terminaler Inversionen stellt ein bis in die jüngste Zeit heftig umstrittenes Problem dar. Da die Terminalinversion für ihr Entstehen nur einen Chromosomenbruch verlangt, wobei ein instabiles Bruchende mit einem stabilen Chromosomenende fusionieren muß, ist zu erwarten, daß sie häufiger als Zweibrucharrangements auftritt. Entgegen allen Erwartungen werden im Strahlenexperiment nur Inversionen vom interkalaren Typ gefunden. Es ergibt sich die Frage, woraus das Fehlen terminaler Inversionen resultiert. Wie bekannt ist, unterscheiden sich die freien Enden der Chromosomenarme von den interkalaren Segmenten durch eine charakteristische strukturelle Organisation und spezifische physiologische Eigenschaften (BROWN 1949, WARTERS und GRIFFEN 1950, BEERMANN 1952, LIMA DE FARIA und SARWELLA 1958). Diese von MULLER (1941, 1954) als Telomeren bezeichneten permanenten Chromosomenendstrukturen sind demnach „Organellen" des Chromosoms; ihr wichtigstes Charakteristikum ist nach MULLER die Monopolarität ihrer Gene. Die Telomeren von *Drosophila* haben nach MULLERs Ansicht eine solche Stabilität ihrer Monopolarität erreicht, daß eine Vereinigung mit Bruchstellen nicht mehr möglich ist. Die bisher bei *Drosophila* bekannt gewordenen terminalen Defizienzen (DEMEREC und HOOVER 1936, SUTTON 1940) erklären MULLER und HERSKOWITZ für interkalar. Sie nehmen an, daß der distale Bruch so dicht am Chromosomenende erfolgte, daß das Endstück lichtoptisch nicht mehr nachgewiesen werden kann (MULLER und HERSKOWITZ 1954, HERSKOWITZ 1954). Das schließt jedoch nach MULLERs Ansicht nicht aus, daß ein Ersatz des Telomers in der Phylogenese des Karyotyps möglich war, zu einem Zeitpunkt, zu dem die Differenzierung in bi- und monopolare Gene noch nicht eingetreten war. Unter diesem Aspekt wird das Heilen terminaler Defizienzen bei *Tradescantia* (SAX 1940), *Trillium* (MATSUURA und HAGA 1950) und im Sporophyten von *Zea mays* (MCCLINTOCK 1941) erklärlich.

Bei vielen Autoren herrscht Übereinstimmung darüber, daß es sich bei den bei *Sciara impatiens* (CARSON 1943, 1946), *Drosophila ananassae* (KAUFMANN 1936, DOBZHANSKY und DREYFUS 1943), *D. robusta* (CARSON und STALKER 1947) und *D. melanica* (WARD 1952) lichtoptisch gefundenen terminalen Inversionen um den interkalaren Typ handelt (BAUER 1939, RILEY 1949, WHITE 1954). PAVAN (1946) fand eine terminale Inversion im 3. Chromosom von *D. nebulosa*, spricht aber die Vermutung aus, daß möglicherweise eine zweite Bruchstelle in unmittelbarer Nähe des Telomers vorhanden sei. Eine von BEERMANN und BAHR (1954) lichtoptisch untersuchte terminale Inversion des Bastards *Camptochironomus tentans* × *C. pallidivittatus* erwies sich unter dem Elektronenmikroskop als interkalar; distal von der letzten lichtmikroskopisch erkennbaren Querscheibe befand sich noch eine heterochromatische Scheibe. Daneben gibt es auch Autoren, die einen interkalaren Einbau des Telomers bei *Drosophila* gelten lassen (CATCHESIDE und LEA 1945).

Inversionstypen, die aus mehreren Inversionen zusammengesetzt sind, werden als *komplexe Inversionen* bezeichnet. Im einfachsten Fall handelt es sich um zwei einfache, durch ein nichtinvertiertes Segment getrennte unabhängige Inversionen ("independent inversions"). Sie können in einem Chromosomenarm oder zu beiden Seiten des Centromers liegen. Als Sonderfall der unabhängigen Inversionen kann die Tandeminversion ("tandem inversion") angesehen werden. Sie verlangt nur drei Brüche und kann entweder als direkte (ABCDEF — A/CB/ED/F) oder umgekehrte Tandeminversion (ABCDEF — A/ED/CB/F) vorkommen. Eine eingeschlossene Inversion ("included inversion") wird dann gebildet, wenn innerhalb einer bestehenden Inversion ein Segment invertiert (ABCDEF — A/EDCB/F — A/E/CD/B/F). Eine übergreifende Inversion ("overlapping inversion, compound inversion") greift nur mit einer Seite in den Bereich der ersten Inversion (ABCDEFGHI — A/EDCB/FGHI — A/E/HGF/BCD/I) und kann zur ersten Inversion distal oder proximal verschoben sein (Abb. 1).

Alle komplexen Inversionstypen wurden in den Riesenchromosomen der Dipteren gefunden. Dagegen ist ihr Nachweis im Pachytän der Meiose oft mit Schwierigkeiten verbunden, doch kann man — wie später gezeigt wird — aus bestimmten Anaphase-Konfigurationen auf ihr Vorhandensein schließen.

b) nach der Beziehung der Inversionspunkte zum Centromer

Hinsichtlich der Beziehung ihrer Inversionspunkte zum Centromer werden parazentrische und perizentrische Inversionen unterschieden. Während die Inversionspunkte parazentrischer Inversionen in einem Chromosomenarm lokalisiert sind (homobrachiale oder intraradiale Inversionspunkte), schließt eine perizentrische Inversion das Centromer ein, d. h. die Inversionspunkte liegen in beiden Chromosomenarmen (heterobrachiale oder interradiale Inversionspunkte). Für para- und perizentrisch gibt es zahlreiche Synonyme wie azentrisch, dyszentrisch, parakinetisch, asymmetrisch bzw. transzentrisch, euzentrisch, transkinetisch und symmetrisch.

Im Tierreich sind perizentrische Inversionen bisher nur in wildlebenden Populationen von Heuschrecken und Dipteren gefunden worden. So wurden für *Trimerotropis sparsa* (White 1951), *Drosophila algonquin* (Miller 1939) und *D. bifurca* (Wharton 1943) je eine perizentrische Inversion und für *D. robusta* (Carson und Stalker 1947) und *D. ananassae* (Freire-Maia 1952, 1953a, 1954) zwei bzw. fünf verschiedene perizentrische Inversionen festgestellt. Im allgemeinen ist die Konzentration perizentrischer Inversionen in natürlichen Populationen sehr gering. Freire-Maia gibt für die perizentrischen Inversionen von *D. ananassae* einen durchschnittlichen Gesamtwert von 0,65% an. Eine Ausnahme macht in dieser Hinsicht die über ein weites Gebiet der Vereinigten Staaten verbreitete perizentrische Inversion 2L—R von *D. robusta*, die am Mt. Vernon in Iowa eine Konzentration von 30,8% erreicht (Carson und Stalker 1947). Nach Levitan (1951a) hat dieselbe Inversion in einer Population von New Jersey eine Frequenz von 38,6%. Für Pflanzen liegen eindeutige Befunde bei *Lilium formosanum* (Brown und Zohary 1955) und *Zea mays* (Morgan 1955, Zohary 1955) vor.

Perizentrische Inversionen haben bei asymmetrischer Verteilung ihrer Inversionspunkte eine Änderung der Längenverhältnisse der Chromosomenarme zur Folge. Metazentrische Chromosomen verwandeln sich in akrozentrische und umgekehrt (Abb. 4). Im recht seltenen Grenzfall kann aus einem metazentrischen Chromosom ein telozentrisches hervorgehen. So fand Seshachar (1939) eine homozygote, perizentrische Inversion in der mitotischen Metaphase der Blindwühle *(Ichthyophis)*, die ein akrozentrisches Chromosom in ein nahezu metazentrisches verwandelt hatte. Über einen umgekehrten Fall, die distale Verlagerung des Centromers, berichteten MILLER (1939) bei *Drosophila algonquin* und COLEMAN (1947) bei *Chorthippus longicornis*. Es besteht wohl kein Zweifel darüber, daß perizentrische Inversionen von Bedeutung für die Evolution des Karyotyps sind, da intraspezifische Differenzierungen zur Ausbildung verborgener Rassen innerhalb einer Art führen können. Einen zusammenfassenden Überblick über den Anteil perizentrischer Inversionen an der Evolution des Karyotyps innerhalb der Gattung *Drosophila* gaben PATTERSON und STONE (1952) und STONE (1955).

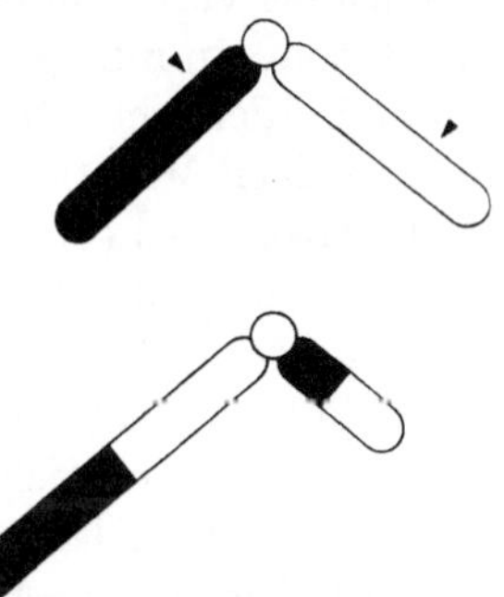

Abb. 4. Verwandlung eines metazentrischen Chromosoms in ein akrozentrisches durch eine perizentrische Inversion

2. Das Entstehen von Inversionen

Die Spontanrate einfacher Inversionen ist in natürlichen Populationen sehr gering. Das erklärt sich aus dem recht seltenen Eintreten von Chromosomenbrüchen und der großen Wahrscheinlichkeit ihrer Restitution. WARTERS (1944) prüfte über 3000 Chromosomen von *Drosophila virilis* ohne eine Inversion zu finden. In 20000 dritten Chromosomen von *D. pseudoobscura* stellten DOBZHANSKY und EPLING (1944) keine einzige neue Inversion fest. MAINX, FIALA und KOGERER (1955) fanden bei der Untersuchung von 10360 Chromosomen von *Liriomyza urophorina* nur zwei verschiedene Inversionen, die nur je in einem Exemplar vorlagen.

Daß die Spontanrate nicht nur Außeneinflüssen unterliegt, sondern als Funktion des Gesamtgenotyps ebenso durch genetische Faktoren modifiziert werden kann, wiesen IVES (1950) und HINTON, IVES und EVANS (1950) speziell für den Ursprung einfacher Inversionen bei *Drosophila melanogaster* nach. Sie entdeckten das Mutatorgen high (hi), das im X-Chromosom bevorzugt einfache Inversionen entstehen läßt. Von Interesse ist, daß innerhalb des X-Chromosoms Reaktionsunterschiede bezüglich des Brucheintritts bestehen, da die proximalen Brüche von zwölf Inversionen in einem heterochromatischen Abschnitt in Nähe des Centromers lokalisiert sind.

Die Wahrscheinlichkeit des zweimaligen Auftretens identischer Inversionen wird unter normalen Bedingungen durch die niedrige Spontanrate und die außerordentlich große Zahl der Rekombinationsmöglichkeiten ausgeschlossen. Setzt man im Idealfall eine zufallsgemäße Verteilung der Chromosomenbrüche voraus, so sind bei einem telozentrischen Chromosom mit n Querscheiben bei Ausschluß von Einband- und Terminalinversionen $^1/_2$ (n—1) (n—2) — (n—2) Inversionen möglich. Das sind bei 1000 Querscheiben schon fast 500000 verschiedene Inversionen.

Selbst dann, wenn eine gleichmäßige Verteilung der Chromosomenbrüche nicht gegeben ist und schon einmal Koinzidenz zweier Brüche eintritt, ist die Freiheit der Rekombination immer noch so groß, daß identische Inversionen praktisch ausgeschlossen sind. Auf Grund dieser Überlegung ist die Annahme gerechtfertigt, daß der hohe Inversionspolymorphismus vieler Dipteren-Arten nicht auf einer hohen Mutationsrate beruht, sondern infolge Akkumulation selektiv begünstigter Inversionen entstand. Identische Inversionen lassen sich letzten Endes immer auf einen einzigen Mutationsschritt zurückführen.

Die Größe des geographischen Areals einer Inversion gibt demnach Aufschluß über die Verbreitung der unmittelbaren Nachkommen der Initialinversion, doch kann sie nicht als Maß für das Alter der Inversion gelten, da besonders in kleinen Populationen der Einfluß der genetischen Drift mathematisch nicht erfaßt werden kann. Nur unter besonders günstigen Umständen läßt sich das phylogenetische Alter des Inversionspolymorphismus eines Areals bestimmen. So konnten MAINX, FIALA und KOGERER (1956) nachweisen, daß es sich bei dem Inversionspolymorphismus von *Liriomyza urophorina* (Agromyzide) in Mitteleuropa um einen relativ jungen Fall handelt. Da die einzige Wirtspflanze dieser Art *(Lilium martagon)* erst nach dem Rückgang der letzten Eiszeit mit der Besiedlung des mitteleuropäischen Raumes begann und dann erst die allmähliche Einwanderung von *Liriomyza* erfolgte, muß der Inversionspolymorphismus jünger als das Diluvium (10000 Jahre) sein.

Übergreifende, eingeschlossene und andere komplexe Inversionen, die durch zeitliche Aufeinanderfolge einfacher Inversionen entstehen, lassen deutlich eine Anhäufung von Inversionspunkten in bestimmten Chromosomenbereichen erkennen. In einigen Fällen wird sogar über Koinzidenz von Inversionspunkten berichtet (KUNZE-MÜHL und SPERLICH 1955). Jede Hypothese, die eine Erklärung für das Entstehen komplexer Inversionen zu geben versucht, muß demnach eine Beantwortung der Frage nach den Ursachen der bevorzugten Lokalisation von Inversionspunkten in bestimmten Chromosomenabschnitten enthalten. Für die Deutung dieses Phänomens gibt es zwei alternative Ausgangspunkte.

Die eine Möglichkeit wäre, daß eine zufallsgemäße Verteilung der Chromosomenbrüche von vornherein gegeben und das Vorhandensein bestimmter für gesteigerten Brucheintritt prädestinierter Chromosomenbereiche ausgeschlossen ist. Diese Annahme stützt sich auf die an *Drosophila* mit Spermienbestrahlung gewonnenen Versuchsergebnisse, die hinsichtlich der Bruchverteilung — abgesehen von einer geringen Anhäufung in heterochromatischen Abschnitten — keine nennenswerte Abweichung von einer gleichmäßigen Streuung im Chromosom erbrachten (BAUER, DEMEREC und KAUFMANN 1938, HELFER 1941, KAUFMANN 1941, FANO 1941, BAKER 1949). Dagegen weisen bestrahlte Pflanzen für spezielle

Chromosomenregionen erhöhte Bruchempfindlichkeit nach Röntgenbestrahlung auf (Câmara 1941, Jackson und Barber 1958). Der Grund mag darin liegen, daß Pflanzen stärker auf den Sekundäreffekt von Strahlen ansprechen, obwohl auch bei *Drosophila* in einem Fall auf erhöhte Bruchempfänglichkeit einzelner Chromosomenabschnitte durch Kaufmann (1940) aufmerksam gemacht wurde. Bei Vernachlässigung geringer Abweichungen von einer gleichmäßigen Bruchverteilung bei *Drosophila* könnte man sich das Zustandekommen einer Anhäufung von Inversionen in einem Chromosomenabschnitt so vorstellen, daß Inversionen mit Brüchen in einer bestimmten Region infolge eines positiven Selektionswertes erhalten blieben, während alle anderen Anordnungen eliminiert würden. Kunze-Mühl und Sperlich (1955) fassen die Möglichkeit ins Auge, daß von Chromosomenbrüchen an bestimmten Stellen Wirkungen im Sinne eines günstigen Positionseffekts ausgehen, was zwangsläufig zu einer Anreicherung von Inversionen in diesen Regionen führen muß. Zur Frage der Koinzidenz von Inversionspunkten wird von den Autoren außerdem die Möglichkeit erwogen, daß eine einmal gebrochene Stelle leichter von einem zweiten Bruch betroffen wird. Für die besondere Empfindlichkeit verheilter Brüche spricht die von Hinton (1950) bei *Drosophila melanogaster* in vier Fällen festgestellte Reinversion der Inversion In(2LR)40d.

Die andere Denkmöglichkeit geht davon aus, daß Chromosomenbrüche nicht gleichmäßig über das ganze Chromosom verteilt sind, sondern an gewissen Stellen bevorzugt eintreten. Allgemein wird jedoch unter Zugrundelegung der oben angeführten Strahlenversuche an *Drosophila* das Vorhandensein von Prädilektionsstellen für den spontanen Chromosomenbruch abgelehnt. Kunze-Mühl, Müller und Sperlich (1958) folgern, daß, wenn schwache Stellen im Chromosom vorhanden wären, diese auch bei Einwirkung von Röntgenstrahlen leichter brechen müßten als andere. Diese Vorstellung unterliegt dem Trugschluß, hauptsächlich Strahlungsmutagene für die spontane Bruchauslösung verantwortlich zu machen bzw. allen Mutagenen gleiche Spezifität zuzuschreiben. Wie gering der Anteil der Strahlung an der Gesamtmutabilität ist, geht aus den Angaben von Muller und Mott-Smith (1930), Timofeeff-Ressovsky (1931) und Dessauer (1954) hervor, die etwa ein Promille der tatsächlich beobachteten Mutationen auf kosmische Strahlung zurückführen. Daraus wird ersichtlich, daß der Spontanrate in überwiegendem Maße andere Ursachen zugrunde liegen müssen. Eine viel größere Rolle, als bisher angenommen wurde, kommt wahrscheinlich automutagenen Stoffen zu, die in der Zelle selbst infolge zellphysiologischer Störungen entstehen und in Abhängigkeit von ihrer Natur, den Unterschieden im chemischen Aufbau der chromosomalen Längselemente und der jeweiligen Aktivitätsphase bestimmter Loci und Chromosomenabschnitte ein ganz spezifisches Muster von Chromosomenbrüchen induzieren. Zwingende Anhaltspunkte sind für keine der beiden Vorstellungen vorhanden; weitere Versuche werden zeigen müssen, inwieweit diese Vorstellungen zu Recht bestehen.

Auf ganz anderen Vorstellungen basiert die Hypothese von NOVITSKI (1946), die auf Grund seiner Untersuchungen an *Drosophila athabasca* entwickelt wurde und das Entstehen eingeschlossener, übergreifender und zusammengesetzter Inversionen und deren Konzentration in einzelnen Chromosomenabschnitten erklärt. Der Autor postuliert als Ursache einfache mechanische Vorgänge in einem inversionsheterozygoten Chromosom während der meiotischen Prophase. Bei der Paarung zweier für eine einfache Inversion heterozygoter Chromosomen wird durch die Schleifenbildung des invertierten oder nichtinvertierten Chromosoms an der Basis der Inversion eine Überkreuzungsstelle gebildet. An dieser Stelle sind die Homologen ungepaart, und der einzelne Chromosomenstrang ist demzufolge größerer mechanischer Belastung ausgesetzt. Bricht das Chromosom an zwei Stellen der Überkreuzung mit anschließender Fusion der Bruchstellen einer Seite, so entsteht eine neue Inversion. Das Ergebnis der Brüche hängt von ihrer Lokalisation ab. Erfolgen sie im nichtinvertierten Chromosom, entsteht eine einfache Inversion. Dagegen verursachen Brüche im invertierten Chromosom eine Reinversion, die bezüglich der Parentalinversion eingeschlossen oder übergreifend sein kann. Wie die Analyse der alten und neuen Inversionspunkte im 3. Chromosom von *D. athabasca* zeigte, sind die Inversionspunkte der Folgeinversion distal verschoben, d. h. der proximale Inversionspunkt der Folgeinversion liegt zwischen den Inversionspunkten und der distale Inversionspunkt außerhalb der Parentalinversion. Die distale Verschiebung der Inversionspunkte erklärt NOVITSKI damit, daß die Paarung der Homologen proximal vollständiger ist.

Auch bei zahlreichen anderen *Drosophila*-Arten sind die Inversionen auf einzelne Chromosomenabschnitte konzentriert, wobei distale Verschiebung der Inversionspunkte überwiegt. So bei *Drosophila pseudoobscura, D. persimilis* (DOBZHANSKY und EPLING 1944), *D. nebulosa* (PAVAN 1946), *D. melanica* (WARD 1952), *D. guaramunu* (SALZANO 1954), *D. subobscura* (KUNZE-MÜHL und SPERLICH 1955, GOLDSCHMIDT 1956) und bei einigen Arten von *Chironomus* (MAINX, KUNZE und KOSKE 1953, ACTON 1955, BEERMANN 1955, ROTHFELS und FAIRLIE 1957).

NOVITSKIs Hypothese verlangt zwei durch mechanische Belastung gleichzeitig entstehende Brüche. Nach ROTHFELS und FAIRLIE (1957) ist das zwar eine mögliche, doch nicht unbedingt notwendige Voraussetzung für das Entstehen einer neuen Inversion. Ein zweiter Bruch kann ebenso später an einer beliebigen Stelle spontan eintreten und die Bildung einer neuen Inversion zur Folge haben. Das setzt voraus, daß die kritische

Zeitspanne, die gerade noch die Restitution oder Rekombination zweier Brüche zuläßt, sehr groß ist. Bei *Drosophila* liegt das insofern günstig, als Chromosomenbrüche während der ganzen Spermatogenese reaktionsfähig bleiben.

III. Die cytologischen Konsequenzen

1. Das Entstehen neuer Chromosomenformen nach Chiasmenbildung

Während homozygote Inversionen einen völlig normalen Meioseablauf gestatten, können sie im heterozygoten Zustand zu Meiosestörungen führen, die mit der Bildung neuer Chromosomenformen verbunden sind. Diese in der Meiose zahlreicher inversionsheterozygoter Pflanzen und Tiere gefundenen abnormen Chromosomenformen durchbrechen die für jeden Organismus charakteristische Formenkonstanz seiner Chromosomen. Im folgenden sollen die wichtigsten Konfigurationen, die nach Chiasmenbildung in heterozygoten, para- und perizentrischen Inversionen entstehen, cytologisch analysiert werden.

a) in heterozygoten, parazentrischen, einfachen Inversionen

Als Folge eines einfachen Chiasmas in einer heterozygoten, einfachen Inversion entstehen eine dizentrische Chromatide und ein azentrisches Fragment (Abb. 5). Die dizentrische Chromatide ist für die distalen Abschnitte, die das Fragment neben dem invertierten Segment dupliziert enthält, defizient. In der Anaphase I wandern die Centromeren der dizentrischen Chromatide zu den entgegengesetzten Polen; es entsteht eine dizentrische Brücke (Abb. 6a), die in der Mehrzahl der Fälle infolge des mechanischen Zugs der Anaphase-Bewegung in der Mittelregion reißt oder während der Zellwandbildung durchgetrennt wird. Sehr kurze Brücken reißen gewöhnlich, worauf sich die beiden Stummel wie elastische Spiralfedern zusammenziehen, oder sie werden in dünne, stellenweise kaum sichtbare Fäden ausgezogen (GEITLER 1937). Brückenfragmentation gibt oft Anlaß zur Bildung von Restitutionskernen und Mikrocyten (Abb. 7). Lange Inversionsbrücken werden weniger in den fortgeschrittenen Stadien der Anaphase gezerrt. Sie reißen zuweilen nicht, bleiben zwischen den Tochterkernen ausgespannt (RICHARDSON 1936, FRANKEL 1937, UPCOTT 1937, DARLINGTON und LA COUR 1941, CARSON 1946) oder nehmen als verschleppte Brücken an der zweiten meiotischen Teilung teil (SMITH 1935, MEURMAN und THERMAN 1939, WOLF 1941, WALTERS 1942). Da die ohnehin schon defizienten Brücken selten median reißen, unterscheiden sich die resultierenden Chromatiden durch zusätzliche Defizienzen bzw. Duplikationen.

Das azentrische Fragment kann nicht aktiv an der Zellteilung teilnehmen; es bleibt in der Regel im Äquator der Spindel zurück und wird eliminiert. Nur bei günstiger Lage im Zellraum bleibt es bis zur Telophase bestehen und kann sich unter Membranbildung als Mikronukleus abkapseln (Geitler 1937). In seltenen Fällen gelangt das Fragment zu einem der Spindelpole, indem es durch ein distal zur Inversion gelegenes nicht terminalisierendes Chiasma mit einer Nichtaustauschchromatide verbunden bleibt (Swanson 1940, Tanaka 1940, Russel und Burnham 1950, Rhoades und Dempsey 1953). Überdauert es beide meiotischen Teilungen, so erscheinen in der nächsten Metaphase nach Reduplikation zwei "acentric rods", die infolge Fusion auch von U-förmiger Gestalt sein können (Giles 1940). Nach erfolgter Tetradenbildung sind nur zwei Zellen im Besitz des vollständigen Chromatidensatzes.

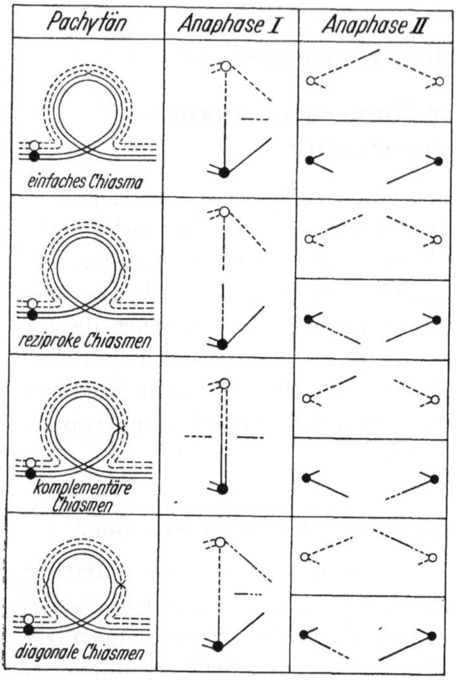

Abb. 5. Meiotische Anaphase-Konfigurationen nach Chiasmenbildung in heterozygoten, parazentrischen, einfachen Inversionen

Bei zwei innerhalb einer heterozygoten Inversion auftretenden Chiasmen wird die Gestalt der Anaphase-Konfigurationen von der Beziehung der Chiasmen untereinander bestimmt. Morphologisch völlig normale Chromatidendyaden mit je einer vom Austausch betroffenen Chromatide sind auf die Bildung reziproker Chiasmen zurückzuführen. Alle vier Zellen

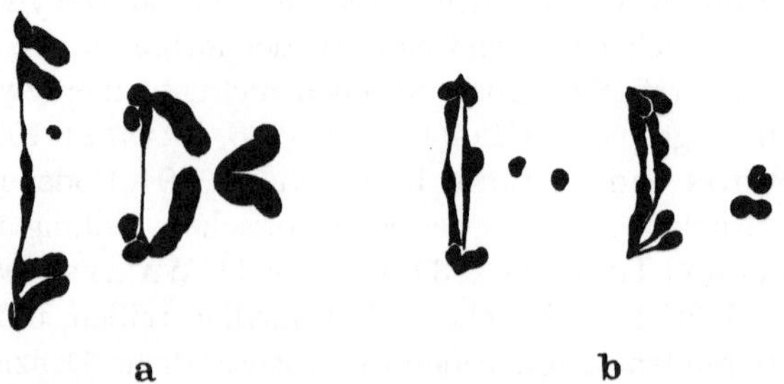

Abb. 6. a) einfache und b) doppelte Inversionsbrücken in der Meiose von *Tulipa* (nach Upcott 1937)

erhalten den vollständigen Chromatidensatz. Eine doppelte Inversionsbrücke und zwei gleichgroße azentrische Fragmente in der Anaphase I haben stets komplementäre Chiasmen als Ursache (Abb. 5 und 6b). Da beide Inversionsbrücken defizient sind, erhält nach der zweiten Teilung keine Tetradenzelle den vollständigen Chromatidensatz. Diagonale Chiasmen ergeben rein äußerlich die gleiche Anaphase-Konfiguration, die nach einem einfachen Chiasma auftritt, nur sind die Defizienzen anderer Art (Abb. 5).

Kompliziertere Konfigurationen entstehen nach gleichzeitiger Chiasmenbildung innerhalb und außerhalb einer heterozygoten, einfachen Inversion. Ein Loop-Univalent, eine morphologisch normale Chromatidendyade mit einer defizienten Austauschchromatide und ein azentrisches Fragment sind das Ergebnis eines Inversionschiasmas und eines Chiasmas proximal zur Inversion. Zwei Loop-Univalente in Verbindung mit zwei azentrischen Fragmenten weisen auf ein zweites Inversionschiasma hin (Abb. 8). Die Loop-Univalente wandern in der Anaphase I zum Spindelpol und geben in der Anaphase II Anlaß zu Brückenbildung (Abb. 9). Im ersten Fall werden zwei, im letzten Fall vier unbalancierte Gameten gebildet.

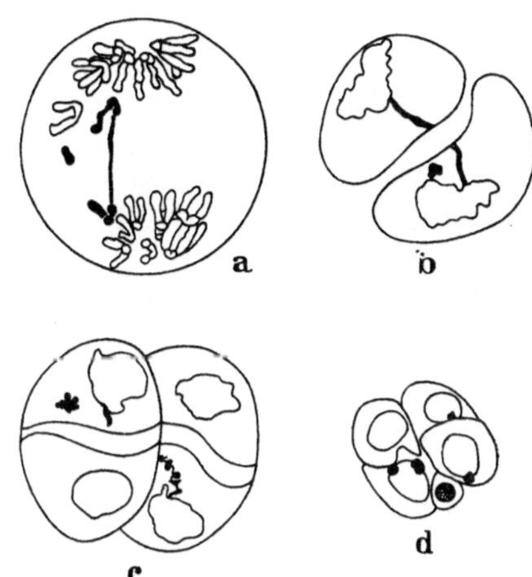

Abb. 7. Konsequenzen nach Brückenbildung in der Mikrosporogenese von *Fritillaria*. a) Inversionsbrücke in der Anaphase I, b) und c) Brückenfragmentation, d) Pollentetrade mit Restitutionskernen und einer Mikrocyte (nach FRANKEL 1937)

Pachytän | Anaphase I | Anaphase II

Abb. 8. Meiotische Anaphase-Konfigurationen nach Chiasmenbildung innerhalb und außerhalb heterozygoter, parazentrischer, einfacher Inversionen

Abb. 9. a) Loop-Univalent in der Anaphase I und b) die daraus entstehende Inversionsbrücke in der Anaphase I der Meiose von *Fritillaria* (nach FRANKEL 1937)

b) in heterozygoten, parazentrischen, unabhängigen Inversionen

Bei der Untersuchung der Meiose von *Tulipa* wurde UPCOTT (1937) auf Konfigurationen in der Anaphase I aufmerksam, die sich nur als die Folge reziproker Chiasmen in zwei in einem Chromosomenarm lokalisierten unabhängigen Inversionen deuten ließen. Sie fand eine dizentrische Chromatide, ein azentrisches Ringfragment und ein kleines azentrisches Fragment (Abb. 10). Die Anaphasebewegung führt zum Zerreißen der Brücke und zu ungleicher Verteilung des Chromatidensatzes; die Fragmente werden eliminiert. Wie UPCOTT fand, können auch Inversionsbrücken in der Anaphase II auftreten. Für ihr Entstehen sind zwei Inversionschiasmen und ein proximales Chiasma notwendig (Abb. 11 und 12). Komplementäre Chiasmen ergeben zwei dizentrische Chromatiden

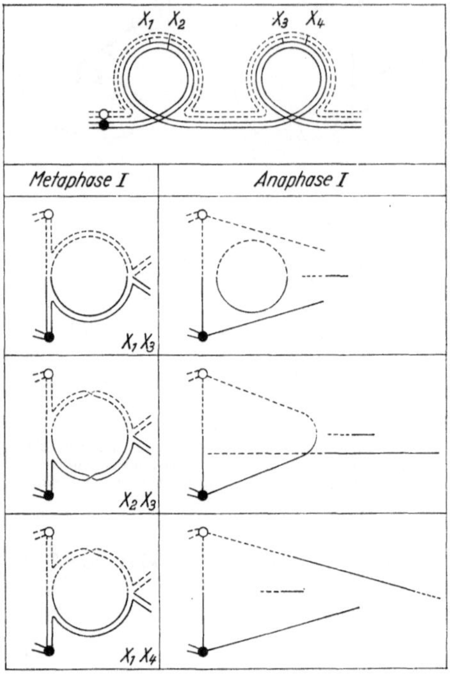

Abb. 10. Konfigurationen nach Chiasmenbildung (reziproke, komplementäre und diagonale Chiasmen) in heterozygoten, parazentrischen, unabhängigen Inversionen

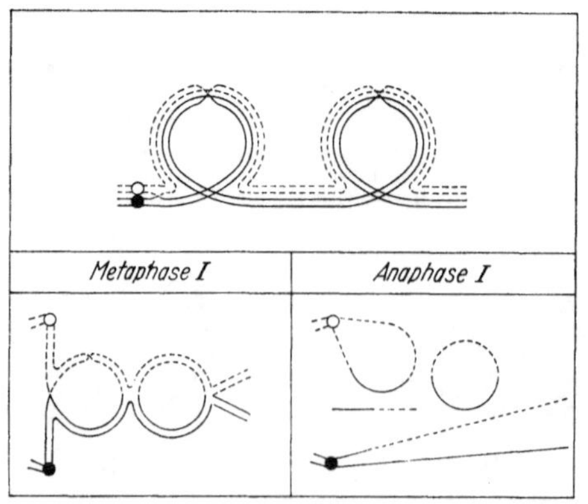

Abb. 11

Abb. 12. Inversionsbrücke und azentrisches Ringfragment in der Anaphase II von *Tulipa* als Folge eines proximalen Chiasmas und reziproker Chiasmen in heterozygoten, parazentrischen, unabhängigen Inversionen (nach UPCOTT 1937)

Abb. 11. Loop-Univalent, azentrisches Ringfragment und einfaches azentrisches Fragment in der Anaphase I nach einem proximalen Chiasma und reziproken Chiasmen in heterozygoten, parazentrischen, unabhängigen Inversionen

von ungleicher Länge und ein großes und kleines azentrisches Fragment. Diese Konfigurationen wurden von SWANSON (1940) in der Anaphase I von *Tradescantia* gefunden. Diagonale Chiasmen lassen eine asymmetrische, dizentrische Chromatide und ein kleines azentrisches Fragment entstehen (Abb. 10).

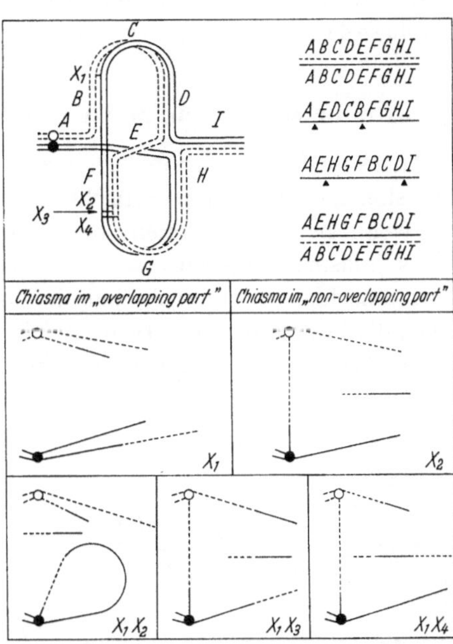

Abb. 13. Konfigurationen in der Anaphase I nach Chiasmenbildung (einfaches Chiasma im "overlapping part" und "non-overlapping part"; reziproke, diagonale und komplementäre Chiasmen) in einer heterozygoten, parazentrischen, übergreifenden Inversion. Rechts oben: Schematische Darstellung des Entstehens einer übergreifenden Inversion

c) in heterozygoten, parazentrischen, übergreifenden und eingeschlossenen Inversionen

Der Erfolg eines einfachen Chiasmas in einer heterozygoten, übergreifenden Inversion hängt von seiner Lokalisation ab. Eine dizentrische Chromatide wird nur dann in der Anaphase I gebildet, wenn das Chiasma im nichtübergreifenden Teil ("non-overlapping part") der Inversion entsteht. Nach Auftreten im übergreifenden Teil ("overlapping part") werden in der Anaphase I zwei Chromatidendyaden verteilt, die je eine abnorme Chromatide enthalten (Abb. 13).

Komplementäre und diagonale Chiasmen lassen symmetrische und asymmetrische Brücken mit einem Fragment und reziproke Chiasmen ein Loop-Univalent, eine Chromatidendyade mit Chromatiden ungleicher Länge und ein Fragment in der Anaphase I entstehen.

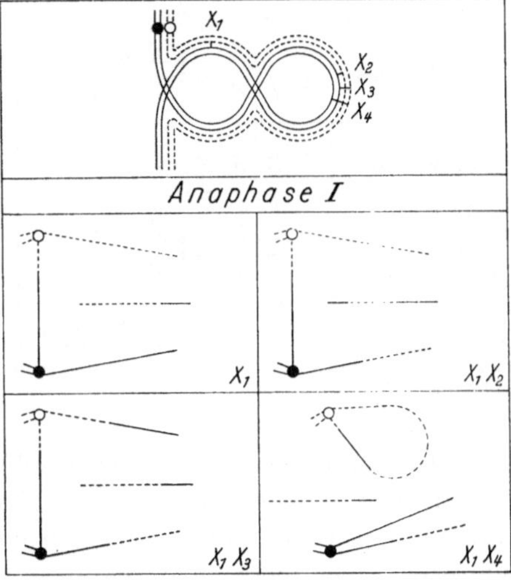

Abb. 14

Abb. 14. Konfigurationen in der Anaphase I nach Chiasmenbildung (einfaches Chiasma, reziproke, diagonale und komplementäre Chiasmen) in einer heterozygoten, parazentrischen, eingeschlossenen Inversion

In eingeschlossenen Inversionen führen ein einfaches Chiasma ebenso wie reziproke und diagonale Chiasmen zu Brückenbildung in der Anaphase I, während komplementäre Chiasmen eine Inversionsbrücke in der Anaphase II zur Folge haben (Abb. 14).

d) in heterozygoten, perizentrischen, einfachen Inversionen

Austauschkonfigurationen heterozygoter, perizentrischer, einfacher Inversionen sind in den meiotischen Anaphasen nicht als solche zu erkennen, wenn die Chromosomen von metazentrischer Gestalt sind. Ein einfaches Chiasma hat ebenso wie reziproke, komplementäre und diagonale Inversionschiasmen morphologisch normale Chromatidendyaden zur Folge, die in cytologischer Hinsicht nichts über ihre Entstehungsursache aussagen (Abb. 15). Die Austauschchromatiden unterscheiden sich durch Defizienz bzw. Duplikation der Endabschnitte. Je zwei normale und anormale Chromatiden sind auf ein einfaches Chiasma oder auf diagonale Chiasmen zurückzuführen. Reziproke Inversionschiasmen bilden keine defekten Chromatiden, dagegen sind nach komplementärer Chiasmenbildung alle Chromatiden defizient.

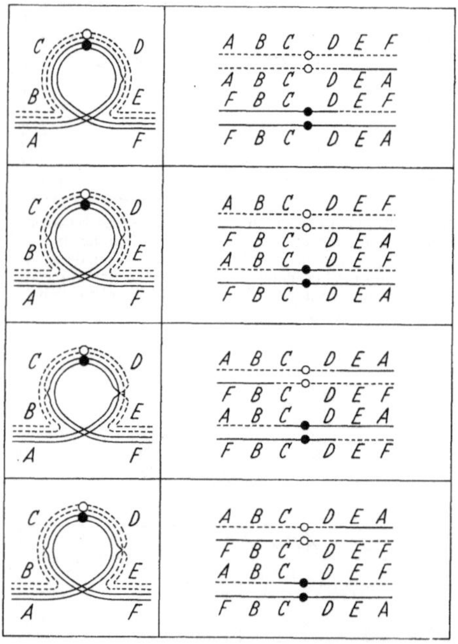

Abb. 15. Cytogenetische Konsequenzen nach Chiasmenbildung in einer heterozygoten, perizentrischen, einfachen Inversion eines metazentrischen Chromosoms. Duplikationen und Defizienzen sind durch die Sequenz der Buchstaben veranschaulicht

Ist dagegen die Inversion von asymmetrischer Gestalt, oder sind die Chromosomen akrozentrisch (Abb. 16), können die Austauschkonfigurationen im cytologischen Bild erkannt und den einzelnen Chiasmenformen zugeordnet werden. Die genetischen Konsequenzen verhalten sich wie oben. Folgende Anaphase-Konfigurationen sind möglich:

Typ A: Ein- und beidseitig reziproke Chiasmen ergeben Chromatidendyaden mit morphologisch normalen Chromatiden.

Typ B: Einfache und einseitig diagonale Chiasmen ergeben zwei Chromatidendyaden mit je einer morphologisch normalen und einer kleinen bzw. großen metazentrischen Chromatide.

Typ C: Einseitig komplementäre Chiasmen ergeben eine große und eine kleine metazentrische Chromatidendyade.

Typ D: Beidseitig komplementäre Chiasmen ergeben zwei Chromatidendyaden mit je einer großen und einer kleinen metazentrischen Chromatide.

Typ E: Beidseitig diagonale Chiasmen ergeben. eine Chromatidendyade mit zwei morphologisch normalen Chromatiden und eine Dyade mit einer großen und einer kleinen metazentrischen Chromatide.

Alle diese Konfigurationen wurden von GILES (1944) in der Meiose von *Gasteria* gefunden. Die quantitative Klassifikation der Typen ergab folgendes Bild: 18,5% der untersuchten Zellen hatten normale Chromatiden, ließen sich also dem Typ A zuordnen, wobei zu bemerken ist, daß in diesem Wert auch die Zellen ohne Austausch erfaßt sind. Dem Typ B fielen 55,4% zu, und der Rest verteilte sich auf die Typen C (8,4%), D (5,9%) und E (11,6%).

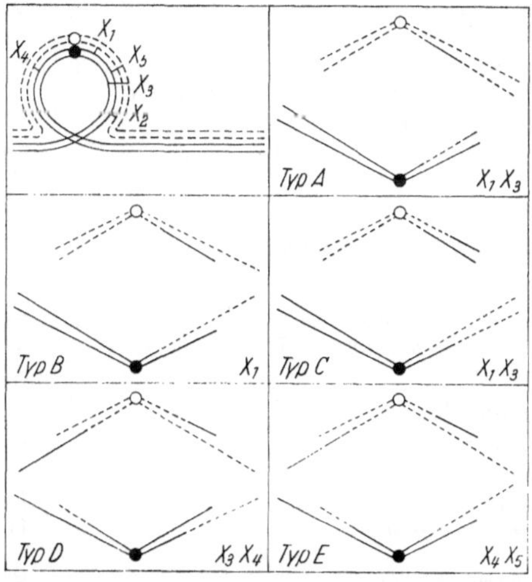

Abb. 16. Morphologisch mögliche Anaphase-Konfigurationen nach Chiasmenbildung in einer heterozygoten, perizentrischen, einfachen Inversion eines akrozentrischen Chromosoms

Wie die angeführten Beispiele zeigen, ist nach Chiasmenbildung in heterozygoten Inversionen eine Vielzahl von Konfigurationen möglich, die nicht in allen Fällen einem Inversionstyp in Verbindung mit einem bestimmten Chiasmentyp zugeordnet werden können. Unterschiedliche Chiasmen haben in Inversionen verschiedenen Typs morphologisch gleiche Anaphase-Konfigurationen zur Folge, die sich nur durch den Grad ihrer Duplikationen und Defizienzen unterscheiden. So kann die einfache Inversionsbrücke — die häufigste Konsequenz und daher ein wichtiges Kriterium der Inversionsheterozygotie — verschiedenen Inversionstypen zugeordnet werden. Da eine Pachytänanalyse am gleichen Objekt vielfach an der Ungunst der Verhältnisse scheitert, ist eine sichere Aussage über die Entstehungsart von Inversionsbrücken oft nicht möglich. Es liegt deshalb nahe, das Auftreten dizentrischer Brücken in Verbindung mit einem azentrischen Fragment nur als einen allgemeinen Hinweis auf die Existenz heterozygoter, parazentrischer Inversionen zu werten.

In der Mehrzahl der Fälle können Inversionsbrücken mit Fragment von einem einfachen Chiasma in einer heterozygoten, parazentrischen,

Tabelle 1

Monocotyledonae

Gramineae	
Triticum turgidum × *Secale cereale* . .	LILJEFORS 1936
T., sp.	MÜNTZING 1939
Dactylis, sp.	MÜNTZING 1937
Lolium loliaceum × *L. rigidum*	JENKINS und THOMAS 1939
L. perenne L.	MYERS 1940
Poa pratensis.	MÜNTZING 1940
Agropyrum junceum × *A. repens* . . .	ÖSTERGREN 1940
A. scabrum.	HAIR 1953
Secale cereale	MÜNTZING und PRAKKEN 1941, LEVAN 1942
Oryza sativa	SAMPATH und MOHANTY 1954
Pennisetum typhoides STAPF. u. HUB. × *P. purpureum* SCHUMACH.	KRISHNASWAMY und RAMAN 1954
Bothriochloa ischaemum	CELARIER 1957
Zea mays × *Teosinte*	TING 1958
Cyperiaceae	
Carex siderosticta HANCE.	TANAKA 1940
Palmae	
Thrinax barbadenis LODD.	SHARMA und SARKAR 1956
Dictyosperma album H. WENDL. und DRUDE	SHARMA und SARKAR 1956
Commelinaceae	
Tradescantia, sp.	SAX 1937, GILES 1940, SWANSON 1940, BHADURI 1942a
T. (4 Arten)	DARLINGTON 1938
Rhoeo discolor	DARLINGTON 1938, STRAUB 1941, BHADURI 1942b, FLAGG 1958
Zebrina pendula	SHARMA 1955
Liliaceae	
Trillium erectum L.	SMITH 1935
Lilium martagon album × *L. hansonii* .	RICHARDSON 1936
L. testaceum	DARLINGTON und LA COUR 1941
Tulipa praecox	UPCOTT 1937
T. silvestris	UPCOTT 1939a
T. gesneriana.	DARLINGTON und LA COUR 1941
Paris quadrifolia	GEITLER 1937, 1938, HEILBORN 1943
Fritillaria (4 Arten)	FRANKEL 1937
Scilla indica BAKER	RAGHAVAN und VENKATASUBBAN 1939
S. sibirica	GOPAL-AYENGAR 1942
Allium (7 Arten)	MENSINKAI 1940
A. odorum	HÅKANSSON und LEVAN 1957
Aloe variegata × *Gasteria verrucosa*, var., *latifolia*	SINOTÔ und SATÔ 1940
A. variegata × *G. gyuzetu*	SINOTÔ und SATÔ 1940
A. sp. × *G.* sp.	RILEY 1947
Tricyrtis hirta × *T. formosana*	SINOTÔ und SATÔ 1941
Amaryllidaceae	
Agava amaniensis.	DOUGHTY 1936
A. sisaliana	DOUGHTY 1936
Narcissus bulbocodius	FERNANDES und NEVES 1941
Musaceae	
Musa, sp.	DODDS 1943

Dicotyledonae

Polygonaceae	
Rumex acetosella	MEURMAN 1925
Ranunculaceae	
Paeonia, sp.	DARK 1936
P. delavayi, var., *lutea* × *P. suffruticosa*	STEBBINS 1938
P. californica.	STEBBINS und ELLERTON 1937
P. tenuifolia	MARQUARD 1953
Clematis viticella	MEURMAN und THERMAN 1939
C. Ville de Lyon	MEURMAN und THERMAN 1939
C. jackmani	MEURMAN und THERMAN 1939
Anemone apenina	BÖCHER 1945
Cruciferae	
Raphanus sativus × *Brassica oleracea* .	HOWARD 1938
Brassica wrightii	SIKKA 1940
B. juncea × *B. campestris*	SIKKA 1940
Rosaceae	
Rubus vitifolius	THOMAS 1940
R. vitifolius × *R. idaeus*	THOMAS 1940
Leguminosae	
Pisum humile × *P. arvense*	HÅKANSSON 1936
P. sativum L.	KOLLER 1938
P. sativum L. × *P. abyssinicum* . . .	ROSEN 1944
Vicia faba L.	ROWLANDS 1958
Hippocastanaceae	
Aesculus hippocastanum × *Ae. carnea* .	UPCOTT 1936
Balsaminaceae	
Impatiens sultani	BHATTACHARIYA 1958
Malvaceae	
Gossypium, sp.	BEASLEY 1942, BROWN und MENZEL 1952
Onagraceae	
Godetia deflexa × *G. bottae*	HÅKANSSON 1943
Clarkia lingulata × *C. biloba australis* .	LEWIS und ROBERTS 1956
C. rubicunda × *C. franciscana*	LEWIS und RAVEN 1958
C. amoena × *C. rubicunda*	LEWIS und RAVEN 1958
Umbelliferae	
Coriandrum sativum	SHARMA und GHOSH 1955
Foeniculum vulgare	SHARMA und GHOSH 1955
Primulaceae	
Primula kewensis	UPCOTT 1939b
Solanaceae	
Nicotiana (Hybriden)	MÜNTZING 1935, KOSTOFF 1943
Solanum douglasii DUNAL × *S. nodiflorum* JAQU.	PADDOCK 1943
S. stenotomum	LAMM 1945
S. tuberarium	MAGOON, COOPER und HOUGAS 1958
S. demissum × *S. goniocalyx*	MAGOON, HOUGAS und COOPER 1958
Cestrum diurnum	SHARMA und SHARMA 1957

Scrophulariaceae	
Calceolaria (3 Arten)	SRINATH 1940
Antirrhinum majus L.	RIEGER 1957
Gesneriaceae	
Saintpaulia WENDL.	EHRLICH 1958
Campanulaceae	
Campanula persicifolia	DARLINGTON und GAIRDNER 1938
Compositae	
Crepis (Hybrid)	MÜNTZING 1934
C. fuliginosa × *C. neglecta*	TOBGY 1943
Notonia grandiflora DC.	GANESAN 1939
Hieracium amplexicaule	GENTSCHEFF und GUSTAFSSON 1940
Hemizonia pungens ssp., *interior* × *H. p.* ssp., *maritima*	VENKATESH 1958

einfachen Inversion hergeleitet werden, bedingt durch die Seltenheit anderer Inversionstypen und die bevorzugte Anlage einfacher Chiasmen.

Inversionsbrücken nach Chiasmenbildung in spontanen und induzierten Inversionen sind in der Meiose einer ganzen Reihe von Pflanzen beobachtet worden (Tab. 1). Schon Jahre vor unserem Wissen um die Existenz von Inversionen fand MEURMAN (1925) in der Meiose von *Rumex* Konfigurationen, die er als "small elongated fragments" und "bridges" beschrieb. Wahrscheinlich hat es sich hierbei um Inversionsbrücken gehandelt. Erst MCCLINTOCK (1933) entdeckte, daß zwischen der Brückenbildung in der Meiose und der Inversionsheterozygotie des Objekts ein kausaler Zusammenhang besteht. Daß die Inversionsheterozygotie in einigen Pflanzen ein großes Ausmaß erreichen kann, mögen einige Beispiele zeigen. MENSINKAI (1940) fand in der Meiose von *Allium* in der ersten Teilung einfache und doppelte und in einer Anaphase II vier Inversionsbrücken. Auch ÖSTERGREN (1940) konnte bei *Agropyrum* mehrere Brücken in einer Anaphase I feststellen. BEASLEY (1942) berichtete sogar über acht bis neun Brücken in einer Anaphase I.

Man wird nicht in allen Fällen damit rechnen können, beim Auffinden von Inversionsbrücken die heterozygoten Inversionen im Pachytän des gleichen Objektes nachweisen zu können; das gelingt nur in seltenen Fällen (UPCOTT 1939, EHRLICH 1958, TING 1958). Entweder läßt sich eine Pachytänanalyse nicht durchführen, oder die Inversion kann wegen ihrer geringen Ausmaße nicht gefunden und lokalisiert werden. Allenfalls gibt das Größenverhältnis von Brücke und Fragment Aufschluß über Größe und Lage der Inversion. Wir gehen von der Überlegung aus, daß Brücke und Fragment immer das invertierte Segment neben den duplizierten proximalen bzw. distalen Chromatidenabschnitten enthalten. Aus diesem Grund ist die Inversion nie größer als Brücke und Fragment zusammen, wenn man von einer Dehnung der Inversionsbrücke in fortgeschrittenen Stadien der Zellteilung absieht. Die Lage der Inversion innerhalb des

Chromosoms ist aus dem Größenverhältnis von Brücke (b) und Fragment (f) zu ersehen. Es ergeben sich die folgenden Beziehungen (Abb. 17):

b > f: distale Lage der Inversion
b = f: mediane Lage der Inversion
b < f: proximale Lage der Inversion.

Diese Beziehung gilt nur für Inversionsbrücken, die sich von der Bildung eines einfachen Chiasmas in einer einfachen, heterozygoten, parazentrischen Inversion herleiten. Bei distaler Lage der Inversion ist das Fragment oft so klein, daß es in vielen Fällen cytologisch nicht nachgewiesen werden kann. Straub (1941) fand bei *Rhoeo discolor* ein sehr kleines Fragment und schloß auf eine distal lokalisierte Inversion, die nur einige „Chromomeren" einschloß.

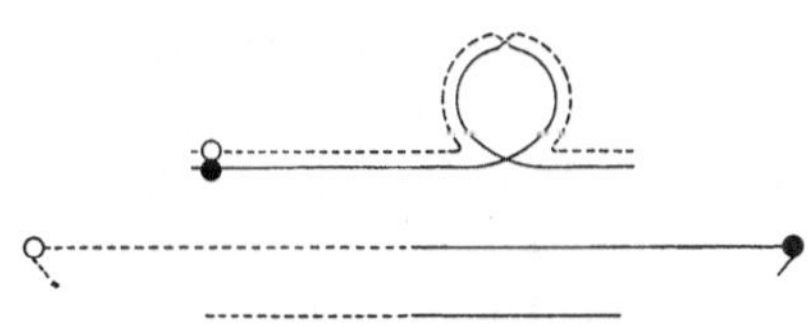

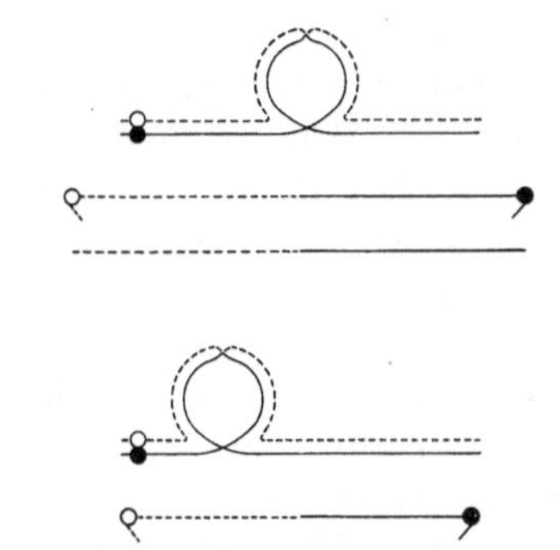

Abb. 17 Abb. 18

Abb. 17. Beziehung zwischen dem Größenverhältnis von Brücke und Fragment und der Lage einer parazentrischen, einfachen Inversion im Chromosom. Der Einfachheit halber ist auf die Zeichnung der Chromatiden verzichtet worden

Abb. 18. Diakinesebivalente mit einem "open" und "reversed chiasma" nach Chiasmenbildung innerhalb und außerhalb einer heterozygoten, parazentrischen, einfachen Inversion (in Anlehnung an Brown u. Zohary 1955)

Interessante Untersuchungen über den Zusammenhang von Chiasmenbildung inner- und außerhalb heterozygoter, parazentrischer Inversionen und den Symmetrieverhältnissen der Diakinese-Konfigurationen stammen von Brown und Zohary (1955). Diese Autoren fanden, daß die Gestalt der Bivalente in der Diakinese von *Lilium formosanum* durch die Lage der Inversionschiasmen bestimmt wird. Bivalente sind nur dann symmetrisch, wenn das Chiasma in der Mitte der Inversionsschleife angelegt wird, d. h. den gleichen Abstand zu den Inversionspunkten hat. Außerdem konnten Brown und Zohary zeigen, daß ein Inversionschiasma in Verbindung mit einem proximalen und distalen Chiasma eine Diakinese-Konfiguration mit einem offenen "reversed chiasma" schafft (Abb. 18). Kommt ein zweites Chiasma in der Inversion hinzu, erfolgt

Verkettung der äußeren Chromatiden ("interlocked reversed chiasma"). Offene "reversed chiasmata" sind im cytologischen Bild gut zu erkennen und gelten als Indizienbeweis für die Hypothese der partiellen Chiasmatypie.

Tiere sind für die Untersuchung auf Meiosestörungen weniger als pflanzliche Objekte geeignet, da die frühen Stadien der Meiose einer cytologischen Analyse schwer zugänglich sind. Angesichts dieser Schwierigkeiten sind die Befunde im Tierreich viel seltener. Die ersten Angaben hat DARLINGTON (1936) für zwei Heuschreckenarten gemacht. Er fand bei *Chorthippus parallelus* und *Stauroderus bicolor* Brücken und Fragmente und konnte auch gleichzeitig Inversionsschleifen im Pachytän nachweisen. Darüber hinaus hat auch KLINGSTEDT (1939) Inversionsbrücken in der Spermatogenese von *Chorthippus*-Bastarden beobachtet. Im Gegensatz zu diesen Beobachtungen stehen allerdings die Angaben von WHITE (1954); er untersuchte nahezu hundert Heuschrecken-Arten, ohne Inversionsbrücken zu finden. WOLF (1941) untersuchte die Spermatogenese einiger Nematoceren und fand bei *Dicranomyia trinotata* Inversionsbrücken in beiden Anaphasen. Ferner wurden Inversionsbrücken in der Spermatogenese von *Sciara impatiens* (CARSON 1946), *Chironomus dorsalis* (ACTON 1956) und *C. tentans* (BEERMANN 1956) festgestellt. Unsere Kenntnisse über das Vorkommen von Inversionen bzw. Inversionsbrücken in anderen Tiergruppen sind noch völlig unzulänglich. CREW und KOLLER (1936) berichteten über Inversionsbrücken in einem Entenbastard *(Cairina × Anas)*. SPURWAY und CALLAN (1950) stellten beim Kammolch *(Triturus cristatus)* und BARIGOZZI (1949) beim Spulwurm *(Ascaris megalocephala)* Inversionsbrücken fest. Über einen besonders interessanten Fall, das Auftreten von Brücken und Fragmenten beim Menschen, berichtete KOLLER (1937). Er fand bei der Untersuchung primärer Spermatocyten einfache Brücken mit einem Fragment und Doppelbrücken mit zwei Fragmenten und nahm an, daß es sich um einen Fall von Inversionsheterozygotie handelte.

Die meisten Autoren vertraten ursprünglich allgemein die Auffassung, daß das regelmäßige Auftreten von dizentrischen Brücken und Fragmenten in der Meiose eine Konsequenz heterozygoter Inversionen sei. Jede Brücke wurde als Beweis für die Inversionsheterozygotie des Objektes angesehen. Nach HAGA (1953) ist die Erklärung der Brückenbildung mit der "inversion-bridge-hypothesis" in vielen Fällen als höchst unwahrscheinlich anzusehen — "the interpretations are highly improbable". Er publizierte als Resultat seiner cytologischen Studien an *Paris verticillata* eine Entstehungshypothese dieser Phänomene. HAGA fand eine große Anzahl spontan auftretender Chromosomenbrüche und beobachtete am gleichen Objekt in den meiotischen Anaphasen freie Chromatiden, dizentrische Brücken, Loop-Univalente und Fragmente. Nach HAGA waren die abnormen Konfigurationen durch Fusion instabiler Bruchenden mit ihresgleichen und Telomeren entstanden. Die Fragmente wurden nicht immer von Brücken begleitet und ließen sich in einzelnen Fällen als Verschmelzungsprodukt ursprünglich selbständiger Bruchstücke interpretieren. HAGA kommt zu dem Schluß, daß die Brückenbildung bei *P. verticillata* nur die Folge von Chromosomenbruch und nachfolgender Reunion sein kann. HAGAs Annahme wird durch weitere Befunde gestützt. So fanden WALTERS (1950), GAUL (1954), REES und THOMPSON (1955) und MARKS (1957) bei anderen Pflanzen Brücken und Fragmente, die infolge eines gleichartigen Bruch-Reunions-Mechanismus entstanden waren.

Es erhebt sich nun die Frage, inwieweit Brücken mit Fragment überhaupt noch eine Beweiskraft hinsichtlich Inversionsheterozygotie zukommt. Charakteristisch für den Bruch-Reunions-Mechanismus ist das unregelmäßige Auftreten einzelner Brücken und Fragmente von wechselnder Gestalt und Größe. Ferner ist zu berücksichtigen, daß infolge ausbleibender oder partieller Reunion eine Reihe „unfertiger" Produkte entsteht — die nicht das Ergebnis von Chiasmenbildung in heterozygoten Inversionen sein kann — wie eine dizentrische Brücke mit zwei Fragmenten, zwei zentrische Fragmente und ein azentrisches Fragment, zwei zentrische und zwei azentrische Fragmente. Regelmäßiges Auftreten von Brücken mit Fragmenten gleicher Gestalt und Größe läßt immer auf Chiasmenbildung in heterozygoten, parazentrischen Inversionen schließen. Es muß allerdings darauf hingewiesen werden, daß eine sichere Aussage über die Entstehungsart von Anaphasebrücken nicht möglich ist; in jedem Fall handelt es sich um Indizienbeweise.

2. Die Persistenz neuer Chromosomenformen

Während des wiederholten Ablaufs seiner identischen Reduplikation kann der Karyotyp Veränderungen der Struktureigentümlichkeit seiner Chromosomen erfahren. Es entstehen abnorme Chromosomenformen, die in der Mehrzahl der Fälle lebensunfähig oder vitalitätsmindernd sind und eliminiert werden. So unterliegen auch die nach Chiasmenbildung in heterozygoten, parazentrischen Inversionen entstehenden neuen Chromosomenformen der Selektion.

In einigen Fällen wird berichtet, daß Neukombinationen des Chromatins zahlreiche Zellteilungen überdauern, auf die nachfolgende Generation übertragen und schließlich konstant werden. Besonderes Interesse verdienen hierbei die Untersuchungen von McCLINTOCK (1938, 1941a, b, 1942a, b, 1943, 1944, 1951) über Ursprung und Verlauf des "breakage-fusion-bridge-cycle" bei *Zea mays*. Dieser Cyclus wird durch den Bruch einer dizentrischen Chromatidenbrücke ausgelöst, die auf folgende Art entstehen kann:

a) durch ein Chiasma in einer langen, heterozygoten, parazentrischen Inversion (Abb. 19) (McClintock 1938),

b) durch ein Chiasma zwischen dem duplizierten Segment einer Chromatide eines Chromosoms und einer normalen Chromatide des homologen Chromosoms (McClintock 1941a).

Die in beiden Fällen entstehenden dizentrischen Brücken weisen Unterschiede hinsichtlich ihres Chromatingehaltes auf. Während im Fall b die Chromatide im Besitz der vollständigen Chromatinmenge ist, wird im Fall a nicht immer der konsequente Ablauf des Cyclus garantiert. Eine Inversionsbrücke ist stets für das distal zur Inversion gelegene Segment defizient und kann u. U. die normale Entwicklung des Gametophyten verhindern. Von der Größe der Defizienz (Lage der Inversion) und ihrer genetischen Wirkung hängt es ab, ob der Gametophyt lebensfähig ist.

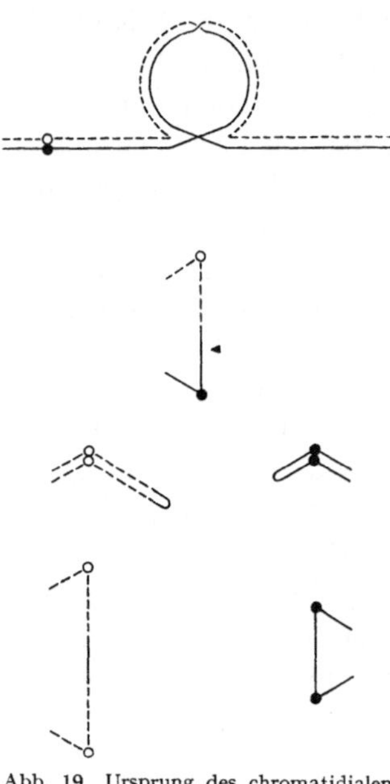

Abb. 19. Ursprung des chromatidialen Teils des Bruch-Fusion-Brücken-Cyclus (in Anlehnung an McClintock 1951)

Durch den Bruch der dizentrischen Chromatide in der Anaphase I entstehen zwei zentrische Fragmente mit je einem instabilen ("unsaturated") Bruchende. Nach erfolgter terminaler Fusion der Subchromatiden eines jeden Fragmentes erscheinen in der ersten mitotischen Anaphase des Gametophyten zwei neue Brückenkonfigurationen, die wiederum nach Bruch und Fusion Anlaß zu Brückenbildung geben. So nimmt der als *Chromatidentyp* bezeichnete Teil des Cyclus seinen Fortgang, bis schließlich alle Zellen des weiblichen und männlichen Gametophyten ein defizientes Chromosom enthalten. Der Cyclus ist nicht nur auf den Gametophyt beschränkt, sondern wird ebenso auf das Endosperm übertragen. Da der Bruch der Brücke wohl in den seltensten Fällen an der Fusionsstelle, sondern weit häufiger submedian oder sogar proximal erfolgt, entstehen Chromatiden, die sich durch größere und kleinere Defizienzen bzw. Duplikationen voneinander unterscheiden. Bestimmte Loci gehen verloren, sind dagegen in den Schwesterkernen dupliziert vorhanden, wo sie auf Grund ihrer Umgruppierung zu neuer Wirksamkeit gelangen können, was sich in einer Mosaikbildung des ausgereiften Endospermgewebes manifestiert.

Der Cyclus hat seinen Ursprung in der Meiose, so daß "unsaturated chromosomes" auch in den Sporophyten gelangen. Hier tritt keine Fusion der instabilen Enden ein ("the newly broken ends heal"), und der Cyclus kommt zum Stillstand. Unklar ist, weshalb sich gleiche Bruchenden in verschiedenen Geweben unterschiedlich verhalten.

Der als *Chromosomentyp* bezeichnet Teil des Cyclus wird eingeleitet, sobald jede Gamete bei der Befruchtung ein Chromosom mit einem instabilen Ende beisteuert. Zum Ursprung des chromosomalen Typs nahm McClintock (1941a) anfangs an, daß die Stabilisierung der Bruchenden sich erst nach Bildung der Zygote vollzieht. Die zentrischen Fragmente sind demnach in der Zygote noch reaktionsfähig und können eine chromosomale Fusion vollziehen. Diese Ansicht ersetzte McClintock (1951) später durch die Annahme, daß erst chromatidiale Fusion der Subchromatiden der zentrischen Fragmente erfolgt und in der nachfolgenden Anaphase zwei Brücken gebildet werden, so daß jeder Telophasekern zwei Chromosomen mit instabilen Bruchenden erhält. Nun kann chromosomale Fusion den Chromosomentyp des Cyclus einleiten. Unklar bleibt nur, weshalb "in the telophase nuclei the fusions now occur between the broken ends of chromosomes rather than between the broken ends of sister chromatids".

Etwas anders liegen dagegen die Verhältnisse im Endosperm. Hier kann der Cyclus dadurch ausgelöst werden, daß entweder der weibliche und der männliche Gametophyt oder nur der weibliche Gametophyt ein Chromosom mit einem instabilen Ende enthalten. Im ersten Fall gelangen drei, im zweiten Fall zwei instabile Chromosomen in das Endosperm und lösen durch chromosomale Fusion den Cyclus aus. Das äußert sich ebenfalls in einer Mosaikbildung im Endospermgewebe. McClintock konnte zeigen, daß der chromosomale Teil des Cyclus in den jungen Pflanzen seinen Fortgang nahm. In 50% der untersuchten Wurzelspitzenmitosen junger Pflanzen wies sie dizentrische Chromosomenbrücken nach. Dagegen blieben in den Mitosen älterer Pflanzen die Brücken aus.

Während der von McClintock beschriebene "breakage-fusion-bridge-cycle" in den jungen Pflanzen schon nach einigen Wochen zum Stillstand kommt, konnten Darlington und Wylie (1953) bei *Narcissus* ein dizentrisches Chromosom vier Jahre lang verfolgen. Der wahrscheinlich ebenfalls durch eine Inversionsbrücke ausgelöste Cyclus unterschied sich von dem Maiscyclus dadurch, daß der chromosomale Teil jederzeit in den chromatidialen Teil übergehen konnte. Wahrscheinlich liegt darin der Grund für die lange Dauer des Cyclus. Andere gleichfalls durch Inversionsbrücken ausgelöste Cyclen sind von Hair (1953) bei *Agropyrum* und Fernandes und Neves (1941) bei *Narcissus bulbocodium* entdeckt worden. Die bisher im Tierreich bekannt gewordenen Bruch-Fusion-Brücken-Cyclen sind nicht die Folge von Inversionsheterozygotie. Den ersten Nachweis eines dizentrischen Chromosoms bei Tieren erbrachten Dalton und Hall (1950). Aus mit Hitzeschock behandelten Eiern vom Axolotl erhielten sie eine sich normal entwickelnde diploide Larve, die im Besitz eines dizentrischen Chromosoms war. Noch nach 64 Tagen wiesen sie in den Mitosen der Schwanzspitze dizentrische Chromosomen nach. KOLLER (1953) entdeckte in einem chemisch induzierten Rattentumor einen Cyclus vom chromosomalen Typ.

Im Verlauf eines Bruch-Fusion-Brücken-Cyclus verändern sich Größe und Gestalt des dizentrischen Chromosoms in Abhängigkeit von der Lokalisation des Brückenbruchs. Dizentrische Chromosomen mit Zentralsegmenten unterschiedlicher Länge sind die Folge submedianen Brückenbruchs; sie stellen den Hauptteil der Variationsprodukte eines Bruch-Fusion-Brücken-Cyclus dar. So fand HAIR (1953) bei *Agropyrum* lange, dizentrische Chromosomen, deren Zentralsegment das 6—8fache der Länge ihres freien Armes hatte, und kurze dizentrische Chromosomen, die nach Verlust des größten Teils des Zentralsegments nur noch ein Viertel seiner Länge besaßen. Durch Eliminierung des Zentralsegments trat bei *Agropyrum* als extremes Variationsprodukt ein "telescoped dicentric" auf.

Die Schwestercentromeren dizentrischer Chromosomen unterliegen in der Anaphase verschiedenen Verteilungsmodi (Abb. 20):

a) Parallelverteilung: Die Centromeren einer Chromatide werden konvergent verteilt, so daß jeder Telophasekern eine dizentrische Chromatide erhält (Abb. 20a).

b) Kreuzverteilung: Bei halber Umwindung der zentralen Chromatidenstränge erscheint in der Anaphase eine Criss-cross-Konfiguration

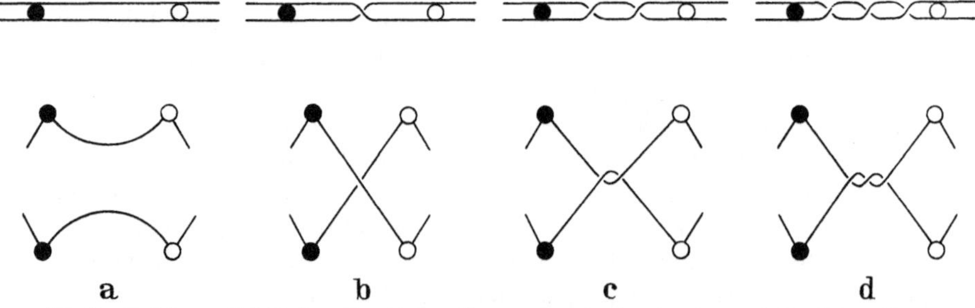

Abb. 20. Verteilungsmöglichkeiten dizentrischer Chromosomen in der Anaphase: a) Parallelverteilung, b) Kreuzverteilung, c) und d) Interlocking (nach KOLLER 1953)

(Abb. 20b). Zerreißen der Zentralsegmente führt zu divergenter Verteilung der Schwestercentromeren. Die zentrischen Fragmente eines Telophasekerns können in Kontakt kommen und fusionieren. Bei Partnermangel ist auch Schwesterchromatiden-Reunion möglich (DARLINGTON und WYLIE 1953, KOLLER 1953). Criss-cross-Konfigurationen gehen oft verloren, da ihre Anaphasebewegung gehemmt ist. Solche Zellen gehen zu Grunde (DARLINGTON und WYLIE 1953).

c) Interlocking: In Abhängigkeit vom Umwindungsgrad der Zentralsegmente zeigen die Anaphase-Konfigurationen singuläres oder multiples Interlocking. Im ersten Fall erfolgt konvergente, im zweiten Fall divergente Verteilung der Schwestercentromeren (Abb. 20c, d).

Die Länge des Zentralsegments dizentrischer Chromosomen bestimmt die Art der Anaphase-Konfiguration und die daraus resultierende Verteilung der Schwestercentromeren. Lange, dizentrische Chromosomen bilden bevorzugt Criss-cross-Konfigurationen oder zeigen Interlocking ihrer Chromatiden. Dagegen nimmt mit Verkürzung des Zentralsegments der Anteil der Parallelverteilung zu, um schließlich bei "telescoped dicentrics" den alleinigen Verteilungsmodus darzustellen (Hair 1953).

Aus Mangel an Fusionspartnern und infolge Stabilisierung der Bruchenden kommt ein Bruch-Fusions-Brücken-Cyclus zum Stillstand. Über die Lebensfähigkeit der dizentrischen Chromosomen und ihrer Derivate entscheiden Grad und genetische Wirksamkeit der Defizienzen bzw. die Größe des zurückgehaltenen Zentralsegments. Bei der von Hair (1953) untersuchten *Agropyrum*-Form hatte der Verlust des größten Teils des Zentralsegments keinen genetischen Nachteil, da die Pflanze durch Polyploidie gegen den schädlichen Einfluß chromosomaler Strukturveränderungen geschützt war. Ein durch den nahezu völligen Verlust des Zentralsegments entstandenes kurzes dizentrisches Chromosom wurde persistent und konnte mit dem Pollen übertragen werden.

IV. Die genetischen Konsequenzen

Durch die Inversion eines Chromosomensegments wird keine quantitative Veränderung im Chromosom hervorgerufen, sondern nur die lineare Gensequenz einer Koppelungsgruppe geändert. Trotzdem haben Inversionen einen genetischen Effekt, wobei die wohl am meisten diskutierte Wirkung die Erscheinung des Positionseffekts darstellt. Positionseffekte von Inversionen sind selten, zumal nicht jede Inversion mit einem nachweisbaren Effekt verbunden ist, und bisher nur für *Drosophila* mit Sicherheit nachgewiesen worden.

1. Die Beeinflussung des Genotyps

Viele als Genmutationen angesehene Veränderungen des Phänotyps haben sich im Laufe der Zeit als Strukturumbauten der Chromosomen erwiesen. Zur Klärung solcher Befunde nahm man an, daß Inversionen mit Beschädigung oder Verlust der den Bruchstellen benachbarten Loci verbunden sind. Die cytologische Analyse ergab, daß zwar in einigen Fällen, aber nicht immer, sichtbare chromosomale Defekte vorlagen. Die Änderung der Reihenfolge der Gene eines Chromosoms kann demnach der Wirkung einer Genmutation gleichkommen, obwohl nicht die Gene und deren Zahl, sondern nur ihre Lage zueinander verändert wird. Diese in der Literatur als Positionseffekt bekannte Erscheinung äußert sich in der Modifikation der Expressivität anderer Gene, die meist herabgesetzt wird, und in der Beeinflussung von Dominanz und Penetranz. Außerdem sind stabile Änderungen des Phänotyps bekannt.

Zum besseren Verständnis soll erst auf die Klassifizierung der mannigfaltigen Erscheinungen des Positionseffekts eingegangen werden. Eine Unterteilung der Phänomene des Positionseffekts in zwei Gruppen schlägt LEWIS (1950) vor. Er postuliert zwei Typen, die er als völlig unterschiedliche Erscheinungen ansieht, den instabilen V-Typ ("variegated type") und den stabilen S-Typ ("stabile type"). Auf Grund aller bisherigen Befunde vertritt LEWIS die Auffassung, daß der S-Typ sehr selten ist, wogegen der V-Typ fast die Gesamtheit der Positionseffekte darstellt. GOLDSCHMIDT (1955) entgegnete darauf, daß diese Beobachtung nicht dem wahren Sachverhalt entspräche. Der V-Typ wird nur leichter gefunden, da das „mottling" eine sehr auffällige Erscheinung darstellt, während der einfache Effekt des S-Typs in der Regel übersehen oder als gewöhnliche Mutation angesehen wird. Beide Typen sind Variationen eines Phänomens; der Unterschied ist nur quantitativer Art.

Der erstgenannte Typ liegt dann vor, wenn ein in die unmittelbare Nähe des Heterochromatins gebrachtes Gen ein instabiles, somatisches Mosaik induziert, das sehr sensibel gegen Umwelteinflüsse ist. So beschrieb HINTON (1949) die Wirkung der Inversion In(2LR)40d im 3. Chromosom von *Drosophila melanogaster* mit jeweils einem Inversionspunkt im Eu- und Heterochromatin, die die Beschaffenheit der Facettenaugen älterer Tiere verändert. Die Inversion bewirkt eine starke Vergröberung der Augen, eine Sprenkelung des Augenpigments und die Verlagerung schwarzen, tumorähnlichen Materials auf die Augenoberfläche. Die Wirkung der Inversion ist sehr variabel und wird von Geschlecht und Alter bestimmt.

Häufig kann man beobachten, daß durch eine Inversion die Expressivität mehrerer durch zahlreiche Querscheiben vom Inversionspunkt getrennter Gene modifiziert wird. Diese als "spreading effect" bekannte Erscheinung wurde von DEMEREC (1940) an *D. melanogaster* eingehend untersucht. Ein langes, invertiertes Segment (N 264—52) im X-Chromosom beeinflußte die Expressivität von fünf Genen (rst, fa, dm, ec, bi) innerhalb der Inversion, die einen mehr oder weniger "variegated phenotype" hervorriefen. Da sich der "spreading effect" bis zum bi-Locus ausbreitete, war er noch 50 Querscheiben vom Inversionspunkt entfernt wirksam.

Über einen interessanten Fall der Beeinflussung von Expressivität und Dominanz durch den Positionseffekt von Inversionen bei *Drosophila* berichteten GOLDSCHMIDT, GARDNER und KODANI (1939). Sie fanden, daß die dominante Mutante Beaded (Bd), die durch die Ausbildung eines geperlten Kammes auf dem Flügelrand gekennzeichnet ist, in ihrer Expressivität und Dominanz durch die Gegenwart der Inversion In(3R)C gesteigert wird. Da der "enhancing effect" dieser Inversion annähernd durch die Payne-Inversion und andere Inversionen des 3. Chromosoms zu ersetzen war und nicht alle Inversionen das gleiche Intensivierungsgen enthalten können, ist anzunehmen, daß diesem Effekt nicht die Existenz eines Dominigens, sondern ein Positionseffekt zugrunde liegt. Ein analoger Wirkungstyp wurde für die Beeinflussung der heterozygoten Mutante vestigial (vg) von GARDNER (1942) und GOLDSCHMIDT und GARDNER (1942) beschrieben. Von der Mutante vg (stummelflügelig) ist eine seriierbare Reihe multipler Allele bekannt, die von GOLDSCHMIDT (1935) im einzelnen beschrieben und nach aufsteigendem Dominanzgrad klassifiziert wurden. Die Autoren fanden, daß einige Inversionen die Dominanz nur bei Männchen, andere dagegen bei Männchen und Weibchen steigerten. Gewöhnlich traten nur die ersten Klassen mit schwach ausgeprägter Dominanz auf ("niched, notched"). Eine Ausnahme machte die Glazed-Inversion In(2R)Gla, die in der Kombination vg/In(2R)Gla die Ausbildung ganz schmaler Flügel ("strap") hervorrief. Kombiniert hatten die Inversionen die stärkste Wirkung.

Selbst kleinste Inversionen am Ende des X-Chromosoms von *D. melanogaster*, die nur drei Querscheiben enthalten, bedingen Abänderungen im Auftreten bestimmter Borsten, für deren Ausbildung die Gene scute und achaete verantwortlich sind (MULLER, PROKOFYEVA und RAFFEL 1935).

Inversionen sind im heterozygoten und homozygoten Zustand oft von Letalfaktoren begleitet. So ist die Inversion In(2LR)40d homozygot letal (HINTON 1950). Die Curly-Inversion In(2L)Cy wirkt im homozygoten Zustand letal, während die heterozygote Inversion nur ein leichtes Aufkrümmen der Flügel verursacht. Die Inversion In(3R)C läßt im späten Puppenstadium die Imago absterben, die nicht die Fähigkeit zum Schlüpfen erlangt (HADORN 1951). Die CLB-Inversion In(1)Cl wirkt im heterozygoten Zustand im Ei- und Larvenstadium letal (BREHME 1939). Über die cytogenetischen Ursachen der Letalität, die in Verbindung mit induzierten Chromosomenmutationen auftritt, wird bis in die neuste Zeit lebhaft diskutiert. Die Frage, ob an den Bruchstellen lokalisierte Defizienzen der Grund des Letaleffekts sind, läßt sich durch eingehende Untersuchung der Struktur der Riesenchromosomen beantworten. Schwieriger ist dagegen eine Entscheidung zwischen Genmutation und Positionseffekt zu treffen, die beide als Ursache in Frage kommen. Nach strahlengenetischen Untersuchungen an *Drosophila* vertreten LEA und CATCHESIDE (1945) die Auffassung, daß Letalität kaum durch Positionseffekt, sondern fast nur durch gleichzeitig mit den Strukturveränderungen in der Nähe der Bruchstellen eintretende Genmutationen und submikroskopische Defizienzen verursacht wird. Verallgemeinernd behaupten sie, daß die meisten Letaleffekte nach Strahlenbehandlung durch Verheilen frischer Bruchstellen ausgelöst werden, wobei das Verheilen zu Restitutionen und Rekombinationen führen kann. Dagegen sieht Herskowitz (1951) die Ursache vieler Letalfaktoren im Positionseffekt und in unabhängig von Chromosomenbrüchen auftretenden Genmutationen. Durch Positionseffekt bedingte Letalfaktoren müssen — der Einteilung von Lewis (1950) folgend — dem stabilen S-Typ zugeordnet werden.

Sollte die durch den Positionseffekt einer Inversion bedingte Änderung der Genaktivität nur auf der neuen Lage des Gens und nicht auf einer Veränderung seiner submikroskopischen Struktur beruhen, so müßte eine in den ursprünglichen Inversionspunkten eintretende Reinversion mit einer Reversion verbunden sein. Tatsächlich fand Grüneberg (1937) bei *D. melanogaster* einige spontane Fälle, die eine völlige Reversion zum Wildtyp in genetischer und cytologischer Hinsicht darstellten. Eine lange, durch Röntgenstrahlen im X-Chromosom induzierte Inversion, deren Inversionspunkte in der Nähe des bb- und rb-Locus lagen, war mit einem Positionseffekt verbunden, der eine Vergröberung der Augenoberfläche bewirkte. Vier Imagines stellten bezüglich der Augenausbildung eine Rückmutation dar. Die cytologische Bestätigung dieses Falles erbrachte Emmens (1937). Es ist offensichtlich, daß dieser Fall, der von größter theoretischer Bedeutung ist, besonderes Interesse fand und teilweise mit Skepsis aufgenommen wurde. Bauer (1938) und Dobzhansky und Sturtevant (1938) sahen in den Ausnahmefliegen Verunreinigungen oder die Folge eines Crossing-over und forderten die exakte Reproduktion des Versuches. Dagegen dürfte gegen die Befunde von Hinton (1950) nichts einzuwenden sein. Hinton fand in verschiedenen, unabhängig voneinander angesetzten Experimenten eines Stammes mit der Inversion In(2LR)40d (Vergröberung der Augen, "mottling" des Augenpigments) vier Ausnahmefliegen, die sich sowohl in cytologischer als auch in genetischer Hinsicht als Rückmutationen erwiesen. Besonderes Interesse verdienen in diesem Zusammenhang die Befunde von Kaufmann (1942); er konnte zeigen, daß die völlige Rekonstruktion der Gensequenz für den Eintritt einer Reversion nicht unbedingt erforderlich ist. Es genügt schon, wenn nur Teilabschnitte des dislozierten Segments umgebaut werden.

Die angeführten Beispiele zeigen, daß Inversionen nicht nur die Mannigfaltigkeit des Karyotyps erhöhen, sondern durch Verlagerung der Gene direkten Einfluß auf die Ausbildung des Genotyps haben. In ihrer Wirkung können sie mit Genmutationen identisch sein. Diese Erscheinung hat Anlaß zu spekulativen Betrachtungen über die Organisation des Chromosoms und den Zusammenhang zwischen Gen- und Chromosomenmutation gegeben. Fest steht, daß das klassische Konzept vom korpuskularen Gen nicht mehr aufrechterhalten werden kann. Das Chromosom muß als Funktionseinheit einzelner spezialisierter Untereinheiten angesehen werden. GOLDSCHMIDT (1946, 1951, 1955) geht jedoch soweit, daß er, ausgehend von der Überlegung, daß Gene nur durch Mutationen bekannt sind und sich alle Mutationen bei cytologischer Analyse als Positionseffekte erwiesen, die Vorstellung vom Gen aufgibt und alle Genmutationen als submikroskopische Chromosomenmutationen erklärt. Dagegen spricht, daß nicht alle Chromosomenmutationen einen nachweisbaren genetischen Effekt haben und z. B. kein Zusammenhang zwischen Länge der Inversion und der Expressivität des Positionseffekts besteht. Die Frage nach der Grenze zwischen Gen- und Chromosomenmutation kann beim heutigen Stand unseres Wissens nur spekulativ beantwortet werden.

2. Der intrachromosomale Effekt heterozygoter Inversionen

In der frühen *Drosophila*-Literatur finden sich zahlreiche Hinweise auf Gene, die auf Grund ihrer Eigenschaft, das Crossing-over lokal herabzusetzen oder zu unterdrücken, als C-Faktoren ("crossing-over suppressor") bezeichnet wurden. Diese Faktoren sind nur im heterozygoten Zustand (Cc) in der Lage, das Crossing-over zwischen den Homologen zu reduzieren. Bei der Untersuchung eines Wildstammes von *Drosophila melanogaster* kam STURTEVANT (1926) zu der Erkenntnis, daß es sich bei dem C-Faktor im 3. Chromosom nicht um ein Gen, sondern um eine Inversion im heterozygoten Zustand handelte, die allem Anschein nach spontanen Ursprungs war. C-Faktoren haben sich in der Mehrzahl der Fälle als heterozygote Inversionen erwiesen.

Besonders umfangreiche Beobachtungen dieses Phänomens sind an *D. melanogaster* gemacht worden. Weitere Mitteilungen stammen von OFFERMANN und MULLER (1932), GRAUBARD (1934) und GRÜNEBERG (1935). STURTEVANT und BEADLE (1936) untersuchten den Effekt verschiedener Inversionen des X-Chromosoms und stellten fest, daß alle Inversionen im heterozygoten Zustand das Crossing-over im Bivalent lokal begrenzt reduzieren, dagegen homozygote Inversionen Grad und Verteilung des Crossing-over nicht verändern. Am markantesten ist dieser Effekt im Inversionsbereich selbst ausgeprägt, wobei eine ungefähre Korrelation zwischen Länge der Inversion und dem Grad der Reduktion besteht, derart, daß die Austauschwerte kurzer Inversionen (In sc-8, In sc-4, In y-4) weit niedriger liegen als die sehr langer Inversionen (In dl-49, In ClB, In sc-7). Die Reduktion des Crossing-over außerhalb der Inversion ist nicht so signifikant und nur in unmittelbarer Nähe der Inversionspunkte vorhanden. Am gleichen Objekt prüften NOVITSKI und

Braver (1954) die Inversion In dl-49 eines "tandem metacentric compound X-chromosome" auf ihre Wirkung. Die Autoren fanden, daß das Crossing-over innerhalb der heterozygoten Inversion bezüglich des Standardwertes um 25% reduziert und zufällig verteilt war. Ausgenommen war eine schwache Depression beiderseits der Inversionspunkte. Dobzhansky und Epling (1948) untersuchten den intrachromosomalen Effekt der heterozygoten Inversionen des 3. Chromosoms einer natürlichen Population von *D. pseudoobscura* und fanden, daß die Rekombinationswerte nicht nur innerhalb der Inversion, sondern auch außerhalb stark herabgesetzt waren. Besonderes Interesse verdienen die Angaben von Komai und Takaku (1940) über zwei unabhängige Inversionen des akrozentrischen X-Chromosoms von *D. virilis*. Beide Inversionen hatten kombiniert folgende Wirkung: Das Crossing-over in der Inversion lag unter 1%; im proximalen Abschnitt (bb-rg) war ebenfalls eine starke Reduktion zu verzeichnen (1,9%), wogegen das Crossing-over im distalen Abschnitt mit zunehmender Entfernung von der Inversion anstieg und schließlich am äußersten Ende (v-y) sogar "a tendency to increase" zeigte. Bemerkenswert ist, daß die Crossing-over-Werte zwischen beiden Inversionen ebenfalls unter 1% lagen. Für die einzelnen Inversionen bestanden bezüglich der Crossing-over-Reduktion keine nennenswerten Unterschiede. Carson (1953) wies nach, daß die Verhältnisse bei *D. robusta* ähnlich liegen. Er berichtete, daß, wenn jeder Arm des metazentrischen X-Chromosoms eine heterozygote Inversion enthält, die Crossing-over-Reduktion im interkalaren Segment absolut ist.

Die Frage nach den Ursachen des intrachromosomalen Effekts heterozygoter Inversionen kann nur hypothetisch beantwortet werden. Es ist anzunehmen, daß der intrachromosomale Effekt aus dem Zusammenspiel verschiedener Faktoren resultiert. Am ältesten ist die von einer Reihe von Autoren geäußerte Vorstellung, derzufolge die Crossing-over-Reduktion nicht eine Folge der Inversionsschleife selbst ist, sondern auf Paarungsunstimmigkeiten während des Zygotäns beruht. Sturtevant und Beadle (1936) wiesen darauf hin, daß "a short inversion may be supposed to have its pairing more interfered with by the uninverted sections than does a long inversion which has short inverted sections". Carson (1953) folgerte: "It is clear that inversions, especially short ones, do interfere to some extent with the actual formation of crossovers within the inverted segment." Es ist einleuchtend, daß bei Abnahme der Inversionslänge aus mechanischen Gründen (begrenzte Flexibilität der Homologen) die Möglichkeit der Schleifenbildung erschwert wird, um schließlich im Extremfall ganz auszubleiben und somit die Anlage von Chiasmen auszuschließen. Unter diesem Aspekt wird die besonders starke Crossing-over-Reduktion in kurzen Inversionen verständlich.

Nach neueren Vorstellungen, die von NOVITSKI und BRAVER (1954) an einem "tandem metacentric compound X-chromosome" gewonnen wurden, soll eine ursächliche Beziehung zwischen der Crossing-over-Reduktion und der unterschiedlichen Paarungstendenz des Eu- und Heterochromatins bestehen. Die Autoren sind der Ansicht, daß die Attraktionskräfte längs der euchromatischen Regionen so schwach sind, daß diese erst dann paaren, wenn nach erfolgter Paarung bestimmter, chromosomenspezifischer Kontaktstellen — in diesem Fall distales Heterochromatin — die Homologen angenähert werden. Enthält ein euchromatischer Abschnitt eine heterozygote Inversion, wird die Euchromatinpaarung mehr oder weniger erschwert und erfolgt teilweise unvollständig. Die Crossing-over-Reduktion ist somit nicht so sehr die Folge der Inversionsschleife, sondern die Konsequenz "of the interruption of the continuity of euchromatic pairing".

Ein anderer Faktor, der besonders für die Crossing-over-Reduktion in Nähe der Inversionspunkte verantwortlich ist, ist die positive Interferenzwirkung der Inversionspunkte, die wohl am ehesten mit dem "spindle-fibre effect" des Centromers verglichen werden kann. Begründen läßt sich diese Erscheinung durch die an den Inversionspunkten ausbleibende Paarung, die die Anlage von Chiasmen in ihrer Nähe erschwert und damit die Möglichkeit eines Crossing-over verhindert. Über eine besonders starke Depression des Crossing-over in unmittelbarer Nähe der Inversionspunkte berichten HOOVER (1938), KOMAI und TAKAKU (1940) und NOVITSKI und BRAVER (1954). Nach RHOADES und DEMPSEY (1953) fehlt bei *Zea mays* diese Depression in der Nachbarschaft der Inversionspunkte.

Die Reichweite des Reduktionseffekts außerhalb der Inversion wird von der Länge der nichtinvertierten Abschnitte bestimmt. Während bei *D. melanogaster* die heterozygoten Inversionen In dl-49 und In M das Crossing-over distal stärker als proximal reduzierten (STONE und THOMAS 1935, STURTEVANT und BEADLE 1936), berichteten KOMAI und TAKAKU (1940) über genau umgekehrte Verhältnisse bei *D. virilis*, wo das Crossing-over im distalen Ende des X-Chromosoms nur schwach reduziert wurde und am äußersten Ende sogar den Standardwert übertraf. Diese Diskrepanz erklärt sich aus der unterschiedlichen Länge der proximalen bzw. distalen Abschnitte des X-Chromosoms von *D. melanogaster* und *D. virilis*. Der distale Abschnitt des X-Chromosoms von *D. melanogaster* hat nur die Länge von 10 Morgan-Einheiten und erlaubt Crossing-over in beschränktem Maße; dagegen erlangt der entsprechende Abschnitt von *D. virilis* bei einer Länge von 45 Morgan-Einheiten einen gewissen Grad von Autonomie bezüglich Paarung und Crossing-over. Das heißt, daß über eine bestimmte Distanz hinaus der Crossing-over reduzierende Effekt heterozygoter Inversionen nicht mehr wirksam ist.

Die Frage nach den Ursachen des intrachromosomalen Effekts kann dahingehend beantwortet werden, daß für die Crossing-over-Reduktion im Bereich der Inversion vornehmlich Paarungsunstimmigkeiten verantwortlich sind und der Effekt außerhalb der Inversion durch ausstrahlende Paarungsstörungen und die positive Interferenzwirkung der Inversionspunkte zustande kommt. Eine zusätzliche Verstärkung erfährt der intrachromosomale Effekt noch dadurch, daß — wie im Abschnitt IV, 4 gezeigt wird — ein Teil der Austauschchromatiden infolge selektiver Eliminierung von den Gameten ferngehalten wird.

3. Der interchromosomale Effekt heterozygoter Inversionen

Bei der sichtbaren Abnahme des Crossing-over in einem inversionsheterozygoten Bivalent steigt in den heterologen Chromosomen das Crossing-over an. Nachdem schon STURTEVANT (1919) und WARD (1923) bei *Drosophila melanogaster* auf diese Erscheinung hingewiesen hatten, wurde sie erst von SCHULTZ in den dreißiger Jahren eingehend untersucht (MORGAN, BRIDGES und SCHULTZ 1930, 1932, 1935) und später von SCHULTZ und REDFIELD (1951) zusammenfassend dargestellt. Dieses in die Literatur unter der Bezeichnung Schultz-Redfield-Effekt eingegangene Phänomen ist bisher außerhalb der Gattung *Drosophila* nicht gefunden worden.

Bei der Analyse des interchromosomalen Effekts der ClB- und Payne-Inversion des 1. und 3. Chromosoms kamen SCHULTZ und REDFIELD zu folgendem Ergebnis: Beide Inversionen hatten annähernd die gleiche Wirkung auf die Verteilung des Crossing-over im 2. Chromosom, indem sie das Crossing-over besonders im zentralen Teil (Centromerenregion) und an den Endabschnitten des Chromosoms erhöhten. Dagegen wurden die Mittelabschnitte der Chromosomenarme weniger betroffen. Die ClB-Inversion war etwas effektiver als die Payne-Inversion; kombiniert erlangten sie größte Wirksamkeit (Abb. 21). In einem anderen Experiment ermittelten die Autoren die Beeinflussung des Crossing-over im 3. Chromosom durch die ClB- und Curly-Inversion. Auch in diesem Fall war die Mittelregion des metazentrischen 3. Chromosoms am stärksten betroffen, doch fehlte jeglicher Effekt an den Chromosomenenden. Anders sah die Verteilung des Crossing-over im X-Chromosom aus. Bedingt durch die terminale Lage des Centromers wiesen beide Chromosomenenden die höchsten Crossing-over-Werte auf (Abb. 22). Kennzeichnend für den Schultz-Redfield-Effekt ist demnach ein Ansteigen des Crossing-over im Bereich des Centromers und an den Chromosomenenden, in Abschnitten, die normalerweise die niedrigsten Crossing-over-Werte aufweisen.

Andere Autoren, die sich ebenfalls mit dem interchromosomalen Effekt heterozygoter Inversionen bei *D. melanogaster* befaßten, kamen zu analogen Resultaten. Interessante Einzelheiten erbrachten die Untersuchungen von STEINBERG. Dieser prüfte den Effekt verschiedener heterozygoter, autosomaler Inversionen auf das Crossing-over im X-Chromosom (STEINBERG 1936). Dabei ergab sich, daß die Curly-Inversion ihren größten Effekt am Ende des X-Chromosoms zwischen y (0) und ec (5,5) erreichte, wo das Crossing-over von 7,4 auf 171,1% anstieg. In Richtung Centromer nahmen die Werte ab und betrugen zwischen cv (13,7) und ct (20,0) nur noch 13% (Standard 8,7%). Die Wirkung der Payne-Inversion war ungefähr gleich, doch distal nicht so stark ausgeprägt. Beide Inversionen kombiniert gaben der Kurve den gleichen Verlauf, doch lagen die Werte in allen Regionen weitaus höher. Die Inversionen ließen sich nach ihrer Wirksamkeit folgendermaßen ordnen: Curly/Payne > Payne > Curly. Von Interesse ist, daß der Crossing-over-Anstieg auf dem Anwachsen des multiplen Crossing-over beruhte und vornehmlich auf Kosten des singulären erfolgte.

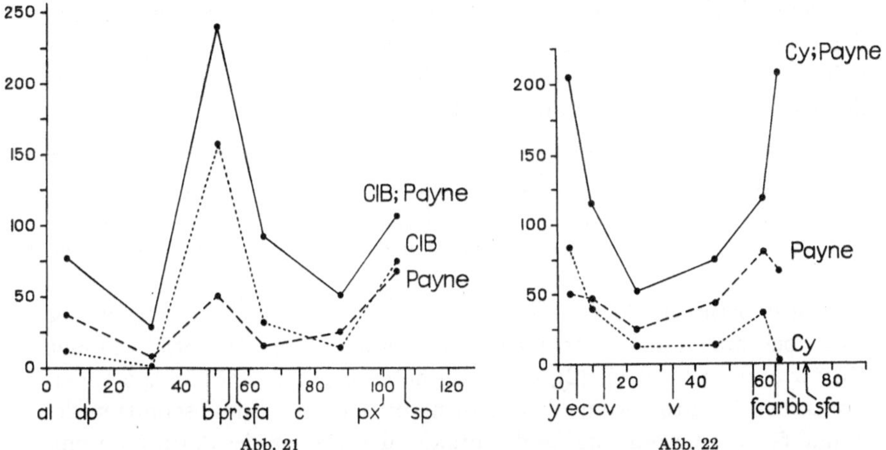

Abb. 21 Abb. 22

Abb. 21. Der Einfluß heterologer, heterozygoter Inversionen des X- und 3. Chromosoms auf das Crossing-over im 2. Chromosom von *Drosophila melanogaster*. Die Ordinate zeigt den prozentualen Anstieg des Crossing-over die Abszisse den genetischen Abstand der Loci (nach SCHULTZ und REDFIELD 1951)

Abb. 22. Der Einfluß heterologer, heterozygoter Inversionen des 2. und 3. Chromosoms auf das Crossing-over im X-Chromosom von *Drosophila melanogaster*. Ordinate und Abszisse wie in vorhergehender Abbildung (nach SCHULTZ und REDFIELD 1951)

Der Einfluß nahezu aller Inversionen des X-Chromosoms auf das Crossing-over im 3. Chromosom wurde von STEINBERG und FRASER (1944) untersucht. Von zwölf Inversionen waren acht effektiv, ließen aber keine direkte Beziehung zwischen dem unterschiedlichen Grad ihrer Wirkung und ihrer Lokalisation und Größe erkennen. STEINBERG (1937) konnte auch zeigen, daß der Wirkungsgrad einer Inversion nicht nur eine Funktion der Inversion selbst ist, sondern ebenso von der

Beschaffenheit der heterologen Chromosomen bestimmt wird. Er wies nach, daß die Wirkung der ClB-, Payne- und Curly-Inversion auf das Crossing-over in den heterologen Chromosomen nicht gleich ist:

Inversion	X-Chr.	2. Chr.	3. Chr.	insgesamt
ClB	—	0,38	0,46	0,84
Curly	0,22	—	0,49	0,71
Payne	0,33	0,55	—	0,88

Beim Vergleich der Wirkung der Inversionen mit der relativen Länge der betroffenen heterologen Chromosomen ergab sich die Gesetzmäßigkeit, daß der Effekt einer Inversion direkt proportional der Metaphaselänge des heterologen Chromosoms ist:

Chromosom	X/2. Chr.	X/3. Chr.	2. Chr./3. Chr.
Größenverhältnis. . .	100/145 = 0,69	100/175 = 0,75	145/175 = 0,83
Wirkungsverhältnis. .	0,33/0,55 = 0,60	0,22/0,49 = 0,45	0,38/0,46 = 0,83

Die experimentelle Bestätigung des Schultz-Redfield-Effekts wurde auch bei anderen *Drosophila*-Arten erbracht. Bei *D. virilis* fanden KOMAI und TAKAKU (1942), daß zwei heterozygote, unabhängige Inversionen des X-Chromosoms das Crossing-over in den Endabschnitten des 3. und 5. Chromosoms und in der proximalen Region des 3. Chromosoms steigerten. Beide Inversionen waren von größter Wirksamkeit.

Obwohl die Analyse durch eine ungenügende Anzahl von Markierungsgenen erschwert war, gelang CARSON (1953) bei *D. robusta* die Feststellung eines interchromosomalen Effekts. Übereinstimmende Ergebnisse liegen auch für *D. pseudoobscura* vor. Schon MCCNIGHT (1937) wies auf den Einfluß heterozygoter, autosomaler Inversionen des Bastards *D. pseudoobscura* × *D. simulans* auf das Crossing-over im X-Chromosom hin, der nach Rückkreuzung verloren ging. Später bestätigten LEVINE und LEVINE (1954) das Vorhandensein des Schultz-Redfield-Effekts in natürlichen Populationen von *D. pseudoobscura*.

Unsere Kenntnisse über die Ursachen der unterschiedlichen Intensivierung des Crossing-over in verschiedenen Chromosomenabschnitten sind noch völlig unzureichend. Die Tatsache, daß z. B. die ClB-Inversion im 2. und X-Chromosom einen unterschiedlichen Verlauf der Crossing-over-Kurve hervorruft, weist darauf hin, daß für die Ausbildung des Effekts neben der Inversion und der relativen Länge der heterologen Chromosomen wahrscheinlich auch die Morphologie der betroffenen Chromosomen ausschlaggebend ist. SCHULTZ und REDFIELD (1951) konnten zeigen, daß das Heterochromatin bei der Modifikation des Crossing-over eine wichtige Rolle spielt. Um den Einfluß des den bb-Locus enthaltenden heterochromatischen Abschnittes auf das Crossing-over zu prüfen, benutzten sie die Inversion In (X)rst[3], die das Heterochromatin aus der Einflußsphäre des Centromers distal verlagerte. Da die Anzahl der Querscheiben zwischen dem y- und ec-Locus im nichtinvertierten X-Chromosom annähernd mit der Querscheibenzahl zwischen dem y- und car-Locus im invertierten Chromosom übereinstimmte, war ein Vergleich der Rekombinationswerte beider Regionen gerechtfertigt. Bei Benutzung der heterozygoten Curly- und Payne-Inversion erwies sich, daß die Crossing-over-Werte für die y-car-Region unter dem Einfluß des Heterochromatins weitaus höher lagen als für die als Standard angenommene y-ec-Region. Das veranlaßte die Autoren in Übereinstimmung mit MATHER (1939a) zu der Annahme, daß Heterochromatin, das sich außerhalb der Interferenzzone des Centromers befindet, einen größeren Crossing-over-Koeffizienten aufweist als Euchromatin.

Die Wechselwirkung des intra- und interchromosomalen Effekts heterozygoter Inversionen modifiziert die Häufigkeit des Crossing-over in den Bivalenten eines Genoms unterschiedlich, ohne seinen absoluten Betrag wesentlich zu ändern. Es tritt nur eine Verschiebung innerhalb des Genoms ein, indem das Crossing-over in Abschnitte heterologer Chromosomen verlagert wird, die im Normalfall die geringsten Rekombinationswerte aufweisen. Zur Deutung dieses Phänomens finden sich in der Literatur zahlreiche Hypothesen, denen teilweise der Mangel anhaftet, daß sie sich entweder auf die Annahme zusätzlicher Hilfshypothesen stützen müssen oder sich nur auf die Erklärung der Inversionseffekte beschränken und außer acht lassen, daß ähnliche interchromosomale Effekte auch für Translokationen (DOBZHANSKY 1931), Ringchromosomen (TROFIMOV und POSTNIKOVA 1935) und überzählige Y-Chromosomen bei *Drosophila* beschrieben wurden.

So geben SCHULTZ und REDFIELD (1951) für den Crossing-over-Anstieg in den heterologen Chromosomen eine Erklärung, der mechanische Vorgänge zugrunde liegen. Sie postulieren, daß die durch eine heterozygote Inversion oder ein überzähliges Y-Chromosom hervorgerufene Paarungsstörung der homologen Chromosomen eine verstärkte Torsion der heterologen Chromosomen nach sich zieht und zu einem Anstieg des Crossing-over führt. Um jedoch die cytomorphologische Verbindung zwischen den Chromosomen herzustellen, wird für die Autoren die Annahme eines polarisierten Paarungsmusters in der Meiose erforderlich.

Eine andere Ansicht ist die, derzufolge das Genom als ein physiologisches System alle quantitativen Verschiebungen seines Funktionsgleichgewichts durch einen Homoeostasismechanismus selbstregulierend abschwächt oder kompensiert. So antwortet das Genom bei Störung seiner spezifischen Crossing-over-Frequenz durch Kompensation dieses

Effekts, indem die Crossing-over-Frequenz in den heterologen Chromosomen ausgleichend ansteigt. Diese Interpretation setzt voraus, daß es für das Genom eine oberste Grenze der Crossing-over-Frequenz gibt und die Bivalente untereinander um die Anlage von Chiasmen wetteifern (MATHER 1936, 1939b).

Es hat auch nicht an dem Bestreben gefehlt, den interchromosomalen Effekt heterozygoter Inversionen als die Folge eines Positionseffekts anzusehen. So äußerten STEINBERG und FRASER (1944) die Vorstellung, daß der Crossing-over-Anstieg durch "an unspecified physiological effect caused by the inversion" hervorgerufen wird. Diese Auffassung, die später auch von GOLDSCHMIDT (1955) aufgegriffen wurde, beschränkt sich jedoch nur auf die Deutung des Effekts heterozygoter Inversionen.

Zu einer etwas anderen Ansicht sind LEVINE und LEVINE (1954) aufgrund ihrer Befunde an einer natürlichen Population von *D. pseudoobscura* gekommen. Sie fanden, daß gewisse Inversionen des 3. Chromosoms nicht nur im heterozygoten, sondern auch im homozygoten Zustand das Crossing-over im X-Chromosom steigerten. Ein Positionseffekt ist nach Meinung der Autoren ausgeschlossen, da zwei in cytologischer Hinsicht identische Arrowhead-Inversionen unterschiedliche Crossing-over-Frequenzen induzierten. Sie postulieren, daß das Crossing-over im X-Chromosom unter der genetischen Kontrolle des 3. Chromosoms steht.

4. Die Beeinflussung der Fertilität durch heterozygote Inversionen

Nur kurze Inversionen besitzen die Fähigkeit, das Crossing-over innerhalb der Inversion weitgehend zu unterbinden. In langen Inversionen kann durchaus Crossing-over stattfinden, was zwangsläufig mit einer ungleichen Verteilung des Chromatins verbunden ist. Theoretisch müßte das zur Bildung unbalancierter Gameten und zu partieller Sterilität führen.

BEADLE und STURTEVANT stellten zuerst die Frage nach den genetischen Konsequenzen des Crossing-over in heterozygoten, parazentrischen Inversionen für die Nachkommenschaft von *Drosophila* (BEADLE und STURTEVANT 1935, STURTEVANT und BEADLE 1936). Sie prüften im Experiment den Einfluß der Inversionen In sc-4, In sc-8, In y-4 und In dl-49 des X-Chromosoms und fanden, daß die Häufigkeit des einfachen Crossing-over in langen Inversionen nahezu der für das uninvertierte Segment spezifischen Frequenz entspricht. Bei *Drosophila* findet Crossing-over nur in der Meiose der Weibchen statt; die Männchen gehören dem achiasmatischen Spermatogenesetyp an und bilden normale Spermien. Es müßten also theoretisch nach einem einfachen Crossing-over in einer heterozygoten Inversion eines weiblichen X-Chromosoms bei zufallsgemäßer Verteilung der Chromatiden einer Chromatidentetrade zur

Hälfte unbalancierte Ootiden entstehen, die nach Befruchtung mit normalen Spermien lebensunfähige Zygoten ergäben. Aus der geringen Anzahl letaler Zygoten, die den Kontrollwert nicht übertraf, schlossen STURTEVANT und BEADLE auf einen Mechanismus, der den Ootiden in einer der meiotischen Teilungen bei gleichzeitiger Elimination der Austauschchromatiden eine Nichtaustauschchromatide zuteilt. Sie nahmen also einen hypothetischen Selektionsmechanismus an, der der zu erwartenden Sterilität entgegenarbeitet.

Die cytologische Bestätigung dieser Hypothese erbrachte CARSON (1946) an einem Stamm von *Sciara impatiens*, der eine lange, heterozygote, parazentrische Inversion im X-Chromosom hatte. Diese Inversion ließ eine hohe Chiasmafrequenz zu, so daß fast alle Weibchen in der Oogenese Inversionsbrücken aufwiesen. *Sciara* verkörpert ebenfalls den achiasmatischen Spermatogenesetyp und bildet normale Spermien. CARSON konnte zeigen, daß die in der Anaphase I entstehende dizentrische Brücke von dem Einschluß in die Ootide ferngehalten wird. Die lineare Orientierung der vier in der Meiose entstehenden Kerne hat zur Folge, daß die Inversionsbrücke zwischen den mittleren Kernen ausgespannt bleibt, während die äußeren Kerne die normalen Chromatiden erhalten (Abb. 23). Aus einem dieser Kerne entsteht die Eizelle. Die lineare Anordnung der Kernspindeln sichert demnach die Eliminierung der Inversionsbrücke, die von den Polkörpern aufgenommen wird und das Ei dadurch vor Sterilität bewahrt. Damit war der cytologische Beweis erbracht, daß ein einfaches Crossing-over in einer heterozygoten, parazentrischen Inversion keinen Einfluß auf die Fertilität der Dipteren-Weibchen hat.

Abb. 23. Orientierung der Inversionsbrücke in der Oogenese von *Sciara impatiens* (nach CARSON 1946)

Dieser Mechanismus kann jedoch nicht bewirken, daß die nach gleichzeitigem Crossing-over innerhalb der Inversion und proximal zu ihr entstehende Inversionsbrücke in die Polkörper ausgestoßen wird. Sie gelangt in der Hälfte der Fälle in die Ootide und führt nach Befruchtung mit normalen Spermien zur Bildung funktionsunfähiger Zygoten. Ebenso beeinträchtigen 3- und 4-Strang-Crossing-over die Fertilität der Weibchen. Nach STURTEVANT und BEADLE (1936) gibt 4-Strang-Crossing-over in einem inversionsheterozygoten X-Chromosom Anlaß zur Bildung einer dizentrischen Doppelbrücke, die nicht an der Zellteilung teilnimmt und verhindert, daß die Eizelle ein X-Chromosom erhält. Nach Befruchtung mit X-tragenden Spermien entstehen zur Hälfte X0-Zygoten. Wie NOVITSKI (1952, 1955) zeigen konnte, stellt diese Gesetzmäßigkeit einen Ausnahmefall dar und hat nur für telozentrische Chromosomen Gültigkeit. Dizentrische Doppelbrücken werden nur dann eliminiert und haben den Verlust des X-Chromosoms zur Folge, wenn das Chromosom von telozentrischer Gestalt ist. Besitzt aber nur eines der X-Chromosomen einen zusätzlichen Schenkel in Form des langen oder kurzen Armes des Y-Chromosoms, ist die Anzahl der X0-Zygoten der F_1 stark reduziert, und sind schließlich beide X-Chromosomen zweischenklig, werden überhaupt keine Zygoten vom X0-Typ gefunden. NOVITSKI, der sich hauptsächlich auf die genetische Seite des Experiments beschränkt, hat folgende Vorstellung von den cytologischen Grundlagen dieser Erscheinung: Eine telodizentrische Doppelbrücke, die sich von zwei telozentrischen Chromosomen herleitet, wird immer vom Einschluß in die Eizelle ferngehalten. Die Eizelle ist für dieses Chromosom hypoploid und normalerweise nicht lebensfähig. Die Chance, daß eine dizentrische Chromatide den Eikern erreicht, steigt, wenn sich die Doppelbrücke von einem telo- und einem akrozentrischen Chromosom herleitet und zusätzlich einen freien Schenkel enthält. Dagegen erhalten beide äußeren Kerne eine dizentrische Chromatide, wenn die Doppelbrücke aus zwei akrozentrischen Chromosomen entstanden ist und beiderseits einen freien Schenkel hat. Die Reduktion des X0-Typs resultiert demnach aus dem Einschluß einer dizentrischen Chromatide in den Eikern. NOVITSKI erklärt diesen Vorgang als "a consequence of unequal centromere 'strength' ". Da symmetrisch-telodizentrischen Brücken dieser Effekt fremd ist, liegt es nahe, den zusätzlichen heterochromatischen Schenkel des Y-Chromosoms oder eine cytologisch nicht nachweisbare unterschiedliche Herkunft des Centromers für diesen Effekt verantwortlich zu machen. "It is questionable wether the observed genetic differences among the various chromosome types are attributable to essentially different centromeres, as such, or to the additional chromosome material present as an extra arm" (NOVITSKI 1955).

3- und 4-Strang-Crossing-over ist bei *D. melanogaster* bei einer Chiasmafrequenz von 1,2 je Chromosomenschenkel ein seltenes Ereignis (WHITE 1954) und dürfte somit keinen großen Einfluß auf die Fertilität der Weibchen nehmen. Das schließt nicht aus, daß bei anderen Dipteren-Arten die Verhältnisse anders liegen.

Um die Beeinflussung der Fertilität im männlichen Geschlecht der Dipteren zu untersuchen, mußten Tiere vom chiasmatischen Spermatogenesetyp gefunden werden. Diese Bedingung erfüllen einige Nematoceren-Arten. In der Spermatogenese inversionsheterozygoter Männchen von *Dicranomyia trinotata* treten Brücken verschiedener Länge in beiden

Anaphasen auf (WOLF 1941). Die langen Inversionsbrücken reißen meistens nicht in der Anaphase I, sondern bleiben zwischen den Tochterkernen ausgespannt. Erst in der Anaphase II wandern die Centromeren der Brücke zu entgegengesetzten Polen und verbinden zwei Kerne miteinander. Die Weiterentwicklung der durch die Brücke verbundenen Spermatidenkerne führt zur Bildung diploider Spermien mit entweder einem Fusionskern oder zwei Kernen ungleicher Größe. Der größere Kern enthält die Inversionsbrücke. Durch Aufnahme des Fragments können Spermien entstehen, die vorübergehend hyperploid sind. Unbalancierte Spermien müßten einen Nachteil bei der Befruchtung haben. Obwohl 95% der Bivalente in der Spermatogenese von *Chironomus tentans* ein Chiasma bilden, zeigten die Kreuzungsversuche von BEERMANN (1956), daß die inversionsheterozygoten Männchen eine lebensfähige und cytologisch normale F_1 hatten. Anzeichen für eine selektive Eliminierung der defekten Chromatiden bzw. der unbalancierten Spermien waren nicht vorhanden. PHILIP (1942) nimmt an, daß unbalancierte Spermien besamungsuntauglich sind. Die Beeinflussung der Fertilität der Dipteren im männlichen Geschlecht infolge Crossing-over in heterozygoten, parazentrischen Inversionen bleibt weiterhin ein ungelöstes Problem. Fest steht nur, daß durch Verschmelzen der defekten Spermatiden die absolute Zahl der anomalen Spermien verringert wird und sich die relative Anzahl der funktionsfähigen Spermien erhöht.

Über die Ausmaße des von STURTEVANT und BEADLE entdeckten Selektionsmechanismus außerhalb der Dipteren und bei Pflanzen ist wenig bekannt. Bei Pflanzen ist ein ähnlicher Mechanismus nur für *Tulipa, Lilium* und *Zea* nachgewiesen worden. Daß sich dizentrische Brücken in der Makrosporogenese von *Tulipa* und *Lilium* ebenso wie in der Oogenese von *Drosophila* verhalten, haben DARLINGTON und LA COUR (1941) zeigen können. Die geringe Anzahl steriler Samenanlagen von nur 4% nach Crossing-over in einer langen, parazentrischen Inversion des 4. Chromosoms von *Zea mays* ließ MORGAN (1950) vermuten, daß die Austauschchromatiden eliminiert wurden. Demnach orientieren sich die Inversionsbrücken in der Anaphase II der Makrosporogenese so, daß die normalen Chromatiden zu den nicht generativen Kernen wandern. Maispflanzen hatten normal gefüllte Kolben, die nach Chiasmenbildung in einer langen, heterozygoten, parazentrischen Inversion des 2. Chromosoms zahlreiche Inversionsbrücken in der Anaphase I aufwiesen (RUSSEL und BURNHAM 1950). Nach RHOADES und DEMPSEY (1953) ist die Eliminierung der Austauschchromatiden nicht in jedem Fall erfolgreich. Statt der zu erwartenden 2,9% fanden sie 12% sterile Samenanlagen. Wahrscheinlich lagen hier 3- und 4-Strang-Crossing-over vor.

Da auch in der Mikrosporogenese von *Zea mays* Crossing-over stattfindet, ist demgemäß in inversionsheterozygoten Pflanzen Pollensterilität zu erwarten. Der Grad der Pollensterilität wird von zwei Faktoren bestimmt: a) von der Art des Crossing-over und b) von seiner Häufigkeit. Über die Abhängigkeit der Sterilität von der Art des Crossing-over gibt nachfolgende Tabelle Aufschluß:

Crossing-over	Anaphase I	Pollenkörner/PMZ	
		steril	fertil
einfaches Crossing-over (einfaches Chiasma)	Brücke Fragment	2	2
2-Strang-Crossing-over (reziproke Chiasmen)	2 Chromatiden-dyaden	—	4
3-Strang-Crossing-over (diagonale Chiasmen)	Brücke Fragment	2	2
4-Strang-Crossing-over (komplementäre Chiasmen)	Doppelbrücke 2 Fragmente	4	—

Bei gleichzeitigem Crossing-over außerhalb (proximal) und innerhalb einer Inversion sind 2 Pollenkörner steril, wenn nur ein Crossing-over in der Inversion stattfindet, dagegen sind alle Pollenkörner steril, wenn ein zweites Crossing-over in der Inversion hinzukommt.

Die Häufigkeit der verschiedenen Crossing-over-Arten in einer heterozygoten, parazentrischen Inversion von *Zea mays* wurde von Rhoades und Dempsey (1953) anhand der Brückenbildung in der Anaphase I festgestellt. Sie kamen zu folgenden Resultaten:

Konfigurationen	PMZ in %
normal	65,2
Brücke und Fragment	35,1
Brücke und assoziiertes Fragment	2,2
Loop-Univalent	5,3
Doppelbrücke und 2 Fragmente	1,2

Aus der Frequenz der verschiedenen Crossing-over-Arten schlossen die Autoren auf den Grad der Pollensterilität. Die reale Sterilität war stets geringer als die mittels der theoretischen Berechnung vorausgesagte. Diese Diskrepanzen erklären Rhoades und Dempsey dadurch, daß in einigen Fällen das durch ein distales Chiasma mit der Brücke verbundene Fragment in den Pollenkern gelangt und, obwohl die Zelle nicht das vollständige Chromatinmaterial enthält, eine normale Entwicklung des Pollenkornes zuläßt. Ferner erhält eine Zahl von Zellen Chromatiden mit genetisch unbedeutenden Defizienzen; es entstehen fertile Pollenkörner, die sich nur durch ihre geringe Größe von dem normalen Pollen unterscheiden. Die genetische Konstitution dieses Pollens ist demnach von der Lokalisation des Brückenbruches abhängig.

Die in Wildpopulationen von Dipteren und Pflanzen gefundenen Inversionen stellen hauptsächlich den parazentrischen Typ dar. Perizentrische Inversionen sind selten, da Chiasmenbildung in diesen zur Produktion defekter Chromatiden führt, die nicht der selektiven Eliminierung in der Meiose unterliegen. Perizentrische Inversionen müßten somit einen selektiven Nachteil haben. Die Beeinflussung der Fertilität von *Drosophila melanogaster* durch zwei strahleninduzierte Inversionen des 2. Chromosoms, die symmetrische Glazed- und die asymmetrische Plum-2-Inversion, untersuchte ALEXANDER (1952). Die Glazed-Inversion, die das doppelte Crossing-over in der Inversion völlig reduzierte, verminderte die Schlüpfrate der Eier um 6,7%; dagegen zeigte die Plum-2-Inversion, die einen gewissen Grad von doppeltem Crossing-over zuließ, eine Verringerung der Schlüpfrate um 9,9%. Daß die Plum-2-Inversion, die sich längenmäßig kaum von der Glazed-Inversion unterschied, diese hinsichtlich Crossing-over-Frequenz und Sterilitätseffekt übertraf, mag dem Umstand zugeschrieben werden, daß sie von asymmetrischer Gestalt war und fast den ganzen rechten Schenkel des 2. Chromosoms einschloß. Demnach bestimmen Länge und Lage der perizentrischen Inversion ihren Einfluß auf die Fertilität. Besonders begünstigt sind kurze perizentrische Inversionen, die infolge der Interferenzwirkung des Centromers vor schädlichem Crossing-over bewahrt werden. Doch auch lange perizentrische Inversionen können zuweilen in natürlichen Populationen von *Drosophila* eine gewisse Verbreitung erlangen, wenn sie in Verbindung mit parazentrischen Inversionen auftreten, die das Crossing-over in ihrer Nähe weitgehend reduzieren (MILLER 1939).

MORGAN (1950) berichtete von einer perizentrischen Inversion im 5. Chromosom von *Zea mays*, die 28,3% des Pollens und 12,5% der Samenanlagen zum Verkümmern brachte. Der prozentuale Unterschied in der Sterilität des männlichen und weiblichen Geschlechts soll auf einer höheren Crossing-over-Frequenz in der Mikrosporogenese beruhen.

Die Untersuchungen über die Beeinflussung der Fertilität von Tier und Pflanze durch Crossing-over in heterozygoten, parazentrischen Inversionen haben ergeben, daß einfaches Crossing-over als häufigster Typ die Fertilität der weiblichen Gameten nicht mindert. Ein bestimmter Selektionsmechanismus sorgt in der Oogenese bzw. Megasporogenese dafür, daß die entstandenen Austauschchromatiden in die Polkörper ausgeschieden werden, und die zukünftige Eizelle eine normale Chromatide erhält. Die Frage nach der Schädlichkeit des Crossing-over für die Spermatogenese inversionsheterozygoter Dipteren (Chironomiden) vom chiasmatischen Typ kann z. Z. nicht zufriedenstellend beantwortet werden. Allem Anschein nach werden die unbalancierten

Spermien von der Befruchtung durch Vitalitätsminderung ferngehalten. Bei Pflanzen spielt partielle Pollensterilität keine nachteilige Rolle; sie wird durch den zahlreich produzierten Pollen ausgeglichen. Infolge besonderer cytologischer Gegebenheiten unterliegen die Austauschchromatiden heterozygoter, perizentrischer Inversionen nicht dem Selektionsmechanismus in der weiblichen Meiose. Perizentrische Inversionen haben einen selektiven Nachteil und werden ausgemerzt. Daraus resultiert ihre Seltenheit in natürlichen Populationen.

V. Die populationsgenetischen Konsequenzen

Eine Gruppe von Individuen, deren Gen-Pool nach außen hin durch generative Isolierung vor Genaustausch geschützt ist und somit ein in sich geschlossenes System darstellt, wird im allgemeinen von Genetikern als biologische Art definiert. Eine Art erscheint in den seltensten Fällen uniform; sie zerfällt in zahlreiche, genetisch offene, nicht scharf abgrenzbare Untergruppen, die untereinander in Wechselbeziehungen stehen und als Populationen bezeichnet werden. Populationen verkörpern die kleinsten evolutionistischen Kategorien einer Art, in denen die Evolutionsfaktoren auf kleinstem Raum zu größter Wirksamkeit gelangen und der Beobachtung besonders gut zugänglich sind. Die Erforschung des populationsdynamischen Geschehens hat zur Formierung eines neuen Zweiges in der Genetik, der Populationsgenetik, geführt.

1. Der balancierte Inversionspolymorphismus und seine Verbreitung in natürlichen Populationen

Die relative Konstanz einer Population beruht auf ihrer Fähigkeit, ihren Gesamtgenotyp fortlaufend zu reproduzieren. Dieser Konstanz wirken Mutabilität und Selektion entgegen und können zu bleibenden Veränderungen im Erbaufbau einer Population führen. Auf mutativem Wege entstehen ständig neue Formen, deren Schicksal vornehmlich von ihrem Selektionswert abhängt. Die meisten Mutationen werden nicht in den Gen-Pool einer Population aufgenommen, da sie vitalitätsmindernd oder letal wirken und somit nur negatives Selektionsmaterial liefern. Erweist sich dagegen eine Mutante dem Wildtyp überlegen, wird das mutierte Allel in den Gen-Pool aufgenommen, und es kommt zu einem gleichzeitigen Auftreten zweier sympatrischer Alternativformen innerhalb einer Population. Diese Erscheinung wird als Polymorphismus bezeichnet.

Solange die Ausbreitung der neuen Form in Fluß ist, spricht man von transientem Polymorphismus. Das Auftreten einer neuen Mutation stellt normalerweise ein einmaliges Ereignis in der Entwicklung einer Population dar. Die Zunahme einer neuen Form im Falle des transienten Polymorphismus kann also nicht das Ergebnis eines sich ständig wiederholenden Mutationsvorganges sein — was auch in den meisten Fällen die niedrige Mutationsrate ausschließt — sondern nur mit einem höheren Selektionswert dieser Form erklärt werden.

Schreitet dieser Prozeß ungehindert fort, kommt es zur Verdrängung und schließlich zum vollständigen Ersatz der ursprünglichen Form durch die neue und damit zum völligen Verschwinden des Polymorphismus. Oft wird jedoch die Ausbreitung der neuen Form beim Erreichen einer bestimmten Häufigkeit gehemmt; es stellt sich ein mehr oder weniger stabiles Gleichgewicht der Alternativformen ein. In diesem Fall handelt es sich nach FORD (1945) um balancierten Polymorphismus. Transienter und balancierter Polymorphismus verkörpern keine Gegensätze, sondern stellen Variationen eines Phänomens dar.

Der balancierte Polymorphismus ist ein verbreitetes Phänomen in natürlichen Populationen von Pflanze und Tier, das sich nicht nur auf phänotypische Verschiedenheiten beschränkt, sondern ebenso für karyologische Unterschiede bekannt ist, wobei es sich hauptsächlich um gleichzeitiges Auftreten alternativer Strukturtypen handelt. Unter den polymorphen Chromosomenstrukturen sind parazentrische Inversionen, die im homo- und heterozygoten Zustand auftreten, weit verbreitet. Hochgradiger Inversionspolymorphismus tritt in natürlichen Populationen der Dipteren — wo er adaptive Funktionen erfüllt — besonders stark in Erscheinung, während er bei Pflanzen seltener ist und keine adaptive Rolle zu spielen scheint.

Eine Zusammenfassung aller bisher bekannten inversionspolymorphen Arten der Dipteren mit Ausnahme der Drosophiliden (ausführliche Übersicht bei MAINX 1956) gibt Tab. 2. Besonders intensiv ist in dieser Hinsicht die Gattung *Drosophila* untersucht. Fast alle Arten zeichnen sich durch den Besitz von Inversionen aus, wobei auffällt, daß der Grad des Inversionspolymorphismus von Art zu Art sehr unterschiedlich ausfällt. Den höchsten Inversionspolymorphismus weist die *willistoni*-Gruppe auf, in der besonders *D. willistoni* (DOBZHANSKY, BURLA und DA CUNHA 1950, DA CUNHA, BURLA und DOBZHANSKY 1950), *D. paulistorum* (dsgl.), *D. nebulosa* (PAVAN 1946, DA CUNHA, BRNCIC und SALZANO 1953) und *D. bocainensis* (CARSON 1954) hervortreten. Für *D. willistoni* sind über 40 Inversionen nachgewiesen worden. In einzelnen Fällen wurden bis zu 16 heterozygote Inversionen in einem Individum gefunden (DOBZHANSKY 1949). Andere Gruppen mit einem auffallend hohen Inversionspolymorphismus sind die *obscura*-Gruppe mit *D. pseudoobscura* (DOBZHANSKY und STURTEVANT 1938, DOBZHANSKY und EPLING 1944), *D. persimilis* (dsgl.), *D. subobscura* (PHILIP, RENDEL, SPURWAY und HALDANE 1944, MAINX, KOSKE, SMITAL 1953, KUNZE-MÜHL und SPERLICH 1955), *D. tristis* (MAINX, KOSKE und SMITAL 1953) und *D. athabasca* (NIVITSKI

Tabelle 2

Nematocera	
Bibionidae	
Bibio marci	MAINX 1950
Limnibiidae	
Dicranomyia trinotata	WOLF 1941
Limnophila ferruginea	MAINX 1950
Ptychopteridae	
Ptychoptera, sp.	MAINX 1950
Culicidae	
Anopheles messeae	FRIZZI 1951, 1952
Simulidae	
Simulium (4 Arten)	KUNZE 1953
Chironomidae	
Lauterbornia, sp.	MAINX, KUNZE und KOSKE 1953
Glyptotendipes defectus	BAUER 1936
G. bardipes STAEGER	BASRUR 1957
Tendipes decorus	ROTHFELS und FAIRLIE 1957
Chironomus dorsalis	BAUER 1936, PHILIP 1942, MAINX, KUNZE und KOSKE 1953, ACTON 1956
C. riparius	BAUER 1936, PHILIP 1942
C., sp.	HSU und LIU 1948
Camptochironomus tentans	BEERMANN 1955, MAINX 1956
C. pallidivittatus	BEERMANN 1955
C. annularius	BEERMANN nach MAINX 1956
Acricotopus lucidus	MECHELKE 1953
Prodamesia olivacea	BAUER 1936, MAINX, KUNZE und KOSKE 1953
Sciaridae	
Sciara impatiens	CARSON 1944, 1946
S. ocellaris	MCCARTHY 1945, ROHM 1947
S. fenestralis	MCCARTHY 1945
S. nacta	MCCARTHY 1945
S. prolificia	MCCARTHY 1945
S., sp.	MCCARTHY 1945
Cecidomyidae	
Lestodiplosis, sp.	WHITE 1946
Lasioptera rubi	KRACZKIEWICZ 1950
Brachycera	
Agromyzidae	
Liriomyza urophorina	MAINX 1951, MAINX, FIALA und KOGERER 1955
Drosophilidae	
Übersicht bei MAINX 1956	

1946), die *melanica*-Gruppe mit *D. melanica* (WARD 1952) und die *guarani*-Gruppe mit *D. guaru* (KING 1947) und *D. guaramunu* (DA CUNHA, BRNCIC und SALZANO 1953, BRNCIC 1953). Dagegen ist in der *repleta*-

Gruppe keine Art bekannt, die nicht frei von Inversionen ist (WHARTON 1942, WARTERS 1944). Dasselbe gilt für einige Arten der *melanogaster-*, *funebris-* und *virilis*-Gruppe.

Der balancierte Inversionspolymorphismus in der Gattung *Drosophila* ist von größter Bedeutung für die Evolution. Die folgenden Abschnitte befassen sich ausschließlich mit der Darstellung dieses Phänomens unter besonderer Berücksichtigung seines adaptiven Charakters.

2. Die Erhaltung des balancierten Inversionspolymorphismus in natürlichen Populationen

Untersuchungen des balancierten Inversionspolymorphismus zahlreicher *Drosophila*-Arten haben ergeben, daß heterozygote Inversionen trotz ihrer fertilitätsmindernden Eigenschaft in natürlichen Populationen weitaus häufiger anzutreffen sind als ihre homozygoten Formen.

So ist eine südkalifornische Population von *D. pseudoobscura* für eine bestimmte Inversion zu 70% heterozygot (DOBZHANSKY 1952a). In einer Population von *D. tropicalis* aus Honduras existieren von einer Inversion des 2. Chromosoms über 70% Heterozygote. Sechs nordbrasilianische über ein weites Gebiet verbreitete Populationen der sehr inversionspolymorphen *D. willistoni* weisen über 60% und zwei andere Populationen über 50% Heterozygote der J-Inversion des 3. Chromosoms auf. Zwei Populationen der gleichen Art haben über 50% Heterozygote der RE-Inversion des 2. Chromosoms. Von *D. paulistorum* gibt es eine peruanische Population, die für die K-Inversion des 3. Chromosoms zu 79,3% heterozygot ist (PAVAN, DOBZHANSKY und DA CUNHA 1957). Im Populationskasten erreichten die Heterozygoten dieser Population sogar 96% (DOBZHANSKY und PAVLOVSKY 1955). DA CUNHA (1953) fand, daß in einigen brasilianischen Populationen von *D. willistoni* und *D. paulistorum* die heterozygoten Inversionen des X-Chromosoms durch die Selektion begünstigt werden und Frequenzen von über 50% erreichen. Da die Männchen für das X-Chromosom haploid sind, manifestiert sich der Selektionsvorteil nur in den Weibchen. LEVITAN (1951a) bestätigte an einem Populationskastenversuch, daß auch bei *D. robusta* inversionsheterozygote Tiere von der Selektion begünstigt werden. Außerdem fand er, daß auch bezüglich des Heterozygotiegrades der Inversionen des X-Chromosoms Unterschiede zwischen den Geschlechtern bestanden (LEVITAN 1951b). Das stimmt mit den Beobachtungen von KERR und KERR (1952) an *D. melanogaster* überein. PHILIP, RENDEL, SPURWAY und HALDANE (1944) berichteten, daß in einer englischen Population von *D. subobscura* die Heterozygoten von drei Inversionen die Homozygoten überwiegen. Wurden inversionsheterozygote Tiere ingezüchtet, trat keine wesentliche Änderung des Gleichgewichts ein. In einer natürlichen Population von *D. subobscura* aus der Nähe von Sotschi stellten SOKOLOV und DUBININ (1941) von zwei Inversionen 66% und 91% Heterozygote fest. Für eine Inversion des 5. Chromosoms konnten sie mit Ausnahmen keine homozygoten Tiere finden. BUZZATI-TRAVERSO (1952) berichtete, daß in einigen europäischen Populationen von *D. subobscura* inversionshomozygote Tiere selten sind. KUNZE-MÜHL, MÜLLER und SPERLICH (1958) und SPERLICH (1958) stellten ebenfalls bei *D. subobscura* im Populationskastenversuch ein Überwiegen der Inversionsheterozygoten fest. Über ein interessantes Freilandexperiment mit *D. funebris* berichteten DUBININ und TINAIKOV (1946a). Sie setzten in der Nähe von Moskau 100000 für eine Inversion des 2. Chromosoms homozygote Fliegen aus, die sich vom Standardwildtyp phänotypisch nicht unterschieden. Der Wildtyp war für die Standardanordnung ebenfalls stark homozygot; nur 0,92% der untersuchten Fliegen besaßen diese Anordnung im heterozygoten Zustand. Die ausgesetzten Fliegen hatten sich mit der Wildform panmiktisch gepaart und zunächst eine Population ergeben, in der die Chromosomen der HARDY-WEINBERG-Formel entsprechend vertreten waren. Schließlich nahmen die inversionsheterozygoten Fliegen zu und überwogen zahlenmäßig die homozygoten. Die Autoren sahen darin einen Fall von Intrapopulationsheterosis. In englischen Populationen von *D. funebris* fanden BERRIE und SANSOME (1948) ein starkes Überwiegen der heterozygoten Inversionsformen von drei kleinen Inversionen. Eine kleine Inversion von *D. macrospina* ist bisher überhaupt nur im heterozygoten Zustand gefunden worden.

Bei panmiktischer Reproduktion einer natürlichen Population liegt das theoretische Maximum der Häufigkeit einer heterozygoten Inversion unabhängig vom anfänglichen Häufigkeitsverhältnis der Alternativformen bei 50%, da nach HARDY-WEINBERG normalhomozygote, inversionsheterozygote und inversionshomozygote Inversionen im Verhältnis 1:2:1 gebildet werden. Dieses Verhältnis wäre selbst dann gegeben, wenn nur die inversionsheterozygoten Tiere zur Fortpflanzung kämen. In Anbetracht der mit Inversionsheterozygotie verbundenen Fertilitätsverluste dürfte demnach nur eine Unterschreitung des theoretischen Höchstwertes von 50% zu erwarten sein.

Das vielfach beobachtete Überschreiten des Maximums kann deshalb nur als die Folge eines günstigeren Selektionswertes der heterozygoten Strukturtypen gewertet werden.

Die Ursache dieser Erscheinung sieht DOBZHANSKY (1950) in einer erhöhten biologischen Eignung heterozygoter Strukturtypen infolge Heterosis. Da es sich hierbei um einen Heterosiseffekt handelt, der an das Vorhandensein heterozygoter, chromosomaler Strukturunterschiede gebunden ist, schlägt DOBZHANSKY die Bezeichnung Euheterosis vor. Die Euheterosis unterscheidet sich von dem gewöhnlichen Heterosiseffekt (Luxurieren) durch ihren adaptiven Charakter. Sie besteht in der

Wirkung balancierter Genkomplexe, die im heterozygoten Zustand heterotisch wirken, und da sie durch Inversionen vor Crossing-over bewahrt werden, erhalten bleiben und als mechanische Einheiten rekombiniert werden.

Wir wissen heute, daß ein Heterosiseffekt schon durch die Heterozygotie eines einzigen Allelenpaares bedingt sein kann. So gelang es STUBBE und PIRSCHLE (1940) und STUBBE (1953), einen mono- und digen bedingten Fall von Heterosis bei *Antirrhinum majus* aufzudecken. L'HERITIER und TEISSIER (1933), TEISSIER (1947) und ELENS (1958) machten die Beobachtung, daß die rezessiven Mutanten sepia und ebony von *D. melanogaster* im heterozygoten Zustand ihre entsprechenden Formen selektiv übertreffen. Diese Fälle sprechen eindeutig für die Heterozygotiehypothese der Heterosis. Es ist deshalb verständlich, daß ein als Supergen fungierender heterozygoter Genkomplex unter bestimmten Bedingungen einen maximalen Heterosiseffekt erlangt und seinem Träger ein positives Selektionsmoment erteilt.

In den letzten Jahren wurde eine Reihe von Arbeiten veröffentlicht, die sich mit den physiologischen Komponenten des Heterosiseffekts befassen. Es konnte nachgewiesen werden, daß die Eliminierung der homozygoten Inversionsformen vornehmlich auf einer erhöhten Mortalität während der Metamorphose beruht. So fanden DOBZHANSKY (1947a) und DOBZHANSKY und LEVENE (1948) bei der Untersuchung von *D. pseudoobscura* unter natürlichen und künstlichen Bedingungen, daß, während die Verteilung der homo- und heterozygoten Inversionen des 3. Chromosoms im Eistadium noch dem Hardy-Weinberg-Gleichgewicht entspricht, im Larven- und Puppenstadium eine unterschiedliche Mortalität der Strukturtypen einsetzt, so daß schließlich im Imaginalstadium die Heterozygoten in der Mehrzahl vorhanden sind. An einer hondurenischen Population von *D. tropicalis*, die hauptsächlich aus den Heterozygoten einer bestimmten Inversion besteht, wiesen DOBZHANSKY und PAVLOVSKY (1955) nach, daß die homozygoten Kombinationen jeder Generation im Ei- und teilweise auch im Larven- und Puppenstadium absterben. Obwohl die Hälfte der Zygoten in jeder Generation geopfert wird, ist die Vitalität der Heterozygoten so groß, daß die Population weiterbesteht. Nach DA CUNHA (1956) werden die Inversionsheterozygoten von *D. willistoni* durch die Selektion begünstigt, indem die Homozygoten im Ei- und Larvenstadium absterben.

Die Ergebnisse der oben geschilderten Untersuchungen verleiten zu einer Deutung im Sinne eines balancierten Letalfaktorensystems, das durch natürliche Selektion infolge Anreicherung rezessiver Subletal- und Letalgene entwickelt werden kann. Zu dieser Auffassung neigt RENDEL (1953); er demonstrierte den Mechanismus der Euheterosis mit dem Cy/Pm-Stamm von *D. melanogaster*, der im Cy- und Pm-Chromosom je einen Letalfaktor enthält und außerdem für die Curly-Inversion heterozygot ist. Die Kombinationen Cy/Cy und Pm/Pm sind überwiegend letal, so daß eine Gleichgewichtsverschiebung zu Gunsten der Heterozygoten eintritt. Doch täuscht dieser Mechanismus nur eine Euheterosiswirkung vor, da das Überwiegen der Heterozygoten durch Opferung der Homozygoten erreicht wird und nur von relativer Art ist. Obwohl das Phänomen der Euheterosis nicht mit einem balancierten Letalfaktorensystem erklärt werden kann, ist sehr wahrscheinlich, daß der Euheterosiseffekt in einigen Fällen auf dem Zusammenwirken mit einem solchen System beruht.

Daß es sich tatsächlich um einen echten Heterosiseffekt handelt, zeigen Befunde, die eindeutig eine Vitalitätssteigerung inversionsheterozygoter Tiere nachwiesen. Nach SPIESS, KETCHEL und KINNE (1952) zeichnen sich die Träger der Whitney/Klamath-Anordnung von *D. persimilis* durch größere Fortpflanzungsfähigkeit — gemessen an der Eilegepotenz — und Lebensdauer aus als ihre entsprechenden Homozygoten. Zu analogen Resultaten gelangte MOOS (1955) bei der vergleichenden Prüfung der heterozygoten Standard/Chiricahua- und der homozygoten Standard/Standard- und Chiricahua/Chiricahua-Anordnung von *D. pseudoobscura*.

Zur Erklärung des Selektionsvorteils heterozygoter Inversionen wird neben der Heterosishypothese oft die Ansicht geäußert, daß der Heterosiseffekt die Folge eines Positionseffekts sei. Nach MAINX (1956) soll dieser Positionseffekt im Sinne der Strukturmusterhypothese von GOLDSCHMIDT verstanden werden, derzufolge nur solche Inversionen Bestandteil einer Population werden, von denen im heterozygoten Zustand eine solche Strukturmusterwirkung ausgeht. Das hieße demnach, einen Positionseffekt zu postulieren, der im heterozygoten Zustand heterotisch und im homozygoten vitalitätsmindernd oder letal wirkt. Prinzipiell bringt diese Hypothese keinen neuen Gedanken, da es sich hierbei letzten Endes nur um eine monofaktorielle Heterosiswirkung handeln würde. Die gesamte Erscheinung der Euheterosis und ihr adaptiver Charakter läßt sich damit nicht erklären. Immerhin wäre es mit WASSERMAN (1954) denkbar, daß es sich hierbei um die erste Phase der Herausbildung des Euheterosiseffekts handelt, indem auf diesem Wege nur bestimmte Inversionen erhalten blieben, die im Laufe der Zeit durch sekundäre Umformung ihres Gengehalts mit der vor Crossing-over geschützten homologen Gensequenz einen optimalen Heterosiseffekt ergäben. In diesem Sinne kann auch der von MAINX (1956) mitgeteilte Befund über den erfolgreichen Einbau einer strahleninduzierten Inversion in eine künstliche Population von *D. subobscura* gewertet werden.

Gegen die alleinige Annahme eines Positionseffekts lassen sich die Versuche von DOBZHANSKY (1948a, 1949, 1950) mit Inversionen verschiedener geographischer Herkunft ins Feld führen. Sie stellen eine eindeutige Stütze der Euheterosishypothese dar. DOBZHANSKY fand, daß heterozygote Inversionen nur dann einen heterotischen Effekt haben, wenn die homologen Chromosomen aus einem einheitlichen Verbreitungsgebiet stammen. Gehören z. B. die Standard- und Chiricahua-Anordnung verschiedenen Populationen an (Mittel- und Südkalifornien; Südkalifornien und Mexiko), so ist der Euheterosiseffekt nur schwach angedeutet oder bleibt völlig aus. SPIESS (1950) bestätigte diese Befunde an Inversionen des 3. Chromosoms von *D. persimilis* aus verschiedenen Höhenlagen der Sierra Nevada.

DOBZHANSKY (1952a) formuliert: "This finding is most compatible with the assumption that the overdominance in fitness observed in the heterozygotes is the property not of a single gene locus, or of a chromosome structure but rather of integrated systems of polygenes. Such polygene systems are coadapted by natural selection to other polygene complexes present in the same populations."

3. Der adaptive Charakter des balancierten Inversionspolymorphismus

Haben die Strukturtypen eines inversionspolymorphen Systems adaptive Eigenschaften, müssen sie auf eine Veränderung der Umweltbedingungen durch Neuregulierung des adaptiven Gleichgewichts antworten. Versuche an künstlichen Populationen von *Drosophila* haben gezeigt, daß das Häufigkeitsverhältnis der Strukturtypen bei konstanten Umweltbedingungen gewahrt bleibt; ihre Selektionswerte befinden sich in einem balancierten Gleichgewicht (LEVINE und BEARDMORE 1959). Unter natürlichen Verhältnissen ist eine Stabilität der Umweltfaktoren nicht gegeben. Ihre Variabilität sorgt dafür, daß Strukturtypen, die unter verschiedenen Bedingungen unterschiedliche Selektionswerte haben, ihr adaptives Häufigkeitsverhältnis ändern. Strukturtypen, die unter gegebenen Verhältnissen ihren Trägern eine erhöhte Eignung verleihen, können unter veränderten Bedingungen eine Eignungssenkung ihres Trägers verursachen und damit eine Verschiebung des adaptiven Gleichgewichts zu Gunsten eines anderen Strukturtyps zur Folge haben. Ein balanciertes Häufigkeitsverhältnis ist demnach unter natürlichen Bedingungen nicht stabil, sondern geht bei Veränderung der Umwelt bis zu seiner Neuregulierung zeitweilig in ein fluktuierendes Häufigkeitsverhältnis über, ohne daß dieser Selektionsprozeß zur völligen Eliminierung eines Strukturtyps führt.

a) Quantitative und qualitative temporale Schwankungen im Auftreten chromosomaler Strukturtypen

DOBZHANSKY fand, daß sich die saisonmäßigen, durch den Wechsel der Jahreszeiten bedingten Schwankungen der Klimafaktoren in der karyotypischen Zusammensetzung kalifornischer Populationen von *D. pseudoobscura* widerspiegeln. Er untersuchte die relative Häufigkeit des Standard-, Arrowhead- und Chiricahua-Strukturtyps der Populationen von Pinon Flats und Andreas Canyon aus dem Gebiet des San Jacinto im Zeitraum von 1939—1946. Es erwies sich, daß sich die Häufigkeit der Strukturtypen regelmäßig und cyclisch änderte und dabei den jährlichen Klimaschwankungen folgte (Abb. 24). Die relative Häufigkeit des Standard-Strukturtyps nahm vom Frühjahr (52%) bis Juni (28%)

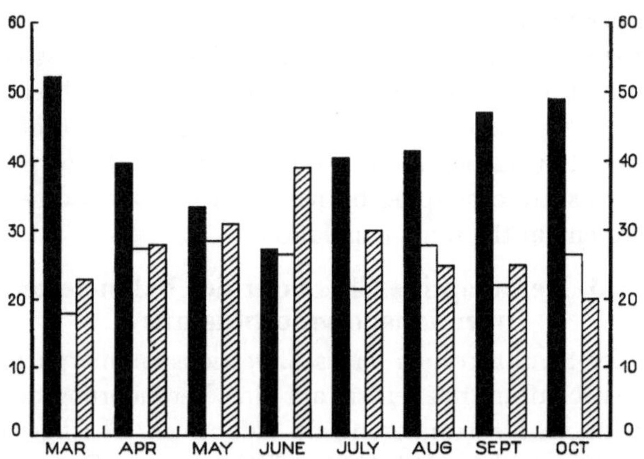

Abb. 24. Jahreszeitliche Häufigkeitsschwankungen der Strukturtypen Standard (schwarz), Arrowhead (weiß) und Chiricahua (schraffiert) des 3. Chromosoms von *Drosophila pseudoobscura*. Die Ordinate gibt die prozentuale Häufigkeit der Strukturtypen an (nach Dobzhansky 1953)

ab, um dann wieder anzusteigen, während sich der Chiricahua-Strukturtyp vom Frühjahr bis zum Sommer an Häufigkeit verdoppelte (39,5%), zum Winter dagegen wieder abnahm (20%). Die Häufigkeitsschwankung des Arrowhead-Strukturtyps war weniger ausgeprägt (Dobzhansky 1943, 1947b). Eine Population aus dem Yosemite Park der Sierra Nevada zeigte ebenfalls cyclische Schwankungen des Standard- und Arrowhead-Strukturtyps. Der Standard-Strukturtyp erreichte im Sommer auf Kosten des Arrowhead-Strukturtyps ein Häufigkeitsmaximum (Dobzhansky 1948b). In der Mather Population der Sierra Nevada waren cyclische Schwankungen nur schwach angedeutet (Dobzhansky 1952b).

Für einige andere *Drosophila*-Arten liegen ebenfalls Angaben über Saisonschwankungen ihrer Strukturtypen vor. So stellten Dubinin und Tinaikov (1945) die Häufigkeit der Strukturtypen einer Moskauer Population von *D. funebris* im Ablauf eines Jahres fest und fanden, daß die Konzentration der Inversionen im Frühjahr kurz nach der Überwinterung am niedrigsten war. Ihre Frequenz stieg an, erreichte im Sommer einen Höchstwert und nahm auf den Winter zu wieder ab. Über weniger stark ausgeprägte Saisonschwankungen in Populationen von *D. robusta* berichteten CARSON und STALKER (1949) und LEVITAN (1951b). Nach LEVITAN traten Konzentrationsschwankungen der Strukturtypen des X-Chromosoms im weiblichen Geschlecht stärker in Erscheinung als im männlichen. Bei *D. persimilis* (SPIESS 1950) und *D. willistoni* (DA CUNHA, BURLA und DOBZHANSKY 1950, TOWNSEND 1958) sind jahreszeitliche Schwankungen nur angedeutet.

Soweit lokale Umweltveränderungen cyclisch-reversible Häufigkeitsschwankungen im chromosomalen Aufbau einer Population auslösen, sind sie für die Evolution bedeutungslos und stellen nach DOBZHANSKY (1943) "a wasted motion" dar. Jedoch haben sie eine gewisse Bedeutung für die Erhaltung einer Population unter extremen Lebensbedingungen, wie sie die jahreszeitlichen Klimaschwankungen mit sich bringen. So gibt es beispielsweise in einer Population einen Strukturtyp, der seinen optimalen Selektionswert im Winter entwickelt, während ein anderer Strukturtyp sein adaptives Optimum in den warmen Sommermonaten hat. Geht in einem strengen Winter die Sommerform zugrunde, wird die Fortdauer der Population durch die Winterform garantiert.

Neben cyclischen Saisonschwankungen chromosomaler Strukturtypen sind auch progressive Veränderungen bekannt geworden, die sich über einen Zeitraum von mehreren Jahren erstrecken und in einigen Fällen wahrscheinlich die Folge größerer Klimaschwankungen sind. DOBZHANSKY (1947a) machte bei *D. pseudoobscura* die Beobachtung, daß sich das Häufigkeitsverhältnis bestimmter Strukturtypen der Population von Keen Camp am Mount San Jacinto von Jahr zu Jahr verschob. Der Standard-Strukturtyp nahm von 1939 (28%) bis 1946 (50%) an Häufigkeit kontinuierlich zu, während im gleichen Zeitraum der Arrowhead-Strukturtyp an Häufigkeit um die Hälfte abnahm (15%). Eine andere Veränderung erfolgte in der Mather-Population von *D. pseudoobscura* und wurde von DOBZHANSKY (1948b, 1952b, 1956, 1958) über einen Zeitraum von 13 Jahren verfolgt. In dieser Population nahm der Arrowhead-Strukturtyp von 1945 (35,7%) bis 1950 (49,8%) ständig zu, während der Standard-Strukturtyp von 35,7 auf 20,3% absank. Mit dem Jahre 1951 trat ein Umschlag in diesem Prozeß ein. Arrowhead verlor an Häufigkeit (33,2%) und Standard nahm zu (45,3%). Eine ähnliche Häufigkeitsverschiebung fand DOBZHANSKY (1952b, 1956) in der Population von *D. persimilis* am gleichen Standort für den Whitney- bzw. Standard- und Klamath-Strukturtyp. Nach DOBZHANSKY besteht eine ursächliche Beziehung zu gleichzeitig während des Beobachtungszeitraumes aufgetretenen Klimaschwankungen. So wurde die in den Jahren 1945—1950 in beiden Populationen beobachtete Häufigkeitsverschiebung durch vorherrschende Trockenheit ausgelöst, schlug jedoch nach dem niederschlagsreichen Winter von 1951 in das Gegenteil um.

Eine von DOBZHANSKY (1944, 1947a, 1948b, 1952b, 1956, 1958) gleichzeitig in der oben genannten Population von *D. pseudoobscura* festgestellte progressive Häufigkeitsverschiebung im Auftreten des Chiricahua- und Pikes-Peak-Strukturtyps ließ sich dagegen nicht mit jenen Klimaschwankungen in Einklang bringen, da der Chiricahua-Strukturtyp während des ganzen Beobachtungszeitraumes (1945—1957) von 17,2 auf 3,8% absank, wogegen der Pikes-Peak-Strukturtyp, der zu Anfang in der Population fehlte, im Jahre 1957 zu 10% vertreten war. Bemerkenswert ist, daß diese Veränderung auf großem Raum stattfand; sie wurde an zehn verschiedenen Orten Kaliforniens festgestellt. DOBZHANSKY vermutet, daß die progressive Veränderung im Auftreten beider Strukturtypen auf einer Ursache beruht.

Die in natürlichen Populationen in Verbindung mit Klimaschwankungen auftretenden Häufigkeitsverschiebungen chromosomaler Strukturtypen lassen erkennen, daß die Umwelt auf den Karyotyp selektionierend einwirkt. Der selektive Einfluß der Umwelt selbst besteht in einem Zusammenspiel verschiedener Faktoren, deren umfassende Analyse noch aussteht. In Populationskastenversuchen konnte eine Reihe von Faktoren erkannt und ihr gestaltender Einfluß auf die karyotypische Konstitution künstlicher Populationen nachgewiesen werden. Über das Ausmaß dieser Faktoren im einzelnen ist jedoch noch wenig bekannt. WRIGHT und DOBZHANSKY (1946) wiesen nach, daß sich der Selektionswert der Strukturtypen des 3. Chromosoms von *D. pseudoobscura* mit der Temperatur ändert. Eine zu gleichen Teilen aus Standard- und Chiricahua-Strukturtypen bestehende künstliche Population zeigte bei 16,5° C keine Veränderung, während bei 25 °C eine Häufigkeitsverschiebung zugunsten des Standard-Strukturtyps eintrat. Die Autoren sehen hierin eine Parallele zu dem in natürlichen Populationen jährlich zu beobachtenden Häufigkeitszuwachs in den warmen Sommermonaten. Nach SPIESS (1950) liegen für *D. persimilis* die Verhältnisse genau umgekehrt, da die Strukturtypen des 3. Chromosoms bei 16,5° C adaptive Eigenschaften entwickeln. Diese Tatsache steht mit den ökologischen Lebensbedingungen beider Arten in Einklang. *D. persimilis* ist bevorzugt in kälteren Gebieten anzutreffen, dagegen bewohnt *D. pseudoobscura* wärmere Gegenden. Neben einer Temperaturabhängigkeit des Selektionswertes der Strukturtypen von *D. pseudoobscura* wies HEUTS (1947, 1948) ebenso eine Abhängigkeit vom Feuchtigkeitsgehalt nach. Es zeigte sich, daß Feuchtigkeit auf die Schlüpfrate der Träger verschiedener Strukturtypen differenzierend wirkt. Der Autor vermutet, daß der jährliche Häufigkeitsanstieg des Chiricahua-Strukturtyps in einer Population am Mount San Jacinto in den Frühlingsmonaten mit der zu dieser Zeit vorherrschenden Feuchtigkeit zusammenhängt.

Von Interesse ist, daß der Selektionswert der Strukturtypen von *D. pseudoobscura* (DA CUNHA 1951, DOBZHANSKY und SPASSKY 1954) und *D. willistoni* (BIRCH und BATTAGUA 1957) von der Art der in der Nahrung enthaltenen Mikroorganismen bestimmt wird. Schwankungen im Nahrungsangebot müssen sich demnach auf die genetische Konstitution einer Population auswirken. Nach neueren Untersuchungen wird der Selektionswert bestimmter Strukturtypen von *D. pseudoobscura* durch das Fehlen oder Vorhandensein anderer Strukturtypen in derselben Population bestimmt (LEVENE, PAVLOVSKY und DOBZHANSKY 1954, 1958).

Die cyclische und progressive Fluktuation chromosomaler Strukturtypen in natürlichen Populationen von *Drosophila* wird von DOBZHANSKY und anderen Autoren als Beweis für den adaptiven Charakter des Inversionspolymorphismus angesehen. Danach bestimmen die Inversionen eines polymorphen Systems die Anpassungswerte ihrer Träger und befähigen die Population, auf Veränderungen ihrer Lebensbedingungen durch Modifikation ihrer genetischen Zusammensetzung zu antworten.

Eine andere Auffassung vertreten EPLING, MITCHELL und MATTONI (1953, 1957). Diese Autoren beobachteten ebenfalls cyclische und progressive Schwankungen chromosomaler Strukturtypen in verschiedenen Populationen von *D. pseudoobscura* aus der Sierra Nevada, die jedoch keine Korrelation zu klimatischen Veränderungen zeigten. Ferner stellten sie Abweichungen von der HARDY-WEINBERG-Formel nach beiden Richtungen fest, was sie zu der Schlußfolgerung veranlaßt, "... that an adaptive disadvantage is not necessarily a property of inversion homozygotes, nor an advantage of heterozygotes, as classes, and that different types of both may differ in their seasonal responses". Nach ihrer Hypothese, die auf dem interchromosomalen Effekt basiert, stellt das Genom bezüglich des Crossing-over eine Einheit dar. Rhythmische Schwankungen im Auftreten der Strukturtypen des 3. und X-Chromosoms sollen dadurch zustande kommen, daß infolge erhöhten Crossing-over in den heterologen Chromosomen Austauschprodukte gebildet werden, die dem Genotyp und jenen Strukturtypen Anpassungsvorteile verleihen. Gestützt wi·d ihre Vorstellung durch die Beobachtung, daß die Saisonschwankungen mit einer gleichzeitigen Zunahme heterozygoter Strukturtypen korreliert sind. Das würde bedeuten, daß die Anpassungsfähigkeit einer Population nicht an das Vorhandensein einzelner Inversionen gebunden ist, sondern eine transitorische Funktion des Gesamtgenotyps darstellt.

Gegen diese Hypothese hat DOBZHANSKY (1958) den Einwand erhoben, daß "the second-order selection which would favour certain third chromosomes because of their effects on recombination in other chromosomes would be altogether too weak to account for the rapidity of the changes actually observed".

b) Der Höhengradient chromosomaler Strukturtypen

Bei der Untersuchung der karyologischen Zusammensetzung einiger Populationen von *D. pseudoobscura* und *D. persimilis* aus verschiedenen Höhenlagen der Sierra Nevada machte DOBZHANSKY (1948b) die Entdeckung, daß die relative Häufigkeit der Strukturtypen des 3. Chromosoms je nach der Höhenlage, die die Population bewohnt, unterschiedlich

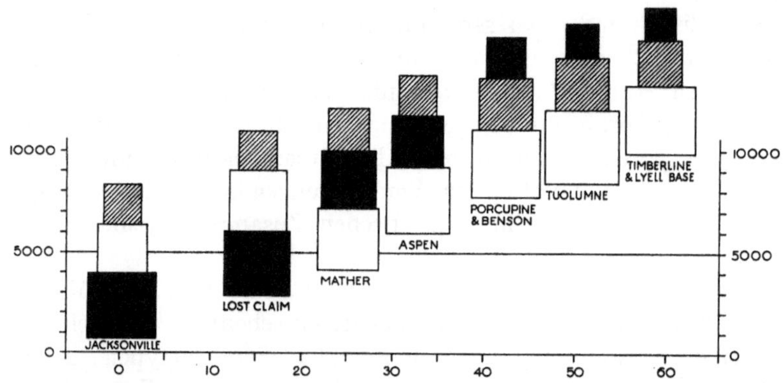

Abb. 25. Häufigkeiten der Strukturtypen Standard (schwarz), Arrowhead (weiß) und Chiricahua (schraffiert) des 3. Chromosoms von *Drosophila pseudoobscura* in verschiedenen Höhenlagen der Sierra Nevada. Die Quadrate entsprechen der prozentualen Häufigkeit der Strukturtypen. Die Ordinate gibt die Höhe in Fuß, die Abszisse die Entfernung in Meilen an (nach DOBZHANSKY 1948)

ist. Der Standard-Strukturtyp von *D. pseudoobscura*, der in den niederen Höhenlagen am häufigsten vertreten ist, nimmt mit zunehmender Höhe ab. Dagegen stellt der Arrowhead-Strukturtyp in den subalpinen Lagen die häufigste Erscheinung dar, ist jedoch in den unteren Lagen nur schwach vertreten (Abb. 25). Mit Ausnahme einer nur schwach angedeuteten Abhängigkeit des Chiricahua-Strukturtyps von der Höhenlage, zeigten alle anderen Strukturtypen des 3. Chromosoms keinen Höhengradienten. Für die Population von *D. persimilis* vom gleichen Standort wies DOBZHANSKY ebenfalls einen Höhengradienten der Whitney-, Standard-, Klamath- und Mendocino-Strukturtypen nach. Whitney ist in den niederen Lagen häufig, während die anderen Strukturtypen in den subalpinen Lagen vorherrschen.

STALKER und CARSON (1948) fanden für Strukturtypen des X-, 2. und 3. Chromosoms von *D. robusta* ebenfalls einen Höhengradienten. Im untersuchten Gebiet der Great Smoky Mountains (Tenessee) existieren in 300 m Seehöhe eine Unterland- und in 1220 m Seehöhe eine Hochlandpopulation, die sich voneinander durch ein spezifisches Verhältnis ihrer Strukturtypen unterscheiden.

Nach DOBZHANSKY stellt das Auftreten abgestufter Häufigkeitsgleichgewichte in verschiedenen Höhenlagen eine Variation des Phänomens der Saisonschwankungen dar. Im Falle des Höhengradienten handelt es sich um die zeitliche Fixierung der Saisonschwankungen eines Ortes in den verschiedenen Höhenlagen eines Gebirges. Die Unterschiede, die sich im Verlauf eines Jahres an einem Orte ergeben, sind mit den Unterschieden identisch, die zwischen Populationen verschiedener Höhenlagen bestehen. Da der Standard-Strukturtyp von *D. pseudoobscura*, der in den subalpinen Lagen selten ist, sein saisonmäßiges Häufigkeitsmaximum in den Sommermonaten und der in den niedrigen Lagen seltene Arrowhead-Strukturtyp sein Maximum in den kalten Monaten erreicht, ist anzunehmen, daß die in gewissen Höhenlagen im Verlauf eines Jahres vorherrschenden klimatischen Bedingungen den an einem Ort zu einer bestimmten Jahreszeit anzutreffenden Witterungsverhältnissen entsprechen.

c) Der territoriale Gradient chromosomaler Strukturtypen

Im Jahre 1950 veröffentlichten DOBZHANSKY und seine Mitarbeiter eine Arbeitshypothese, derzufolge der Grad des Inversionspolymorphismus einer Population von der Mannigfaltigkeit des ökologischen Charakters der Umwelt bestimmt wird (DA CUNHA, BURLA und DOBZHANSKY 1950, DOBZHANSKY, BURLA und DA CUNHA 1950).

Die ersten Hinweise dafür ergaben sich aus den umfassenden Untersuchungen der teilweise sehr inversionspolymorphen Arten der *willistoni*-Gruppe von *Drosophila*. Schon auf den ersten Blick fällt auf, daß *D. willistoni* und *D. paulistorum*, die zu den häufigsten und verbreitetsten Arten dieser Gruppe gehören und im tropischen und subtropischen Südamerika anzutreffen sind, einen stärkeren Inversionspolymorphismus haben als die selteneren und nur lokal vorkommenden Arten *D. tropicalis* und *D. equinoxialis*. Die Häufigkeit der meisten Inversionen von *D. willistoni* schwankt von Population zu Population und läßt eine positive Korrelation zu der Anzahl der von einer Population ausgefüllten ökologischen Nischen ihres Wohngebietes erkennen. Hochpolymorphe Populationen von *D. willistoni* bewohnen vorzugsweise die Savannen des zentral- und nordbrasilianischen Berglandes, die mit einer abwechslungsreichen "fruit-bearing" Flora bestanden sind und den Populationen zahlreiche ökologische Nischen mit ausreichenden und vielseitigen Nahrungsquellen bieten. Hier erreichen die Populationen ihre größte Dichte und weisen durchschnittliche Inversionswerte von 9,7 Inversionen/Weibchen und 6,6/Männchen auf. Diese Werte liegen im weiblichen Geschlecht höher, da ihnen hauptsächlich die heterozygoten Inversionen des X-Chromosoms zugrunde liegen. In der in floristischer Hinsicht artenarmen Gegend um Bahia belaufen sich die Werte auf 0,8 Inversionen im weiblichen und 0,4 im männlichen Geschlecht (DOBZHANSKY 1949). Da sich die Drosophiliden von den auf bestimmten Früchten lebenden Mikroorganismen ernähren, muß ein Gebiet mit einer vielseitigen "fruit-bearing" Flora, das mehr und vielseitigere Nischen enthält als ein Gebiet mit einer artenarmen Pflanzenwelt, seiner Population Anlaß zur Entwicklung eines hochgradigen Inversionspolymorphismus geben, wogegen in Gebieten mit einseitigem ökologischen Charakter nur ein geringer Inversionspolymorphismus ausgebildet wird. Eine Klassifizierung des Biotyps zahlreicher Populationen von *D. willistoni* unter besonderer Berücksichtigung von Klima und Pflanzengemeinschaft ließ eine direkte Korrelation des Ausmaßes des Inversionspolymorphismus zur Heterogenität des Standortes erkennen (DA CUNHA und DOBZHANSKY 1954).

Vergleichende Untersuchungen zahlreicher Populationen von *D. willistoni* aus dem gesamten Verbreitungsgebiet dieser Art erbrachten, daß der Grad des Inversionspolymorphismus gemessen an der Zahl der heterozygoten Inversionen pro Individuum mit zunehmender Entfernung vom Zentrum des Verbreitungsgebietes abnimmt (Abb. 26). So betragen die Inversionswerte der nördlichen Randpopulationen von Florida und Kuba pro Individuum 2 bzw. 1,5 Inversionen. Außerdem ist das Inversionsspektrum peripherer Populationen sehr verarmt; es werden nur noch $^1/_4$ der im Zentrum vorhandenen Inversionen gefunden. Ganz ähnlich liegen die Verhältnisse an der südlichen Randzone in Argentinien (DOBZHANSKY, BURLA und DA CUNHA 1950, TOWNSEND 1952, DA CUNHA und DOBZHANSKY 1954).

Der zentripetale Gradient des Inversionspolymorphismus stellt ein Analogon des von VAVILOV (1926, 1927) formulierten Gesetzes von den Genzentren dar, demzufolge die genische Mannigfaltigkeit einer Art in ihrem Entstehungszentrum am größten ist. Diese von VAVILOV als Genzentren bezeichneten Areale sind der Ausgangspunkt der natürlichen Verbreitung einer Art. Das Ursprungsgebiet von *D. willistoni* liegt demnach in den tropischen Regenwäldern des Amazonas-Beckens und den zentralbrasilianischen Savannen. Von hier aus erfolgte die Besiedlung der angrenzenden Randgebiete bei gleichzeitiger Verarmung des chromosomalen Mannigfaltigkeitsspektrums. Da einerseits im Verlauf dieses Prozesses wirksame Bestandteile des ursprünglichen adaptiven Systems verlorengingen, andererseits in Anbetracht der Kürze der Zeit, die seit der Besiedlung der Randgebiete verstrichen ist, kein neues adaptives System entwickelt werden konnte, haben Randpopulationen nur einen schwach entwickelten Inversionspolymorphismus. Solche Populationen sind wahrscheinlich erst nach längerer Zeit in der Lage, alle sich ihnen bietenden ökologischen Nischen ihres Wohngebietes zu erobern.

Im Verlauf weiterer vergleichender Untersuchungen an Insel- und Festlandpopulationen ergaben sich die gleichen Gesetzmäßigkeiten

(DOBZHANSKY 1957). Die Inselpopulationen der Großen und Kleinen Antillen sind ebenso wie Randpopulationen hinsichtlich ihrer Chromosomenstruktur verarmt, was besonders auf den Kleinen Antillen zum

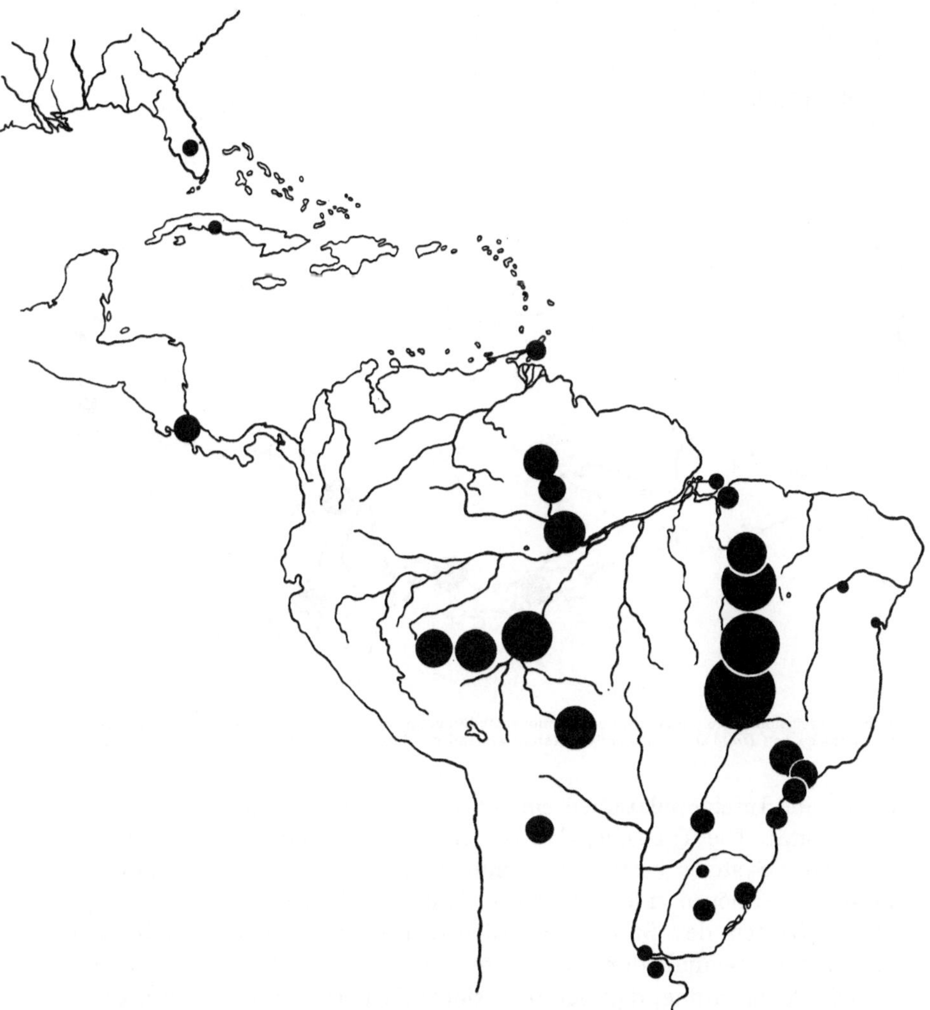

Abb. 26. Inversionswerte (heterozygote Inversionen/Individuum) mittel- und südamerikanischer Populationen von *Drosophila willistoni* dargestellt anhand der Kreisdurchmesser (nach TOWNSEND 1952)

Ausdruck kommt (Abb. 27). Das Inselmilieu wirkt differenzierender als das Festland, so daß die Unterschiede zwischen Populationen verschiedener Inseln stärker als zwischen benachbarten Festlandpopulationen in Erscheinung treten. Wahrscheinlich spielt hierbei die geographische Isolierung der Inseln eine entscheidende Rolle. Die niedrigste

Konzentration an heterozygoten Inversionen hat die Population der kleinen Insel St. Kitts mit Werten von 0,2 Inversionen pro Individuum; das ist die niedrigste Konzentration, die bisher bei *D. willistoni* festgestellt wurde. Die höchste Konzentration hat die kubanische Population mit Werten von 2,7 im weiblichen und 2,9 im männlichen Geschlecht. Mit Ausnahme der dem südamerikanischen Festland vorgelagerten Insel Trinidad (3,2 Inversionen/Weibchen; 2,6 Inversionen/Männchen)

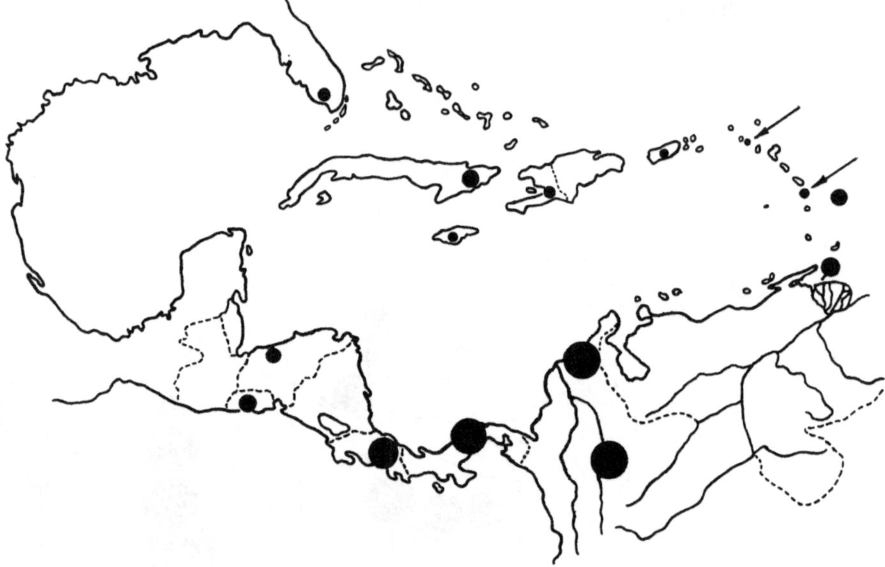

Abb. 27. Inversionswerte (heterozygote Inversionen/Individuum) mittelamerikanischer Festland- und Inselpopulationen von *Drosophila willistoni* dargestellt anhand der Kreisdurchmesser (nach Dobzhansky 1957)

zeigen alle Inselpopulationen eine extreme Reduktion ihrer genetischen Variabilität. Die Tatsache, daß die Größe der Inseln das Ausmaß des adaptiven Systems bestimmt, stimmt mit der These überein, daß das Leistungsvermögen eines adaptiven Systems mit der Zahl der ökologischen Nischen des Standortes wächst. Eine allgemeine Bestätigung fanden diese Befunde durch die Untersuchungen von Townsend (1958).

Die Beobachtung, daß der Inversionspolymorphismus eine Funktion der ökologischen Gegebenheiten ist, wurde auch auf andere Arten von *Drosophila* ausgedehnt. Die ebenfalls zur *willistoni*-Gruppe gehörende Art *D. nebulosa* wurde von Da Cunha, Brncic und Salzano (1953) und Freire-Maia (1955c) untersucht. Die Autoren bringen übereinstimmend zum Ausdruck, daß Populationen in ökologisch mannigfaltigen Gebieten inversionspolymorpher sind als geographische und ökologische Randpopulationen. Nach Dubinin und Tinaikov (1946b, 1947) besteht bei *D. funebris* ein ausgesprochener Stadt-Land-Gradient chromosomaler

Strukturtypen. Sie fanden in dem Gebiet um Moskau eine gesetzmäßige Abnahme der Konzentration heterozygoter Inversionen vom Stadtzentrum (89%) ausgehend über die Stadtgrenze (17%) zu Gebieten mit ländlichem Charakter (1,5%). In einigen weit entfernten Dörfern wurden überhaupt keine Inversionen gefunden. Eine hohe Konzentration bestand auch in Saratov (33,9%) und Ivanov (37,2%). Auffallend ist, daß in Städten, die ihren Stadtcharakter infolge Kriegseinwirkungen verloren haben, die Konzentration von Inversionen ebenso wie auf dem Lande sehr gering ist. Dasselbe gilt für Klein- und Gartenstädte.

Ausnahmen der Nischenhypothese stellen offenbar *D. melanogaster*, *D. simulans*, *D. ananassae*, *D. pararepleta*, *D. immigrans*, *D. kikkawai*, *D. hydei*, *D. repleta*, *D. virilis* und *D. busckii* dar. Bei diesen Arten war es bisher nicht möglich, eine klare Beziehung zwischen dem teilweise schwach entwickelten Inversionspolymorphismus und dem ökologischen Charakter ihres Standortes zu finden (CARSON 1955a, FREIRE-MAIA 1953b, 1955a, b). Bei diesen Arten handelt es sich um Kosmopoliten, die vornehmlich in der Nähe menschlicher Siedlungen anzutreffen sind. Es ist anzunehmen, daß im Adaptationsprozeß kosmopolitischer Arten frei kombinierbare Einzelgene eine größere Rolle spielen als die Polygene der Inversionen.

4. Die Bedeutung des balancierten Inversionspolymorphismus für die Evolution

Der balancierte Inversionspolymorphismus der Dipteren stellt ein adaptives System dar, das eine Population befähigt, sich Umweltveränderungen anzupassen, indem die verschiedenen Strukturtypen des Systems unter spezifischen Umweltbedingungen einen optimalen Anpassungswert entwickeln. Erhalten wird das adaptive System durch Heterosis (Euheterosis), die auch Inversionen, die homozygot mit einem Selektionsnachteil verbunden sind, im strukturheterozygoten Zustand erhält und damit das adaptive Leistungsvermögen einer Population erhöht.

Je stärker der Inversionspolymorphismus einer Population, desto stärker ist die Fixierung ihrer genetischen Variabilität, da Crossing-over weitgehend eingeschränkt und nur noch in nichtinvertierten Chromosomenabschnitten und in homologen Inversionen möglich ist. Solche Populationen sind zwar in adaptiver Hinsicht außerordentlich spezialisiert, haben aber an Flexibilität ihres Genotyps verloren. Hierbei handelt es sich um endemische Populationen und solche, die im Entstehungs- bzw. Verbreitungszentrum einer Art anzutreffen sind. Populationen mit einem nur schwach ausgebildeten Inversionspolymorphismus sind infolge eines höheren Grades der freien Rekombination weniger spezialisiert, passen sich dagegen besser einer veränderten Umwelt an. Zu dieser

Gruppe gehören Randpopulationen und Kosmopoliten. Eine Mittelstellung nehmen nach Carson (1955a, b) solche Arten ein, deren Inversionspolymorphismus auf ein Chromosom beschränkt ist *(D. melanogaster, D. nebulosa)*; sie stellen ein relativ offenes Rekombinationssystem dar.

Die grundlegende evolutionistische Bedeutung von Inversionen besteht in ihrem Einfluß auf das Crossing-over. Durch weitgehende Reduktion des Crossing-over in heterozygoten Inversionen entstehen zusammenhängende Genkomplexe (Polygene), die als fixierte Einheiten rekombiniert werden. Einzelgene und Genkomplexe stellen demnach die eigentliche Grundlage des Adaptationsprozesses dar; durch heterozygote Inversionen werden sie vor ihrer Elimination bewahrt.

Literatur

Acton, A. B.: Larval groups in the subgenus *Chironomus* Meigen. Arch. Hydrobiol. **50**, 64—75 (1955).

— Crossing over within inverted regions in *Chironomus*. Amer. Naturalist **90**, 63—65 (1956).

Alexander, M. L.: The effect of two pericentric inversions upon crossing over in *Drosophila melanogaster*. Univ. Tex. Publ. **5204**, 219—226 (1952).

Baker, W. K.: The production of chromosome interchanges in *Drosophila virilis*. Genetics **34**, 167—193 (1949).

Barigozzi, C.: La struttura e il numero dei chromosomi di *Ascaris megalocephala* cloquet durante la spermatogenesi. Caryologia **1**, 131—143 (1949).

Basrur, V. R.: Inversion polymorphism in the midge *Glyptotendipes barbipes* (Staeger). Chromosoma **8**, 597—608 (1957).

Bauer, H.: Beiträge zur vergleichenden Morphologie der Speicheldrüsenchromosomen. Zool. Jb. **56**, 239—276 (1936).

— Fortschr. Zool. N. F. **3**, 434—460 (1938).

— Die Chromosomenmutationen. Z. indukt. Abstamm.- u. Vererb.-Lehre **76**, 309—322 (1939).

—, M. Demerec and B. P. Kaufmann: X-ray induced chromosomal alterations in *Drosophila melanogaster*. Genetics **23**, 610—630 (1938).

Beadle, G. W., and A. H. Sturtevant: X-chromosome inversions and meiosis in *Drosophila melanogaster*. Proc. nat. Acad. Sci. (Wash.) **21**, 384—390 (1935).

Beasley, J. O.: Meiotic chromosome behavior in species, species hybrids, haploids, and induced polyploids of *Gossypium*. Genetics **27**, 25—54 (1942).

Beermann, W.: Chromosomenkonstanz und spezifische Modifikation der Chromosomenstruktur in der Entwicklung und Organdifferenzierung von *Chironomus tentans*. Chromosoma **5**, 139—198 (1952).

— Cytologische Analyse eines *Camptochironomus*-Artbastards. Chromosoma **7**, 198—259 (1955).

— Inversionsheterozygotie und die Fertilität der Männchen von *Chironomus*. Chromosoma **8**, 1—11 (1956).

— and G. Bahr: The submicroscopic structure of the balbiani-ring. Exp. Cell Res. **6**, 195—201 (1954).

Berrie, G. K., and E. W. Sansome: Wild population studies; *Drosophila funebris* near Manchester. J. Genet. **49**, 151—152 (1948).

BHADURI, P. N.: Application of new technique to cytological reinvestigation of the genus *Tradescantia*. J. Genet. **44**, 87—127 (1942a).
— Cytological analysis of structural hybridity in *Rhoeo discolor*. J. Genet. **44**, 73—85 (1942b).
BHATTACHARIYA, S.: Die Wirkung von Röntgenstrahlen auf Kerne mit verschiedener heterochromatischer Konstitution. Chromosoma **9**, 305—318 (1958).
BIRCH, L. C., and BATTAGUA: Selection in *Drosophila willistoni* in relation to food. Evolution **11**, 94—105 (1957).
BÖCHER, T. W.: Meiosis in *Anemone apenina* with special reference to chiasma localization. Hereditas (Lund) **31**, 221—237 (1945).
BREHME, K.: The time of action of the ClB lethal in *Drosophila melanogaster*. Amer. Naturalist **71**, 567—574 (1937).
BRNCIC, D.: Chromosomal variation in natural populations of *Drosophila guaramunu*. Z. indukt. Abstamm.- u. Vererb.-Lehre **85**, 1—11 (1953).
BROWN, M. S., and M. Y. MENZEL: Polygenomic hybrids in *Gossypium*. I. Cytology of hexaploids, pentaploids, and hexaploid combinations. Genetics **37**, 242—263 (1952).
BROWN, S. W.: The structure and meiotic behavior of differentiated chromosomes of tomato. Genetics **34**, 437—461 (1949).
— and D. ZOHARY: The relationship of chiasmata and crossing over in *Lilium formosanum*. Genetics **40**, 850—873 (1955).
BUZZATI-TRAVERSO, A.: Heterosis in population genetics. In GOWEN: Heterosis. Iowa State College Press, 149—160 (1952).

CÂMARA, A.: O problema da fragmentacao chromosomica, operada pelos raios X, estudado no *Triticum monococcum*. Agronomia Lusitana **3**, 341—359 (1941).
CARSON, H. L.: Cytological analysis of natural populations of *Sciara impatiens*. Genetics **28**, 71—72 (1943).
— An analysis of natural chromosome variability in *Sciara impatiens* JOHANSEN. J. Morphol. **75**, 11—59 (1944).
— The selective elimination of inversion dicentric chromatids during meiosis in the eggs of *Sciara impatiens*. Genetics **31**, 95—113 (1946).
— The effects of inversions on crossing over in *Drosophila robusta*. Genetics **38**, 168—186 (1953).
— Infertile sibling species in the *willistoni* group of *Drosophila*. Evolution **8**, 148—165 (1954).
— The genetic characteristics of marginal populations of *Drosophila*. Cold Spr. Harb. Symp. quant. Biol. **20**, 276—287 (1955a).
— Variation in genetic recombination in natural populations. J. cell. comp. Physiol. Suppl. 1/2, **45**, 221—236 (1955b).
— and H. D. STALKER: Gene arrangements in natural populations of *Drosophila robusta* STURT. Evolution **1**, 131—133 (1947).
— — Seasional variation in gene arrangement frequencies over a three-year period in *Drosophila robusta* STURT. Evolution **3**, 322—329 (1949).
CATCHESIDE, D. C., and D. E. LEA: Dominant lethals and chromosome breaks in ring X-chromosome of *Drosophila melanogaster*. J. Genet. **47**, 25—40 (1945).
CELARIER, R. P.: The cyto-geography of the *Bothriochloa ischaemum* complex. II. Chromosome behavior. Amer. J. Bot. **44**, 729—738 (1957).
COLEMAN, L. C.: Chromosome abnormalities in an individual of *Chorthippus longicornis* (Acrididae). Genetics **32**, 435—447 (1947).
CREW, F. A., and P. C. KOLLER: Genetical and cytological studies of the intergenetic hybrids of *Cairina moschata* and *Anas platyrhyncha*. Proc. roy. Soc. Edinburgh **56**, 210—241 (1936).

CUNHA, A. B. DA: Modification of the adaptive values of chromosomal types in *Drosophila pseudoobscura* by nutritional variables. Evolution **5**, 395—404 (1951).

— Chromosomal inversions with sex-limited effect. Nature (Lond.) **172**, 815—816 (1953).

— Differential variability favoring inversion heterozygotes in *Drosophila willistoni.* Evolution **10**, 231—234 (1956).

—, D. BRNCIC and F. M. SALZANO: A comparative study of chromosomal polymorphism in certain South American species of *Drosophila.* Heredity **7**, 193—202 (1953).

—, H. BURLA and TH. DOBZHANSKY: Adaptive chromosomal polymorphism in *Drosophila willistoni.* Evolution **4**, 212—235 (1950).

— and TH. DOBZHANSKY: A further study of chromosomal polymorphism in *Drosophila willistoni* in its relation to the environment. Evolution **8**, 119—134 (1954).

DALTON, H. C., and J. HALL: Gene action in Axolotl. Carnegie Inst. Wash. Yb. **49**, 181—188 (1950).

DARK, S. O.: Meiosis in diploid and tetraploid *Paeonia* species. J. Genet. **32**, 353—372 (1936).

DARLINGTON, C. D.: Crossing over and its mechanical relationship in *Chorthippus* and *Stauroderus.* J. Genet. **33**, 465—500 (1936).

— Chromosome behavior and structural hybridity in the Tradescantiae. II. J. Genet. **35**, 259—280 (1938).

— and A. E. GAIRDNER: The variation system in *Campanula persicifolia.* J. Genet. **35**, 97—128 (1938).

— and L. LA COUR: The genetics of embryo-sac development. Ann. Bot. **5**, 547—562 (1941).

— and A. P. WYLIE: A dicentric cycle in *Narcissus.* Heredity **6**, Suppl., 195—213 (1953).

DEMEREC, M.: Genetic behavior of euchromatic segments inserted into heterochromatin. Genetics **25**, 618—627 (1940).

— and M. E. HOOVER: Three related X-chromosome deficiencies in *Drosophila.* J. Hered. **27**, 207—212 (1936).

DESSAUER, F.: Quantenbiologie. Einführung in einen neuen Wissenszweig. Berlin: Springer 1954.

DOBZHANSKY, TH.: The decrease of crossing over observed in translocations and its probable explanation. Amer. Naturalist **65**, 214—232 (1931).

— Genetics of natural populations. IX. Temporal changes in the composition of populations of *Drosophila pseudoobscura.* Genetics **28**, 162—186 (1943).

— Chromosomal races in *Drosophila pseudoobscura* and *persimilis.* Carnegie Inst. Wash. Publ. **554**, 47—144 (1944).

— Genetics of natural populations. XIV. A response of certain gene arrangements in the third chromosome of *Drosophila pseudoobscura* to natural selection. Genetics **32**, 142—160 (1947a).

— A directional change in the genetic constitution of a natural population of *Drosophila pseudoobscura.* Heredity **1**, 53—64 (1947b).

— Genetics of natural populations. XVIII. Experiments on chromosomes of *Drosophila pseudoobscura* from different geographic regions. Genetics **33**, 588—602 (1948a).

— Genetics of natural populations. XVI. Altitudinal and seasonal changes produced by natural selection in certain populations of *Drosophila pseudoobscura* and *Drosophila persimilis.* Genetics **33**, 158—176 (1948b).

DOBZHANSKY, TH: Genetic structure of natural populations. Carnegie Inst. Wash. Yb. **48**, 201—212 (1949).
— Genetics of natural populations. XIX. Origin of heterosis through natural selection in populations of *Drosophila pseudoobscura*. Genetics **35**, 288—302 (1950).
— Nature and origin of heterosis. In GOWEN: Heterosis. Iowa State College Press, 218—223 (1952a).
— Genetics of natural populations. XX. Changes induced by drought in *Drosophila pseudoobscura* and *Drosophila persimilis*. Evolution **6**, 234—243 (1952b).
— Genetics of natural populations. XXV. Genetic changes in populations of *Drosophila pseudoobscura* and *Drosophila persimilis* in some localities in California. Evolution **10**, 82—92 (1956).
— Genetics of natural populations. XXVI. Chromosomal variability in island and continental populations of *Drosophila willistoni* from Central America and the West Indies. Evolution **11**, 280—293 (1957).
— Genetics of natural populations. XXVII. The genetic changes in populations of *Drosophila pseudoobscura* in the American Southwest. Evolution **12**, 385—401 (1958).
—, H. BURLA and A. DA CUNHA: A comparative study of chromosomal polymorphism in sibling species of the *willistoni* group of *Drosophila*. Amer. Naturalist **84**, 229—246 (1950).
— and A. DREYFUS: Chromosomal aberrations in Brazilian *Drosophila ananassae*. Proc. nat. Acad. Sci. (Wash.) **29**, 301—305 (1943).
— and C. EPLING: Contributions to the genetics, taxonomy, and ecology of *Drosophila pseudoobscura* and its relatives. Carnegie Inst. Wash. Publ. **554**, 1—183 (1944).
— — The suppression of crossing over in inversion heterozygotes of *Drosophila pseudoobscura*. Proc. nat. Acad. Sci. (Wash.) **34**, 137—141 (1948).
— and H. LEVENE: Genetics of natural populations. XVII. Proof of operation of natural selection in wild populations of *Drosophila pseudoobscura*. Genetics **33**, 537—547 (1948).
— and O. PAVLOVSKY: An extreme case of heterosis in a Central American population of *Drosophila tropicalis*. Proc. nat. Acad. Sci. (Wash.) **41**, 289—295 (1955).
— and D. SOKOLOV: Structure and variation of the chromosomes in *Drosophila azteca*. J. Genet. **30**, 3—19 (1939).
— and N. SPASSKY: Environmental modification of heterosis in *Drosophila pseudoobscura*. Proc. nat. Acad. Sci. (Wash.) **40**, 407—414 (1954).
— and A. H. STURTEVANT: Inversions in the chromosome of *Drosophila pseudoobscura*. Genetics **23**, 28—64 (1938).
DODDS, K. S.: Genetical and cytogenetical studies of *Musa*. V. Certain edible diploids. J. Genet. **45**, 113—138 (1943).
DOUGHTY, L. R.: Chromosome behavior in relation to genetics of Agave. I. Seven species of fibre Agave. J. Genet. **33**, 197—205 (1936).
DUBININ, N. P., and G. G. TINAIKOV: Seasonal cycles and the concentration of inversions in populations of *Drosophila funebris*. Amer. Naturalist **79**, 570—572 (1945).
— — Inversion gradients and natural selection in ecological races of *Drosophila funebris*. J. Hered. **37**, 39—44 (1946a).
— — Structural chromosome variability in urban and rural populations of *Drosophila funebris*. Amer. Naturalist **80**, 393—395 (1946b).
— — Inversionen an den Grenzen der ökologischen Rassen von *Drosophila funebris*. C. R. (Doklady) Acad. Sci. USSR. Nouv. Sér. **55**, 643—645 (1947) (russ.).
— — Stadtökologie und Verbreitung der Inversionen bei *Drosophila funebris*. C. R. (Doklady) Acad. Sci. USSR. Nouv. Sér. **56**, 865—867 (1947) (russ.).

EHRLICH, H. G.: Cytological studies in *Saintpualia* WENDL. (Gesneriaceae). Amer. J. Bot. **45**, 177—182 (1958).
ELENS, A. A.: Le rôle de l'hétérosis dans la competition entre ebony et son allèle normal. Experientia (Basel) **14**, 274—276 (1958).
EMMENS, C.: Salivary gland cytology of roughest^3 inversion and reinversion, and roughest^2. J. Genet. **34**, 191—202 (1937).
EPLING, C., D. F. MITCHELL and R. H. T. MATTONI: On the role of inversions in wild populations of *Drosophila pseudoobscura*. Evolution **7**, 342—365 (1953).
— — — The relation of an inversion system to recombination in wild populations. Evolution **11**, 225—247 (1957).
FANO, U.: On the analysis and interpretation of chromosomal changes in *Drosophila*. Cold Spr. Harb. Symp. quant. Biol. **9**, 113—120 (1941).
FERNANDES, A., and J. DE B. NEVES: Sur l'origine des formes de *Narcissus bulbocodium* L. à 26 chromosomes. Bol. Soc. Brot. **15**, 43—132 (1941).
FLAGG, R. O.: A mutation and an inversion in *Rhoeo discolor*. J. Hered. **49**, 185—188 (1958).
FORD, E. B.: Polymorphism. Biol. Rev. Cambridge Phil. Soc. **20**, 73—88 (1945).
FRANKEL, O. H.: Inversions in *Fritillaria*. J. Genet. **34**, 447—462 (1937).
FREIRE-MAIA, N.: Pericentric inversions in Brazilian populations of *Drosophila ananassae*. D. I. S. **26**, 100—101 (1952).
— New data on the incidence of pericentric inversions in Brazilian populations of *Drosophila ananassae*. D. I. S. **27**, 91—92 (1953a).
— Chromosome variation in *Drosophila immigrans*. Dusenia **4**, 303—311 (1953b).
— Pericentric inversions in *Drosophila*. D. I. S. **28**, 118—119 (1954).
— In Discussion. Cold Spr. Harb. Symp. quant. Biol. **20**, 270 (1955a).
— Chromosome mutations in natural populations of *Drosophila ananassae*. D. I. S. **29**, 116—117 (1955b).
— Inversions in wild and domestic populations of *Drosophila nebulosa*. D. I. S. **29**, 117—118 (1955c).
FRIZZI, G.: Dimorfismo cromosomico in *Anopheles maculipennis messeae*. Sci. Genet. **4**, 79—93 (1951).
— Nuovi contributi e prospettive di ricerca nel gruppo *Anopheles maculipennis* in allo studio del dimorfismo cromosomico (ordinamento ad X invertito e tipico) nel *messeae*. Symp. Genet. **3**, 231—265 (1952).
GANESAN, D.: Cytological studies in a chromosome ring-forming diploid *Notonia grandiflora* DC. J. Genet. **39**, 493—516 (1937).
GARDNER, E. J.: A further study of genetic modification of dominance, especially by position effects. Univ. Calif. Publ. Zool. **49**, 85—99 (1942).
GAUL, H.: Über meiotische Fragment- und Brückenbildung der Bastarde *Secale* und *Triticum* × *Agropyrum*. Chromosoma **6**, 314—329 (1954).
GEITLER, L.: Cytogenetische Untersuchungen an natürlichen Populationen von *Paris quadrifolia*. Z. indukt. Abstamm.- u. Vererb.-Lehre **73**, 182—197 (1937).
— Weitere cytogenetische Untersuchungen an natürlichen Populationen von *Paris quadrifolia*. Z. indukt. Abstamm.- u. Vererb.-Lehre **75**, 161—190 (1938).
GENTSCHEFF, G., and Å. GUSTAFSSON: The balance system of meiosis in *Hieracium*. Hereditas (Lund) **26**, 209—256 (1940).
GILES, N. H.: Spontaneous chromosome aberrations in *Tradescantia*. Genetics **25**, 69—87 (1940).
— A pericentric inversion in *Gasteria* resulting in apparent iso-chromosomes at meiosis. Proc. nat. Acad. Sci. (Wash.) **30**, 1—5 (1944).
GOLDSCHMIDT, E.: Chromosomal polymorphism in a population of *Drosophila subobscura* from Israel. J. Genet. **54**, 474—496 (1956).

GOLDSCHMIDT, R. B.: Gen und Außeneigenschaft (Untersuchungen an *Drosophila*). II. Z. indukt. Abstamm.- u. Vererb.-Lehre **69**, 70—131 (1935).

— Position effect and the theory of the corpuscular gene. Experientia (Basel) **2**, 1—40 (1946).

— Chromosomes and genes. Cold Spr. Harb. Symp. quant. Biol. **16**, 1—12 (1951).

— Theoretical genetics. Berkeley and Los Angeles: Calif. Univ. Press **1955**.

— and E. J. GARDNER: A further contribution to the analysis of scalloped wings in *Drosophila melanogaster*. Univ. Calif. Publ. Zool. **49**, 103—123 (1942).

— — and M. KODANI: A remarkable group of position effects. Proc. nat. Acad. Sci. (Wash.) **25**, 314—317 (1939).

— and A. HANNAH: One-band inversion. Proc. nat. Acad. Sci. (Wash.) **30**, 299—301 (1944).

GOPAL-AYENGAR, A. R.: Structural hybridity in *Scilla* species. Genetics **27**, 143-144 (1942).

GRAUBARD, M. A.: Temperature effect on interference and crossing over. Genetics **19**, 83—94 (1934).

GRÜNEBERG, H.: A new inversion of the X-chromosome in *Drosophila melanogaster*. J. Genet. **31**, 163—184 (1935).

— The position effect proved by a spontaneous reinversion of the X-chromosome in *Drosophila melanogaster*. J. Genet. **34**, 169—189 (1937).

HADORN, E.: Developmental action of lethal factors in *Drosophila*. Advanc. Genet. **4**, 53—85 (1951).

HAGA, T.: Meiosis in *Paris*. II. Spontaneous breakage and fusion of chromosomes. Cytology **18**, 50—66 (1953).

HAIR, J. B.: The origin of new chromosomes in *Agropyron*. Heredity **6**, Suppl. Symp. 215—233 (1953).

HÅKANSSON, A.: Die Reduktionsteilung in einigen Artbastarden von *Pisum*. Hereditas (Lund) **21**, 215—226 (1936).

— Meiosis in a hybrid with one set of large and of small chromosomes. Hereditas (Lund) **29**, 461—474 (1943).

— and A. LEVAN: Endo-duplicational meiosis in *Allium odorum*. Hereditas (Lund) **43**, 179—200 (1957).

HEILBORN, O.: Inversions in *Paris quadrifolia*. Hereditas (Lund) **29**, 498—499 (1943).

HEITZ, E., u. H. BAUER: Beweise für die Chromosomenstruktur der Kernschleifen in den Knäuelkernen von *Bibio hortulans* L. (Cytologische Untersuchungen an Dipteren, I.). Z. Zellforsch. **17**, 67—82 (1933).

HELFER, R. G.: A comparison of X-ray induced and naturally occuring chromosomal variations in *Drosophila pseudoobscura*. Genetics **26**, 1—22 (1941).

HERSKOWITZ, I. H.: The genetic basis of X-ray induced recessive lethal mutations. Genetics **36**, 356—363 (1951).

— The relation between X-ray dosage and the frequency of simulated healing of chromosome breakages in *Drosophila melanogaster*. Proc. nat. Acad. Sci. (Wash.) **40**, 576—585 (1954).

HEUTS, J.: Influence of humidity on the survival of different çhromosomal types in *Drosophila pseudoobscura*. Proc. nat. Acad. Sci. (Wash.) 33, 210—213 (1947).

— Adaptive properties of carriers of certain gene arrangements in *Drosophila pseudoobscura*. Heredity **2**, 63—75 (1948).

HINTON, T.: The modification of the expression of a position effect. Amer. Naturalist **83**, 69—94 (1949).

— A correlation of phenotypic changes and chromosomal rearrangements at the two ends of an inversion. Genetics **35**, 188—205 (1950).

Hinton, T., P. T. Ives and A. Evans: Changing the gene order and number in natural populations. Evolution **6**, 19—28 (1952).

Hoover, M. E.: Cytogenetic analysis of nine inversions in *Drosophila melanogaster*. Z. indukt. Abstamm.- u. Vererb.-Lehre **74**, 420—434 (1938).

Horton, I.: A comparison of the salivary gland chromosomes of *Drosophila melanogaster*. Genetics **24**, 234—243 (1939).

Howard, H. W.: The fertility of amphidiploids from the cross *Raphanus sativus* × *Brassica oleracea*. J. Genet. **36**, 239—273 (1938).

Hsu, T. C., and T. T. Liu: Microgeographic analysis of chromosomal variation in a Chinese species of *Chironomus* (Diptera). Evolution **2**, 49—57 (1948).

Ives, P. T.: The importance of mutation rate genes evolution. Evolution **4**, 236—252 (1950).

Jackson, W. D., and H. N. Barber: Patterns of chromosome breakage after irradiation and agening. Heredity **12**, 1—25 (1958).

Jenkins, T. J., and P. D. Thomas: Interspecific and intergenetic hybrids in herbage grasses. III. *Lolium loliaceum* and *L. rigidum*. J. Genet. **37**, 255—286 (1939).

Kaufmann, B. P.: A terminal inversion in *Drosophila ananassae*. Proc. nat. Acad. Sci. (Wash.) **22**, 591—594 (1936).

— Induced changes in chromosomes carrying inverted sections. Genetics **25**, 124—125 (1940).

— Induced chromosomal breaks in *Drosophila*. Cold Spr. Harb. Symp. quant. Biol. **9**, 82—91 (1941).

— Reversion from roughest to wild type in *Drosophila melanogaster*. Genetics **27**, 537—549 (1942).

Kerr, W. E., and L. S. Kerr: Concealed variability in the X-chromosome of *Drosophila melanogaster*. Amer. Naturalist **86**, 405—408 (1952).

King, J. C.: A comparative analysis of the chromosome of the *guarani* group of *Drosophila*. Evolution **1**, 48—62 (1947).

Klingstedt, H.: Taxonomic and cytological studies on grasshopper hybrids. I. Morphology and spermatogenesis of *Chorthippus bicolor* Charp. × *Ch. biguttulus* L. J. Genet. **37**, 389—420 (1939).

Koller, P. C.: Origin of variations within species. Nature (Lond.) **135**, 69—70 (1935).

— The genetical and mechanical properties of sex chromosomes. III. Man. Proc. roy. Soc. Edinburgh **57**, 194—214 (1937).

— Asynapsis in *Pisum sativum*. J. Genet. **36**, 275—305 (1938).

— Dicentric chromosomes in a rat tumour induced by an aromatic nitrogen mustard. Heredity **6**, Suppl. Symp. 181—196 (1953).

Komai, T., and T. Takaku: Two independent inversions in the X-chromosome of *Drosophila virilis* and their effects on crossing over and disjunction. Cytologia **11**, 245—260 (1940).

— — On the effect of the X-chromosome inversions on crossing over in *Drosophila virilis*. Cytologia **12**, 357—265 (1942).

Kostoff, D.: Cytogenetics of the genus *Nicotiana*. Sofia: State Printing House 1943.

Kraczkiewicz, Z.: Recherches cytologiques sur les chromosomes de *Lasioptera rubi* Heeg. (Cecydomyidae). Zool. Polon. **5**, 73—117 (1950).

Krishnaswamy, N., and V. S. Raman: Studies on the interspecific hybrids of *Pennisetum typhoides*, Stapf and Hub. × *P. purpureum*, Schumach. III. The cytogenetics of the colchicine induced amphidiploid. Genetica **27**, 253—272 (1954).

Kunze, E.: Artenunterschiede im Bau der Riesenchromosomen in der Gattung *Simulium* Latr. Öst. zool. Z. **4**, 23—32 (1953).

KUNZE-MÜHL, E., E. MÜLLER u. D. SPERLICH: Qualitative, quantitative und jahreszeitliche Untersuchungen über den chromosomalen Polymorphismus natürlicher Populationen von *Drosophila subobscura* COLL. in der Umgebung von Wien. Z. indukt. Abstamm.- u. Vererb.-Lehre **89**, 636—646 (1958).

— u. D. SPERLICH: Inversionen und chromosomale Strukturtypen bei *Drosophila subobscura* COLL. Z. indukt. Abstamm.- u. Vererb.-Lehre **87**, 65—84 (1955).

LAMM, R.: Cytogenetic studies in *Solanum*, sect. *Tuberarium*. Hereditas (Lund) **31**, 1—128 (1945).

LEA, D. E., and D. C. CATCHESIDE: The relation between recessive lethals, dominant lethals, and chromosome aberrations in *Drosophila*. Genetics **47**, 10—24 (1945).

LEVAN, A.: Studies on the meiotic mechanism of haploid rye. Hereditas (Lund) **28**, 177—211 (1942).

LEVENE, H., O. PAVLOVSKY and TH. DOBZHANSKY: Interaction of the adaptive values in polymorphic experimental populations of *Drosophila pseudoobscura*. Evolution **8**, 335—349 (1954).

— — — Dependence of the adaptive values of certain genotypes in *Drosophila pseudoobscura* on the composition of the gene pool. Evolution **12**, 18—23 (1958).

LEVENE, L., and I. A. BEARDMORE: A study of an experimental *Drosophila* population in equilibrium. Amer. Naturalist **93**, 35—40 (1959).

LEVINE, R. P., and E. E. LEVINE: The genotypic control of crossing over in *Drosophila pseudoobscura*. Genetics **39**, 677—691 (1954).

LEVITAN, M.: Experiments on chromosomal variability in *Drosophila robusta*. Genetics **36**, 285—305 (1951a).

— Selective differences between males and females in *Drosophila robusta*. Amer. Naturalist **85**, 385—388 (1951b).

LEWIS, E.: The phenomenon of position effect. Advanc. Genet. 3, 73—116 (1950).

LEWIS, H., and P. H. RAVEN: Rapid evolution in *Clarkia*. Evolution **12**, 319—336 (1958).

— and M. R. ROBERTS: The origin of *Clarkia lingulata*. Evolution **10**, 126—138 (1956).

L'HERITIER, P., et G. TEISSIER: Elimination des formes mutants dans les populations de Drosophiles. Cas des Drosophiles «ebony». C. R. Soc. Biol. (Paris) **124**, 882—885 (1933).

LILJEFORS, A.: Zytogenetische Studien über den F_1-Bastard *Triticum turgidum* × *Secale cereale*. Hereditas (Lund) **21**, 240—262 (1936).

LIMA DE FARIA, A., and P. S. SARVELLA: The organisation of telomeres in species of *Solanum, Salvia, Secale, Agapanthus*, and *Ornithogalum*. Hereditas (Lund) **44**, 337—346 (1958).

MAGOON, M. L., D. C. COOPER and R. W. HOUGAS: Cytogenetic studies of some diploid Solanums sect. tuberarium. Amer. J. Bot. **45**, 207—221 (1958).

—, R. W. HOUGAS and D. C. COOPER: Cytogenetic studies of tetraploid hybrids in *Solanum*. J. Hered. **49**, 171—178 (1958).

MAINX, F.: Die Bedeutung der Chromosomenaberrationen in natürlichen Populationen. Port. Acta Biol., Ser. A, GOLDSCHMIDT-Vol., 563—592 (1950).

— Die Verbreitung von Chromosomendislokationen in natürlichen Populationen von *Liriomyca urophorina* MIK. Chromosoma **4**, 521—534 (1951).

— Der chromosomale Struktur-Polymorphismus natürlicher Populationen als Problem der Genetik und Evolution. Novant'anni delle leggi Mendeliane (Roma) 425—453 (1956).

—, Y. FIALA u. E. v. KOGERER: Die geographische Verbreitung der chromosomalen Strukturtypen von *Liriomyca urophorina* MIK. Chromosoma **8**, 18—29 (1956).

MAINX, F,, TH. KOSKE u. E. SMITAL: Untersuchungen über die chromosomale Struktur europäischer Vertreter der *Drosophila obscura*-Gruppe. Z. indukt. Abstamm.- u. Vererb.-Lehre **85**, 354—372 (1953).

—, E. KUNZE u. T. KOSKE: Cytologische Untersuchungen an Lunzer Chironomiden. Öst. zool. Z. **4**, 33—44 (1953).

MARKS, G. E.: The cytology of *Oxalis dispar* (BROWN). Chromosoma **8**, 650—670 (1957).

MARQUARD, H.: Über die spontanen Aberrationen in der Anaphase der Meiose von *Paeonia tenuifolia*. Chromosoma **5**, 81—112 (1953).

MATHER, K.: Competition between bivalents during chiasma formation. Proc. roy. Soc. B **120**, 208—227 (1936).

— Competition for chiasmata in diploid and triploid maize. Chromosoma **1**, 119-129 (1939a).

— Crossing over and heterochromatin in the X-chromosome of *Drosophila melanogaster*. Genetics **24**, 413—435 (1939b).

MATSUURA, H., and T. HAGA: Chromosome studies on *Trillium kamtchaticum* PALL. and its allies. IX. Chromosome aberrations induced by X-ray treatment. Cytology **16**, 37—47 (1950).

McCARTHY, M.: Chromosome studies on eight species of *Sciara* (Diptera) with special reference to chromosome changes of evolutionary significance. Amer. Naturalist **79**, 228—245 (1945).

McCNIGHT, R. H.: Crossing over in the sex chromosome of racial hybrids of *Drosophila pseudoobscura*. Genetics **22**, 249—256 (1937).

McCLINTOCK, B.: Cytological observations of deficiencies involving known genes, translocations and an inversion in *Zea mays*. Univ. Missouri Res. Bull. **163**, 1—30 (1931).

— The association of non-homologous parts of chromosomes in the mid-prophase of meiosis of *Zea mays*. Z. Zellforsch. **19**, 191—237 (1933).

— The fusion of broken ends of sister half-chromatids following chromatid breakage at meiotic anaphasis. Univ. Missouri Res. Bull. **290**, 1—48 (1938).

— The stability of broken ends of chromosomes in *Zea mays*. Genetics **26**, 234—282 (1941a).

— Spontaneous alterations in chromosome size and form in *Zea mays*. Cold Spr. Harb. Symp. quant. Biol. **9**, 72—80 (1941b).

— Maize genetics. Carnegie Inst. Wash. Yb. **41**, 181—186 (1942).

— The fusion of broken ends of chromosomes following nuclear fusion. Proc. nat. Acad. Sci. (Wash.) **28**, 458—463 (1942).

— Maize genetics. Carnegie Inst. Wash. Yb. **42**, 148—152 (1943).

— Maize genetics. Carnegie Inst. Wash. Yb. **43**, 127—135 (1944).

— Chromosome organisation and genic expression. Cold Spr. Harb. Symp. quant. Biol. **16**, 13—47 (1951).

MECHELKE, F.: Reversible Strukturmodifikationen der Speicheldrüsenchromosomen bei *Acricotopus lucidus*. Chromosoma **5**, 511—543 (1953).

MENSINKAI, S.: Cytogenetic studies in the genus *Allium*. J. Genet. **39**, 1—45 (1940).

MEURMAN, O.: The chromosome behavior of some dioecious plants and their relatives with special reference to the sex chromosomes. Soc. Sci. Fennica, Comm. Biol. **2**, 2—104 (1925).

— and E. THERMAN: Studies on the chromosome morphology and structural hybridity in the genus *Clematis*. Cytologia **10**, 1—14 (1939).

MILLER, D. D.: Structure and variation of the chromosomes in *Drosophila algonquin*. Genetics **24**, 699—708 (1939).

MOOS, J. R.: Comparative physiology of some chromosomal types in *Drosophila pseudoobscura*. Evolution **9**, 141—151 (1955).

MORGAN, D.: A cytogenetic study of inversions in *Zea mays*. Genetics **35**, 153—174 (1950).

MORGAN, T. H., C. B. BRIDGES and J. SCHULTZ: The constitution of the germinal material in relation to heredity. Carnegie Inst. Wash. Yb. **29**, 325—359 (1930).

— — — The constitution of the germinal material in relation to heredity. Carnegie Inst. Wash. Yb. **31**, 303—307 (1932).

— — — Constitution of the germinal material in relation to heredity. Carnegie Inst. Wash. Yb. **34**, 284—291 (1935).

MULLER, H. J.: Induced mutations in *Drosophila*. Cold Spr. Harb. Symp. quant. Biol. **9**, 151—165 (1941).

— The nature of the genetic effects produced by radiation. Rad. Biol. **1**, 351—474 (1954).

— and I. H. HERSKOWITZ: Concerning the healing of chromosome ends produced by breakage in *Drosophila melanogaster*. Amer. Naturalist **88**, 177—208 (1954).

— and S. M. MOTT-SMITH: Evidence that natural radioactivity is inadequate to explain the frequency of natural mutations. Proc. nat. Acad. Sci. (Wash.) **16**, 277—285 (1930).

—, A. PROKOFYEVA and D. RAFFEL: Minute intergenic rearrangements as a cause of apparent "gene mutations". Nature (Lond.) **135**, 253—255 (1935).

MÜNTZING, A.: Chromosome fragmentation in a *Crepis* hybrid. Hereditas (Lund) **19**, 284—302 (1934).

— Chromosone behavior in some *Nicotiana* hybrids. Hereditas (Lund) **20**, 251—272 (1935).

— The effect of chromosomal variation in *Dactylis*. Hereditas (Lund) **23**, 113—235 (1937).

— Chromosomenmutationen bei Pflanzen und ihre genetische Wirkung. Z. indukt. Abstamm.- u. Vererb.-Lehre **76**, 309—351 (1939).

— Further studies on apomixis and sexuality in *Poa*. Hereditas (Lund) **26**, 115—190 (1940).

— and R. PRAKKEN: Chromosomal aberrations in rye populations. Hereditas (Lund) **27**, 273—308 (1941).

MYERS, W. M.: Variations in chromosomal behavior during meiosis among plants of *Lolium perenne* L. Cytologia **11**, 388—406 (1940).

NOVITSKI, E.: Chromosome variation in *Drosophila athabasca*. Genetics **31**, 508—524 (1946).

— The genetic consequences of anaphase bridge formation in *Drosophila*. Genetics **37**, 270—287 (1952).

— Genetic measures of centromere activity in *Drosophila melanogaster*. J. cell. comp. Physiol., Suppl. 1/2, **45**, 151—169 (1955).

— and G. BRAVER: An analysis of crossing over within a heterozygous inversion in *Drosophila melanogaster*. Genetics **39**, 197—209 (1954).

OFFERMANN, C. A., and H. J. MULLER: Regional differences in crossing over as a function of the chromosome structure. Proc. 6th Int. Congr. Genet. 143—145 (1932).

ÖSTERGREN, G.: Cytology of *Agropyrum junceum*, *A. repens* and their spontaneous hybrids. Hereditas (Lund) **26**, 304—316 (1940).

PADDOCK, E. F.: The backcross transmission of an inversion in an interspecific hybrid of *Solanum*. Genetics **28**, 85—86 (1943).

Painter, T. S.: A new method for the study of chromosome rearrangements and the plotting of chromosome maps. Science **78**, 585—586 (1933).

— A new method for the study of chromosome aberrations and the plotting of chromosome maps in *Drosophila melanogaster*. Genetics **19**, 175—188 (1934).

— The morphology of the X-chromosome in salivary glands of *Drosophila melanogaster* and a new type of chromosome map for this element. Genetics **19**, 448—469 (1934).

— The structure of salivary gland chromosomes. Amer. Naturalist **73**, 315—338 (1939).

Patterson, J. T., and W. S. Stone: The evolution in the genus *Drosophila*. New York: Macmillan 1952.

Pavan, C.: Chromosomal variation in *Drosophila nebulosa*. Genetics **31**, 546—557 (1946).

—, Th. Dobzhansky and A. B. Da Cunha: Heterosis and the elimination of weak homozygotes in natural populations of three related species of *Drosophila*. Proc. nat. Acad. Sci. (Wash.) **43**, 226—234 (1957).

Philip, U.: An analysis of chromosomal polymorphism in two species of *Chironomus*. J. Genet. **44**, 129—142 (1942).

—, J. M. Rendel, H. Spurway and J. B. S. Haldane: Genetics and Karyology of *Drosophila subobscura*. Nature (Lond.) **145**, 260—262 (1944).

Raghavan, T. S., and K. R. Venkatasubban: Studies in the Indian Scilleae. II. The cytology of *Scilla indica* Baker. Cytologia **10**, 189—204 (1939).

Rees, H., and J. B. Thompson: Localisation of chromosomal breakage at meiosis. Heredity **9**, 399—408 (1955).

Rendel, J. M.: Heterosis. Amer. Naturalist **87**, 129—138 (1953).

Rhoades, M., and E. Dempsey: Cytogenetic studies of deficient-duplicate chromosomes derived from inversion heterozygotes in maize. Amer. J. Bot. **40**, 405—424 (1953).

Richardson, M.: Structural hybridity in *Lilium Martagon album* × *L. Hansonii*. J. Genet. **32**, 411—450 (1936).

Rieger, R.: Inhomologenpaarung und Meioseablauf bei haploiden Formen von *Antirrhinum majus* L. Chromosoma **9**, 1—38 (1957).

Riley, H. P.: Chromosome studies in a hybrid between *Gasteria* and *Aloe*. Genetics **32**, 102 (1947).

— Genetics and Cytogenetics. London/New York: Wiley/Chapman 1949.

Rohm, P. B.: A study of evolutionary changes in *Sciara* (Dipters), chromosome C in the salivary gland cell of *Sciara ocellaris* and *Sciara reynoldsi*. Amer. Naturalist **81**, 5—29 (1947).

Rosen, G. v.: Artkreuzungen in der Gattung *Pisum*. Hereditas (Lund) **30**, 201—400 (1944).

Rothfels, K., and T. Fairlie: The not random distribution of inversion breaks in the midge *Tendipes decorus*. Canad. J. Zool. **35**, 221—263 (1957).

Rowlands, D. G.: The nature of the breeding system in the field bean (*V. faba* L.) and ist relationship to breeding yield. Heredity **12**, 113—126 (1958).

Russel, W., and C. Burnham: Cytogenetic studies of an inversion in maize. Scient. Agric. **30**, 93—111 (1950).

Salzano, F. M.: Chromosomal relations in two species of *Drosophila*. Amer. Naturalist **88**, 399—405 (1954).

Sampath, S., and H. K. Mohanty: Cytology of semisterile rice hybrids. Current Science **6**, 182—183 (1954).

Sax, K.: Chromosome behavior and nuclear development in *Tradescantia*. Genetics **22**, 523—533 (1937).

SCHULTZ, J., and H. REDFIELD: Interchromosomal effect on crossing over in *Drosophila*. Cold Spr. Harb. Symp. quant. Biol. **16**, 175—195 (1951).
SESHACHAR, B. R.: Inversion of a central segment in a chromosome of *Ichthyophis glutinosus* (LINN.). Cytologia **10**, 15—17 (1939).
SHARMA, A. K.: Cytology of some members of Commelinaceae and its bearing on the interpretation of phylogeny. Genetica **27**, 323—363 (1955).
— and C. GHOSH: Cytogenetics of some of the Indian Umbellifers. Genetica **27**, 17—44 (1955).
— and S. K. SARKAR: Cytology of different species of palms and its bearing on the solution of the problems of phylogeny and speciation. Genetica **28**, 361—488 (1956).
— and A. SHARMA: Karyotype studies in *Cestrum* as an aid to taxonomy. Genetica **29**, 83—100 (1957).
SIKKA, S. M.: Cytogenetics of *Brassica* hybrids and species. J. Genet. **40**, 441—509 (1940).
SINOTÔ, Y., and D. SATÔ: Basikaryotype and its analysis. Cytologia **10**, 529—538 (1940).
— — Cyto-genetical studies on *Tricyrtis*. IV. Basikaryotype analysis in hybrids *T. hirta* and *T. formosana*. Cytologia **12**, 302—312 (1941).
SMITH, S. G.: Chromosome fragmentation produced by crossing over in *Trillium erectum* L. J. Genet. **30**, 227—232 (1935).
SOKOLOV, N. N., and N. P. DUBININ: Permanent heterozygosity in *Drosophila*. D. I. S. **15**, 39—40 (1941).
SPERLICH, D.: Modellversuche zur Selektionswirkung verschiedener chromosomaler Strukturtypen von *Drosophila subobscura* COLL. Z. Vererb.-Lehre **89**, 422—436 (1958).
SPIESS, E. B.: Experimental populations of *Drosophila persimilis* from an altitudinal transect of the Sierra Nevada. Evolution **4**, 14—33 (1950).
—, M. KETCHEL and B. P. KINNE: Physiological properties of gene arrangement carriers in *Drosophila persimilis*. I. Egg laying capacity and longevity of adults. Evolution **6**, 208—215 (1952).
SPURWAY, H., and H. G. CALLAN: Hybrids between members of the Rassenkreis *Triturus cristatus*. Experientia (Basel) **6**, 95—96 (1950).
SRINATH, K. V.: Morphological and cytological studies in the genus *Calceolaria*. II. Meiosis in diploid and aneuploid Calceolarias. Cytologia **10**, 467—491 (1940).
STALKER, H. D., and H. L. CARSON: An altitudinal transect of *Drosophila robusta* STURTEVANT. Evolution **2**, 295—305 (1948).
STEBBINS, G. L.: Cytogenetic studies in *Paeonia*. II. The cytology of the diploid species and hybrids. Genetics **23**, 83—110 (1938).
— and S. ELLERTON: Structural hybridity in *Paeonia californica* and *P. Brownii*. J. Genet. **38**, 1—36 (1939).
STEINBERG, A. G.: The effect of autosomal inversions on crossing over in the X-chromosome of *Drosophila melanogaster*. Genetics **21**, 615—624 (1936).
— Relations between chromosome size and effects of inversions on crossing over in *Drosophila melanogaster*. Proc. nat. Acad. Sci. (Wash.) **23**, 54—56 (1937).
— and F. C. FRASER: Studies on the effect of X-chromosome inversions on crossing over in the third chromosome of *Drosophila melanogaster*. Genetics **29**, 83—101 (1944).
STONE, W. S.: Genetic and chromosomal variability in *Drosophila*. Cold Spr. Harb. Symp. quant. Biol. **20**, 256—269 (1955).
— and I. THOMAS: Crossover and disjunctional properties of X-chromosome inversions in *Drosophila melanogaster*. Genetica **17**, 170—184 (1935).

STRAUB, J.: Untersuchungen über die cytologischen Grundlagen der Komplexheterozygotie. Chromosoma **2**, 64—76 (1941).

STUBBE, H.: Über mono- und digen bedingte Heterosis bei *Antirrhinum majus* L. Z. indukt. Abstamm.- u. Vererb.-Lehre **85**, 450—478 (1953).

— u. K. PIRSCHLE: Über einen monogen bedingten Fall von Heterosis bei *Antirrhinum majus* L. Ber. bot. Ges. **58**, 546—558 (1940).

STURTEVANT, A. H.: Contributions to the genetics of *Drosophila melanogaster*. III. Inherited linkage variations in the second chromosome. Carnegie Inst. Wash. Publ. **278**, 305—341 (1919).

— A crossover reducer in *Drosophila melanogaster* due to an inversion of a section of the third chromosome. Biol. Zbl. **46**, 697—702 (1926).

— and G. W. BEADLE: The relations of inversions in the X-chromosome of *Drosophila melanogaster* to crossing over and disjunction. Genetics **21**, 554—604 (1936).

SUTTON, E.: Terminal deficiencies in the X-chromosome of *Drosophila melanogaster* Genetics **25**, 628—635 (1940).

SWANSON, C. P.: The distribution of inversions in *Tradescantia*. Genetics **25**, 438—446 (1940).

TANAKA, N.: Chromosome studies in Cyperiaceae. VIII. Meiosis in diploid and tetraploid forms of *Carex siderosticta* HANCE. Cytologia **11**, 282—310 (1940).

TEISSIER, G.: Variation de la fréquence du géne sepia dans une population stationaire de Drosophiles. C. R. Acad. Sci. (Paris) **224**, 676—677 (1947).

— Variation de la fréquence du géne ebony dans une population stationaire de Drosophiles. C. R. Acad. Sci. (Paris) **224**, 1788—1789 (1947).

THOMAS, P. T.: The origin of new forms in *Rubus*. III. The chromosome constitution of *R. loganobacchus* BAILAY, its parents and derivates. J. Genet. **40**, 141—156 (1940).

TIMOFEEFF-RESSOVSKY, N. W.: Einige Versuche an *Drosophila melanogaster* über die Art der Wirkung der Röntgenstrahlen auf den Mutationsprozeß. Roux' Arch. Entw.-Mech. **124**, 644—665 (1931).

TING, Y. C.: Inversions and other characteristics of *Teosinte* chromosomes. Cytologia **23**, 239—250 (1958).

TOBGY, H. A.: A cytogenetical study of *Crepis fuliginosa, C. neglecta* and their F_1 hybrids, and its bearing on the mechanism of phylogenetic reduction in chromosome number. J. Genet. **45**, 67—110 (1943).

TOWNSEND, J. I.: Genetics of marginal populations of *Drosophila willistoni*. Evolution **6**, 428—442 (1952).

— Chromosomal polymorphism in Caribbean island populations of *Drosophila willistoni*. Proc. nat. Acad. Sci. (Wash.) **44**, 38—42 (1958).

TROFIMOV, I. E., and E. D. POSTNIKOVA: The interaction of non-homologous chromosomes during meiosis in *Drosophila melanogaster*. Biol. Z. **4**, 731—734 (1935).

UPCOTT, M.: The parents and progeny of *Aesculus carnea*. J. Genet. **33**, 135—148 (1936).

— The genetic structure of *Tulipa*. II. Structural hybridity. J. Genet. **34**, 339—398 (1937).

— The genetic structure of *Tulipa*. III. Meiosis in polyploids. J. Genet. **37**, 303—339 (1939a).

— The nature of tetraploidy in *Primula kewensis*. J. Genet. **39**, 79—100 (1939b).

Vavilov, N.: Über den Ursprung der Kulturpflanzen. Bull. appl. Bot., Genet. and Plant Breed. **16**, 1—138 (1926) (russ.).

— Geographical regularities in the distribution of the genes of cultivated plants. Bull. appl. Bot., Genet. and Plant Breed. **17**, 411—428 (1927) (russ. u. englisch).

Venkatesh, C. S.: A cytogenetic and evolutionary study of *Hemizonia*, sect. *Centromadia*. Amer. J. Bot. **45**, 77—84 (1958).

Walters, J. L.: Distribution of structural hybrids in *Paeonia californica*. Amer. J. Bot. **29**, 270—275 (1942).

Walters, M. S.: Spontaneous breakage and reunion of meiotic chromosomes in the hybrid *Bromus trinii* × *B. maritimus*. Genetics **35**, 11—37 (1950).

Ward, L.: The genetics of curly wing in *Drosophila* another case of balanced lethal factors. Genetics **8**, 276—300 (1923).

Ward, C. L.: Chromosomal variation in *Drosophila melanica*. Univ. Texas Publ. **5204**, 137—157 (1952).

Warters, M.: Chromosomal aberrations in wild populations of *Drosophila*. Univ. Texas Publ. **4445**, 129—174 (1944).

— and A. B. Griffen: The telomeres of *Drosophila*. J. Genet. **41**, 183—190 (1950).

Wasserman, M.: Cytological studies of the *repleta*-group. Univ. Texas Publ. **5422**, 130—152 (1954).

Wharton, L. T.: Analysis of the *repleta*-group of *Drosophila*. Univ. Texas Publ. **4228**, 23—52 (1942).

— Analysis of the metaphase and salivary gland chromosome morphology within the genus *Drosophila*. Univ. Texas Publ. **4313**, 282—319 (1943).

White, M. J. D.: The cytology of Cecidomyidae (Diptera). I. Polyploidy and polyteny in salivary gland cells of *Lestodiplosis* ssp. J. Morph. **78**, 201—219 (1946).

— A cytological survey of wild populations of *Trimerotropis* and *Circotettix* (Orthoptera, Acrididae). II. Racial differentiation in *T. sparsa*. Genetics **36**, 31—53 (1951).

— Animal cytology and evolution. Cambridge: Univ. Press 1954.

Wolf, E.: Die Chromosomen in der Spermatogenese einiger Nematoceren. Chromosoma **2**, 192—246 (1941).

Wright, S., and Th. Dobzhansky: Genetics of natural populations. XII. Experimental reproduction of some of the changes by natural selection in certain populations of *Drosophila pseudoobscura*. Genetics **31**, 125—156 (1946).

Zohary, D.: Chiasmata in a pericentric inversion in *Zea mays*. Genetics **40**, 874—877 (1955).

Pheromones

By PETER KARLSON, München

Physiologisch-Chemisches Institut der Universität München, Germany

Contents

I. Introduction and definitions

The concept of internal secretion may be traced back to CLAUDE BERNARD who, about a century ago, first used the term to describe some aspects of the metabolic activity of the liver. The first experimental demonstration of hormonal action (of the gonads) had been given ten years earlier, however, by BERTHOLD, in 1849. At the end of the nineteenth century, the new discipline of endocrinology gradually emerged, and BAYLISS and STARLING, in their famous lecture in 1905, coined the term „hormone" for chemical messengers between different tissues. They demonstrated that the pancreas is stimulated to secretion by a substance, "secretin", produced in the duodenal mucosa.

This concept has been very fruitful; however, their original examples are no longer included among typical endocrine processes. Most workers agree today that the terms coined by BERNARD, and by BAYLISS and STARLING, should be restricted to products of special glands or cell groups controlling specific processes in other tissues, the target organs. A detailed, critical discussion of the use of the term "hormone" in modern physiology may be found in the textbook of VERZAR (1948).

In 1932, A. BETHE drew attention to a number of chemical messengers which act between individuals, not within the organism. His classical example is the sex attractant of a moth. It is produced in a special gland and secreted not internally, into the blood, but externally, into the open air: it is an air-borne substance which stimulates the antennae of the males, indicating the presence of a female.

Going back to the original definition of STARLING, BETHE termed these substances „ectohormones" (chemical messengers secreted not internally but externally). He included any substance with attracting or repelling activity into his system. This was unfortunate for several reasons: (i) it seems to be unjustified to bring food attractants of plant origin, for example, under the same heading as hormones; (ii) the use of the term "hormone" has been generally restricted to the glandular hormones acting via the blood stream; and (iii) if hormones are understood to mean internal secretions, then the term "ectohormone" is self-contradictory.

It was for these reasons that the present author wrote, in 1956: "The rather unfortunate term "ectohormone" comprises those active substances which are given off externally by the organism and effect other individuals of the same species. The author considers that these substances do not fall into the category of hormones, and that it would be better to create a neutral name for them."

In a subsequent discussion with LÜSCHER, a neutral name was selected: Pheromone (derived from the greek pherein, to carry, and horman, to excite)[1]. It was intended to choose a name with some resemblance to "hormone", "gamone", "termone", since these compounds bear some relationship in their physiological function; they are all substances of humoral correlation within a species. On the other hand, it should not be easily confused with "hormone", or with "ectohormone", for which it should substitute. In the definition given below, it was restricted to substances acting within a species.

The definition was first intended only as part of the review of "Pheromones (Ectohormones) in Insects" (KARLSON and BUTENANDT 1959). As a result of a discussion with several authors actively engaged in research in this field, it was finally published (KARLSON and LÜSCHER 1959) and defined as follow: "Pheromones are defined as substances which are secreted to the outside by an individual and received by a second individual of the same species, in which they release a specific reaction, for example, a definite behaviour or a developmental process. The principle of minute amounts being effective holds. The pheromones, messenger substances between individuals, thus take their place as a group beside the hormones, the gamones, and the termones."

[1] The term Pheromone has been criticized as being incorrectly derived from its Greek origin: It should correctly read „pherormone". It is, however, felt more important to have a word which is easily pronounced and well defined even if it is an artificial one.

The meaning of this concept may be explained with reference to the classical example of a pheromone, the sex attractant of the common silk moth, *Bombyx mori* (or other Lepidoptera). It is produced in special glands, the sacculi laterales, at the tip of the abdomen; it has been extracted and isolated, its chemical formula is known, as will be discussed later. Its action can easily be demonstrated: male moths react to minute traces of the substance with excitation, movement of antennae and wings, and finally attempts of copulation. Obviously, the pheromone is responsible for the onset of a sequence of innate, instinctive reactions of the animal; it is, in the terminology of behaviourists, a chemical releaser. Its fundamental role in the biology of the species is clear: it informs the male about the proximity of a female, finally aiding in reproduction and propagation of the species. In one respect, the pheromone is unique among the numerous chemical releasers: namely, it is produced by a member of the same species. This is an essential part of the definition given above, which should not be overlooked. Thus, we exclude from the pheromones not only food attractants but also a number of other scents affecting the behaviour or physiology of the animal.

Stimuli acting as releasers may be very complex; chemical releasers (scents) may turn out to be a mixture of several compounds. In the classical case of *Bombyx*, we are dealing with a single substance which has been studied by chemical and biochemical methods, and has recently been isolated. It is a challenging problem for biochemists to identify substances of high biological activity (in this case, less than 10^{-16} g are effective), and it is believed that other chemical messengers of this kind will be investigated in the near future. It would be convenient for the biochemist studying the substance, as well as for the biologist studying its effect, to have a simple term such as „pheromone" to classify it.

However, the pheromone concept is not restricted to behavioural response. In another case, a pheromone acts as a developmental stimulus. In termites, pheromones which are produced by the king and the queen are essential for the caste differentiation, especially inhibition of sexual differentiation. This substance is ingested by the workers; details will be discussed later. For review purpose, the pheromones may be subdivided into those acting olfactorily, like the *Bombyx* sex attractant, and orally, like some active substances occurring among social insects. In the first group, sensory cells and the central nervous system play a part in the mode of action. This may be true for the second group as well, through chemoreceptors of taste.

It is not intended to give a complete list of pheromones or phenomena in which the participation of pheromones is suggested, but rather to discuss some of them in detail. Most of the examples will be taken from the insect kingdom; this reflects not only the main interest of the present author, but is also due to the fact that insect pheromones are much better studied than any others. Pheromones mediate social correlations; it is not surprising that the best examples are found in those animals with the highest social organisation, the social insects.

II. Olfactorily acting pheromones

1. The Bombyx sex attractant

The fact that female moths attract males has been known to entomologists for a long time. A detailed account is given by FABRE (see A. BUTENANDT 1955). Field studies, however, were not reliable enough to allow a detailed study of the active substance on the basis of a bioassay. As a laboratory animal, the commercial silk worm, *Bombyx mori*, has considerable advantages over nondomestic insects; it was therefore chosen by A. BUTENANDT (1939) for a chemical study of the sex attractant.

The bioassay of the sex pheromone is based on the behaviour of the males. A glass rod is dipped into the solution to be assayed and held in front of a male moth. Excited movements, fluttering of the wings and eventually attempted copulation are the signs of positive response. The biological unit (attractant unit) is defined as the amount of substance per ml solution giving positive reaction with 50% or more of the animals tested (not less than 20, preferably more). The behaviour assay is not very accurate, as it indicates only concentration differences of 1:10. However, it is simple, and proved convenient as a guide during chemical purification of the pheromone. The isolation of the pheromone turned out to be a long-term program carried out with several collaborators and a large number of insects as starting material. The active substance was extracted from the tips of the abdomens of female *Bombyx* moths; suitable solvents were petroleum ether or (preferable) alcohol-ether-mixtures. The crude lipid extract was freed from acids and saponified; the active material was found in the non-saponifiable matter. It is an alcohol and can be further purified by separating the alcohol fraction via the semisuccinates. After saponification, a marked increase in activity is observed, presumably due to the fact that a large portion of the pheromone is bound to some fatty acid; this is believed to be the storage form of the pheromone.

The alcohol fraction was converted to the p-nitro-azobenzenecarboxylic esters. The ester is physiologically inactive; the sex attractant can be recovered by saponification and then assayed. The further purification of the ester mixture included fractional precipitation with water from acetone and chromatographic procedures. Considerable difficulties were encountered in chemical work with this highly active substance, which was present only in minute amounts: contamination of solvents or nearly inactive fractions with traces of active material gave misleading results, and instability in some solvents led to great losses of active material. Finally 12 mg of pure crystalline pheromone ester were obtained from a total of 500000 *Bombyx* glands (BUTENANDT et al. 1959).

Two other papers on the chemistry of the *Bombyx* pheromone should be mentioned. MAKINO et al. (1956) described a purification procedure including paper chromatography. Their final product Bombyxin represents a purified material containing less than 0,1% of the pure pheromone; the analytical data for this mixture are of little value. Another claim for isolation (AMIN 1957) must be rejected as entirely wrong. The isolation was not guided by critical bioassay; no data are given regarding the biological activity of the isolated dimethylamine. Tests performed by BUTENANDT and HECKER (1958) showed dimethylamine to be nearly inactive; the attractant unit is 10,000 μg/ml or 10^{14} times higher than that of the true pheromone.

The empirical formula of the ester isolated by BUTENANDT et al. has been determined as $C_{29}H_{37}O_4N_3$, which leaves as the parent alcohol, the pheromone proper, $C_{16}H_{30}O$. The elucidation of the structure by BUTENANDT et al. (1959) is a masterpiece of organic microchemistry; the crucial degradation, done with less than a milligram of material, yielded butyric acid, oxalic acid and the p-nitro-azobenzenecarboxylic ester of ω-hydroxydecenoic acid. Since cleavage occured at the double bonds, the formula of the *Bombyx* sex pheromone may be written as

$$CH_3\!-\!CH_2\!-\!CH_2\!-\!CH\!=\!CH\!-\!CH\!=\!CH\!-\!(CH_2)_8\!-\!CH_2OH$$

cleavage products:

$$CH_3\!-\!CH_2\!-\!CH_2\!-\!COOH \quad HOOC\!-\!COOH$$

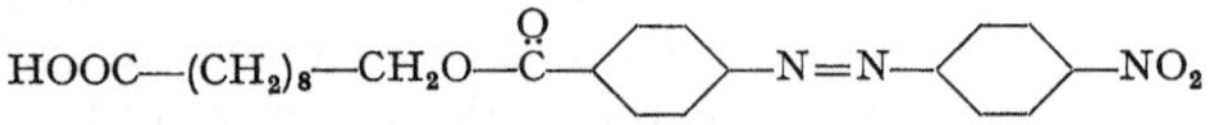

$$HOOC\!-\!(CH_2)_8\!-\!CH_2O\!-\!\overset{O}{\overset{\|}{C}}\!-\!C_6H_4\!-\!N\!=\!N\!-\!C_6H_4\!-\!NO_2$$

The isolation and identification of this compound has been given in some detail since it is the first pheromone that has been obtained in pure form. Its biological activity is very high: one attractant unit as defined above is equal to 10^{-16} g/ml. Taking into account that only a fraction of a ml. adheres to the glass rod, a safe estimate of the number of molecules present on the rod is about 1000; the number finally reaching the antenna of the male must be much less. On the other hand, there are more than a thousand olfactory sensillae per antenna (SCHNEIDER 1957, SCHNEIDER and KAISSLING 1956, 1957, 1959), and it seems highly probable that one or very few molecules per receptor are sufficient to stimulate the receptor, and that only a small number of responding receptors is necessary for a positive reaction in the behavioural assay.

The electrophysiology of the antennal reaction has been studied by SCHNEIDER (1955, 1957) and SCHNEIDER and HECKER (1956). Micro-electrodes introduced into the antenna were used to detect electrical nerve signals (spikes) and a slow change in receptor potential (electro-antennogram). Though only pheromone concentrates (not the pure substances) were available to that time, the slow potential in response to the pheromone was found to be very characteristic. It could not be mistaken for antennograms obtained with such substances as sorbinol, cyclohexanone or cycloheptanone which are active (in high concentrations) in the behavioural assay. Thus, the electrophysiological assay can differentiate between the pheromone and synthetic substances of similiar behavioural action.

The basis of the moth's behaviour is the excitatory effect of the pheromone. The direction of flight is then determined by air current rather than by a concentration gradient (SENGÜN 1954, SCHWINCK 1954, 1955). This will obviously guide the males to the females and accounts also for the fairly long distances of attraction that have been observed by entomologists (cf. DUFAY 1957).

2. Other sex attractants

Although sex attractants are widespread among the insect kingdom, chemical investigations are scanty. The sex pheromone of the gipsy moth, *Lymantria dispar* L., has been extracted from abdomen tips and purified to some extent (HALLER, ACREE and POTTS 1944; ACREE 1953, 1954). Like the *Bombyx* pheromone, it seems to be an unsaturated alcohol, but some esters are apparently active. Mainly due to difficulties in obtaining starting material and in the bioassay, the substance has not been isolated in pure form. — The sex attractant of another insect pest, the cotton leafworm, *Prodenia litura*, has been studied by FLASCHENTRÄGER and AMIN (1950), but some of their findings have not been confirmed by a later thorough investigation (JARCZIK, personal communication).

In the common roach, *Periplaneta americana*, a sex pheromone is produced in certain pouch-like organs in the female (ROTH and WILLIS 1952). It adheres to filter-paper on which the females are sitting and has been extracted therefrom; the extracts act as excitant to males and can be assayed with better accuracy than the *Bombyx* sex attractant (WHARTON et al. 1954a, b, 1957). Its purification is under way; it will be interesting to compare the chemical structure of this substance with the *Bombyx* pheromone.

One other substance has been isolated and identified which is believed to be a sex attractant of a tropical bug, *Belostoma indica*. The substance is secreted in special ducts which occur only in males; its flavour prompted the natives to use it as spice for fatty dishes. It has been identified as Δ^2-hexenol-acetate (BUTENANDT and TAM 1957):

$$H_3C—CH_2—CH_2—CH=CH—CH_2O \cdot COCH_3$$

Synthetic hexenol-acetate is identical in all respects with the isolated material. The biological role of this substance in *Belostoma* has not been investigated.

It is of considerable interest that a very similar compound, Δ^2-hexenal, has been detected by ROTH et al. (1956) in the roach *Eurycotis* (in both sexes).

Sex pheromones are generally believed to be species-specific. In moth, however, some cross-reactions have been observed (SCHWINCK 1954, KETTLEWELL 1946). KULLENBERG (1956) found a close relationship between bumble-bee pheromones and blossom-scents, but they have not been chemically isolated.

Several secretions of vertebrates may be classified as sex pheromones, e. g. the well-known musk, which is produced in a special gland of the musk-deer.

3. Marking scents and alarm substances

It is a well-known fact that ants generally migrate on trails. This is an essential mechanism of colony organization, and is mediated by some marking scent or trail pheromone laid down by the workers. The substance is perceived by chemoreceptors of the antennae and acts as chemical releaser and orienter. The chemical nature and the source of the pheromone may vary. In *Formica rufa*, formic acid is apparently used as trail pheromone (STUMPER, personal communication); in the myrmicine species *Solenopsis saevissima* the trail substance comes from the accessory gland of the sting, as has been found only recently by E. O. WILSON (1959a). Trails laid down artificially with the contents of this gland were followed quite regularly, though their effectiveness lasted only for a few minutes. The substance produced by the accessory glands is steam-volatile and soluble in petroleum ether. It is not impossible that it is identical with the toxic principle of the venom, which has been studied by BLUM et al. (1958). — A volatile substance isolated from *Iridomyrmex detectus* and identified as 2-methylhept-2-en-6-one acts as chemical releaser for trail following (CAVILL and FORD 1953).

The Nassanov gland of the honeybee, *Apis mellifica*, has been shown to produce a marking scent (FREE and BUTLER 1955; M. RENNER 1955, 1959; LECOMTE 1957). No chemical data are available on this pheromone. Similar scents from bumblebees have been analysed olfactorily but not chemically by KULLENBERG (1956). Finally, the territory-marking substances of carnivorous beasts may be mentioned here.

A characteristic behaviour of some gregarious fishes, especially the minnow (*Phoxinus phoxinus* L.), has been studied by v. FRISCH (1941). Small injuries to one individual cause a rapid flight of the swarm. This is due to some substance released from the skin, an alarm substance. The latter has been tentatively classified as a pheromone (KARLSON and LÜSCHER 1959).

Better examples of alarm pheromones may be found in ants. E. O. WILSON (1959b) presents evidence that a secretion of the mandibular gland serves to communicate a state of excitement (alarm behaviour) in *Pogonomyrmex badius* Latr. The substance has not been identified, but a similar phenomenon has been observed in *Atta sexdens rubropilosa* (BUTENANDT, LINZEN and LINDAUER 1959). In this case, the substance has been extracted from the head and shown to be a mixture, somewhat like many essential oils of plants, with citral as the main component.

III. Orally acting pheromones

1. The queen substance of the honeybee

This is a substance which informs the colony about the presence of the queen. Removal of the queen from a hive produces among the worker bees, besides general restlessness, two characteristic effects: brood cells containing young worker larvae are enlarged to emergency queen cells, and the normally atrophied ovaries of the workers begin to develop. Responsible for both effects is the withdrawal of a substance, as has been found by three different laboratories (CHAUVIN and PAIN 1956, PAIN 1956; DE GROOT and VOOGD 1954, VOOGD 1955, 1956, VERHEIJEN-VOOGD 1959; BUTLER 1954, 1957a, b, c, BUTLER and GIBBONS 1958). The queen substance can be extracted from queens, even from museum specimens; the active principle is distributed over the cuticle, but is presumably produced in one of the head glands. A dead queen impregnated with the extract can substitute for a living queen in respect to the above-mentioned effects. This clearly demonstrates the mediation of a substance and can be used as an assay for the queen pheromone.

The chemical nature of the queen substance is not known. It is soluble in alcohol, acetone, and chloroform. According to BUTLER, it may be a steroid, but it has not yet been isolated in pure form. Of considerable chemical and biological interest is the observation of CARLISLE and BUTLER (1956) that extracts containing the ovary-inhibiting hormone of crustaceans can substitute for the queen pheromone and vice versa. It is amazing that the production of a single individual, the queen, is sufficient to affect several tens of thousands of worker bees, and that the lack of pheromone becomes effective within a few hours. The latter fact has been explained by BUTLER (1956). He detected the pheromone in the content of the honeystomach and concluded that worker bees redistribute the substance with regurgitated food. This mechanism, however, creates a considerable dilution; the pheromone must be active in very small amounts, and this could easily be accounted for by the assumption that the pheromone affects the chemoreceptors of taste and acts via the central nervous system. The brain would control the behaviour pattern directly, the ovary development indirectly, i. e. by true hormones. If this assumption holds, injection of the pheromone should be ineffective.

A pheromone similar to the queen substance of the honeybee has been demonstrated in ants (STUMPER 1956). For a detailed discussion the reader is referred to the review of KARLSON and BUTENANDT (1959).

2. Pheromones of termites

The social structure of a termite colony differs considerably from that of a bee hive (for review see BRIAN 1957, BIER 1958). One main feature is, that the workers are larvae (pseudergates) rather than specialized adults; they are capable of metamorphosis into either soldiers or (winged or unwinged) reproductives which can substitute for the king and the queen. Removal of the reproductives, king and queen, results in the production of supplementary reproductives which are wingless and arise from pseudergates by a special molt. The inhibition of the formation of supplementary reproductives is — besides egg production — the main function of the sexual caste in maintaining the social structure of the colony.

Experiments with wire screens separating two colonies with and without a reigning pair indicated that the inhibition effect is due to pheromone production (LÜSCHER 1952, GRASSÉ 1949). Final proof stems from experiments of LIGHT (1944) as interpretated by LÜSCHER (1956); feeding of aequous or methanolic extracts made from heads of functional females inhibited the production of supplementary reproductives. The pheromone is presumably produced in the head and secreted through the gut along with the feces which are an important food factor in a termite colony. Experiments with animals mounted on wire screens in a way that the head belongs to one colony, the abdomen to another one suggest that the active substance passes through the intestine of the worker larvae and is redistributed among the colony members. This would account for a rapid spread of information to several thousand individuals. Also, it can be concluded that both sexes — king and queen — produce specific pheromones, which must be administered simultaneously to ensure full activity. The female substance alone is slightly inhibitory to production of functioning females; the male pheromone alone is without effect.

Production of supplementary reproductives as induced by lack of pheromone takes place through molting and differentiation. The molt is an induced one, occurring much earlier in an orphan colony than in controls (LÜSCHER 1952). Since molting and metamorphosis in insects is controlled by hormones (for review see KARLSON 1956), it is likely that the hormonal system is involved in the mechanism of pheromone action. One special hypothesis, namely a stimulation of only the prothoracic gland, has been tested experimentally (LÜSCHER and KARLSON 1958). Administration of ecdysone (crystalline prothoracic gland hormone, BUTENANDT and KARLSON 1954) to termite larvae induced molting, as expected; but even with large doses, no adult forms could be produced. Administration of ecdysone plus juvenile hormone (the hormone of the corpora allata) resulted in soldier formation (LÜSCHER 1958). The molt to a supplementary reproductive is presumably due to a special hormonal balance (lack of juvenile hormone?) which could not be imitated experimentally. Nevertheless, it seems highly probable that the pheromone influences in some way the hormonal glands, especially the neurosecretory system. But it is open to question if this influence is exerted by the pheromone proper, or if the pheromone is perceived by chemoreceptors of taste, and the nervous stimuli then activate the neurosecretory cells (for correlation of nervous and neurosecretory activity in insects, see VAN DER KLOOT 1955). As in the case of the queen substance of the honeybee, it might be argued that in the latter case successful spread of information to thousands of individuals is easier to imagine; but this is of course only speculation.

IV. General considerations

A number of biological phenomena have been discussed — some long known, others only recently elucidated — which have one feature in common: they are mediated by substances produced and secreted by an animal, and acting upon other members of the species. It has been proposed that these substances be termed pheromones. Only few of them have been chemically characterized; there is need for much more biochemical research in this field. But it is believed that the number of examples will grow in the near future, and that it will prove useful to have a scientific term to classify these substances.

The pheromones seem to be highly active, only a few molecules of the sex pheromone of *Bombyx* being needed for a positive reaction. In this case, however, the individual provides an amplifying system in its chemoreceptors and nerve cells, and this is true for the other olfactorily acting pheromones as well. We are dealing with substances with a well defined biological effect; it can be used as basis of a bioassay involving the very complex system of innate behavioural responses to certain stimuli. But it should be borne in mind that hormones, too, can evoke instincts (e. g. breeding behaviour), and that nervous stress causes hormone release (e. g. corticoid production).

Another point needs discussion: the mode of action of the second group of pheromones. A substance taken up orally is usually assumed to be absorbed and act as such, i. e. biochemically rather than through nervous correlation. In the case of pheromones, no experimental proof is available for this assumption. The opposite hypothesis mentioned above, namely a stimulation of taste receptors and action through the central nervous system, would complete the analogy to the olfactory pheromones and would account for activity in high dilution (distribution among many individuals). But it is today mere hypothesis; this point needs clarification by clear-cut experiments.

The pheromones are substances of social correlation rather than intercellular organisation. In this respect they are unlike hormones, and it is not surprising that they affect the central nervous system. A development of instincts, behaviour patterns etc. is a prerequisite of social life in animals. If a system of pheromones as a method of social correlation has evolved in evolution, it is reasonable that a direct influence upon behaviour via the nervous system is the most likely. Pheromones are links between biochemistry, nerve physiology, and behavioural science — and a challenging field for further research.

References

Acree, F., Jr.: The isolation of gyptol, the sex attractant of the female gypsy moth. J. econ. Entomol. **46**, 313—15 (1953).

— The chromatography of gyptol and gyptyl ester. J. econ. Entomol. **47**, 321—26 (1954).

Amin, El. S.: The sex-attractant of the silkworm moth *(Bombyx mori)*. J. chem. Soc. **1957**, 3764.

Bethe, A.: Vernachlässigte Hormone. Naturwissenschaften **20**, 177—81 (1932).

Bier, K. H.: Die Regulation der Sexualität in den Insektenstaaten. Ergebn. Biol. **20**, 97—126 (1958).

Blum, M. S., J. R. Walker, P. S. Callahan and A. F. Novak: Chemical, insecticidal, and antibiotic properties of fire ant venom. Science **128**, 306—307 (1958).

Brian, M. V.: Caste determination in social insects. Ann. Rev. Entomol. **2**, 107—120 (1957).

— The evolution of queen control in the social hymenoptera. Proceed. X. Intern. Congr. Entomol. 1956, **2**, 497—502 (1958).

Butenandt, A.: Zur Kenntnis der Sexual-Lockstoffe bei Insekten. Jb. preuß. Akad. Wiss. **1939**, 97.

— Über Wirkstoffe des Insektenreiches. Naturwiss. Rdsch. **12**, 457—464 (1955).

—, R. Beckmann, D. Stamm u. E. Hecker: Über den Sexuallockstoff des Seidenspinners *Bombyx mori*, Reindarstellung und Konstitution. Z. Naturforsch. **14b**, 283—84 (1959).

— u. E. Hecker: Dimethylamine, the supposed sex-attractant of the silkworm moth *(Bombyx mori)*. Proc. chem. Soc. **1958**, 53.

— u. P. KARLSON: Über die Isolierung eines Metamorphose-Hormons der Insekten in kristallisierter Form. Z. Naturforsch. **9b**, 389—91 (1954).

— ,B. LINZEN u. M. LINDAUER: Über einen Duftstoff aus der Mandibulardrüse der Blattschneiderameise *Atta sexdens rubropilosa* Forel. Arch. d'Anatomie Microscopique (Emil Witschi-Festschrift), im Druck.

— u. N.-D. TAM: Über einen geschlechtsspezifischen Duftstoff der Wasserwanze *Belostoma indica* Vitalis. Z. physiol. Chem. **308**, 277—283 (1957).

BUTLER, C. G.: The method and importance of the recognition by a colony of honeybees *(A. mellifera)* of the presence of its queen. Trans. roy. entomol. Soc. (London) **105**, 11—29 (1954).

— Some further observations on the nature of "queen substance" and of its role in the organisation of a honey bee *(Apis mellifera)* community. Proc. roy. entomol. Soc. (Lond.) **31**, 12—16 (1956).

— The control of ovary development in worker honeybees *(Apis mellifera)*. Experientia (Basel) **13**, 256—257 (1957a).

— The process of queen supersedure in colonies of honeybees *(Apis mellifera* Linn.*)* Ins. Soc. **4**, 211—223 (1957b).

— Some work at Rothamsted on the social behaviour of honeybees. Proc. roy. Soc. B **147**, 275—288 (1957c).

— and D. A. GIBBONS: The inhibition of queen rearing by feeding queenless worker honeybees *(A. mellifera)* with an extract of "queen substance". J. Insect Physiol. **2**, 61—64 (1958).

CARLISLE, D. B., and C. G. BUTLER: The "queen-substance" of honeybees and the ovary-inhibiting hormone of crustaceans. Nature (Lond.) **177**, 276—277 (1956).

CAVILL, G. W. K., and D. L. FORD: The chemistry of ants. Chem. and Ind. **1953**, 351.

CHAUVIN, R., and J. PAIN: Le développement des ovaires des ouvrières des abeilles et l'ectohormone des reines. Experientia (Basel) **12**, 354—355 (1956).

DUFAY, C.: Sur l'attraction sexuelle chez *Lasiocampa quercus* L. Bull. Soc. entomol. France **62**, 61—64 (1957).

FLASCHENTRÄGER, B., and EL. S. AMIN: Chemical attractants for insects: sex- and food-odours of the cotton leaf worm and the cut worm. Nature (Lond.) **165**, 394 (1950).

FREE, J. B., and C. G. BUTLER: An analysis of the factors involved in the formation of a cluster of honeybees. Behaviour **7**, 304—316 (1955).

FRISCH, K. v.: Über einen Schreckstoff der Fischhaut und seine biologische Bedeutung. Z. vergl. Physiol. **29**, 46—145 (1941).

GRASSÉ, P. P.: Ordre des Isoptères ou termites, in Traité de Zoologie, Tome IX, pp. 443—473. Paris: Masson et Cie. 1949.

GROOT, A. P. DE, and S. VOOGD: On the ovary development in queenless worker bees *(Apis mellifica* L.*)*. Experientia (Basel) **10**, 384—385 (1954).

HALLER, H. L., F. ACREE Jr., and S. F. POTTS: The nature of the sex attractant of the female gypsy moth. J. Amer. chem. Soc. **66**, 1659—1662 (1944).

KARLSON, P.: Biochemical studies on insect hormones. Vitam. and Horm. **14**, 227—266 (1956).

— and A. BUTENANDT: Pheromones (ectohormones) in insects. Ann. Rev. Entomol. **4**, 39—58 (1959).

— and M. LÜSCHER: „Pheromones": A new term for a class of biologically active substances. Nature (Lond.) **183**, 55—56 (1959).

— — „Pheromone", ein Nomenklaturvorschlag für eine Wirkstoffklasse. — Naturwissenschaften **46**, 63—64 (1959).

KETTLEWELL, H. B. D.: Female assembling scents, with reference to an important paper on the subject. Entomologist **79**, 8—14 (1946).

KLOOT, W. G. VAN DER: The control of neurosecretion and diapause by physiological changes in the brain of the Cecropia silkworm. Biol. Bull. **109**, 276—294 (1955).

KULLENBERG, B.: Field experiments with chemical sexual attractants on aculeate hymenoptera males. Zool. Bidr. fran Uppsala **31**, 253—354 (1956).

LECOMTE, J.: Sur le marquage olfactif des sources de nourriture par les abeilles butineuses. C. R. Acad. Sci. (Paris) **245**, 2385—2387 (1957).

LIGHT, S. F.: Experimental studies on ectohormonal control of the development of supplementary reproductives in the termite genus *Zootermopsis* (formerly *Termopsis*). Univ. Calif. (Berkeley) Publ. Zool. **43**, 413—454 (1944).

LÜSCHER, M.: Die Produktion und Elimination von Ersatzgeschlechtstieren bei der Termite *Kalotermes flavicollis* Fabr. Z. vergl. Physiol. **34**, 123—141 (1952).

— Die Entstehung von Ersatzgeschlechtstieren bei der Termite *Kalotermes flavicollis* Fabr. Ins. Soc. **3**, 119—128 (1956).

— Experimentelle Erzeugung von Soldaten bei der Termite *Kalotermes flavicollis* (Fabr.). Naturwissenschaften **45**, 69—70 (1958).

— u. P. KARLSON: Experimentelle Auslösung von Häutungen bei der Termite *Kalotermes flavicollis* (Fabr.). J. Ins. Physiol. **1**, 341—345 (1958).

MAKINO, K., K. SATOH and K. INAGAMI: Bombixin, a sex attractant discharged by female moth, *Bombyx mori*. Biochim. biophys. Acta **19**, 394—395 (1956).

PAIN, J.: Sur L'Ectohormone des Reines D'Abeilles. Ins. Soc. **3**, 199—202 (1956).

RENNER, M.: Neue Untersuchungen über die physiologische Wirkung des Duftorganes der Honigbiene. Naturwissenschaften **42**, 589 (1955).

— Die physiologische Bedeutung des Duftorgans der Honigbiene. Z. vergl. Physiol. 1959 (in Vorbereitung).

ROTH, L. M., and E. R. WILLIS: A study of cockroach behavior. Amer. Midland Naturalist **47**, 66—129 (1952).

— A. D. NIEGISCH and W. H. STAHL: Occurrence of 2-hexenal in the cockroach *Eurycotis floridana*. Science **123**, 670 (1956).

SCHNEIDER, D.: Mikro-Elektroden registrieren die elektrischen Impulse einzelner Sinnesnervenzellen der Schmetterlingsantenne. Industrie-Elektronik **3/4**, 3—7 (1955).

— Elektrophysiologische Untersuchungen von Chemo- und Mechanorezeptoren der Antenne des Seidenspinners *Bombyx mori* L. Z. vergl. Physiol. **40**, 8—14 (1957).

— u. E. HECKER: Zur Elektrophysiologie der Antenne des Seidenspinners *Bombyx mori* bei Reizung mit angereicherten Extrakten des Sexuallockstoffes. Z. Naturforsch. **11 b**, 121—124 (1956).

— u. K. E. KAISSLING: Der Bau der Antenne des Seidenspinners *Bombyx mori* L. I. Architektur und Bewegungsapparat der Antenne sowie Struktur der Cuticula. Zool. Jahrb. Abt. Anat. u. Ontogenie Tiere **75**, 287—310 (1956).

— — Der Bau der Antenne des Seidenspinners *Bombyx mori* L. II. Sensillen, cuticulare Bildungen und innerer Bau. Zool. Jb. Abt. Anat. u. Ontogenie Tiere **76**, 223—250 (1957).

— — Der Bau der Antenne des Seidenspinners *Bombyx mori* L. III. Das Bindegewebe und das Blutgefäß. Zool. Jb. Abt. Anat. u. Ontogenie Tiere **77**, 111—132 (1959).

SCHWINCK, I.: Experimentelle Untersuchungen über Geruchsinn und Strömungswahrnehmung in der Orientierung bei Nachtschmetterlingen. Z. vergl. Physiol. **37**, 19—56 (1954).

— Weitere Untersuchungen zur Frage der Geruchsorientierung der Nachtschmetterlinge: Partielle Fühleramputation bei Spinnermännchen, insbesondere am Seidenspinner *Bombyx mori* L. Z. vergl. Physiol. **37**, 439—458 (1955).

SENGÜN, A.: Über die biologische Bedeutung des Duftstoffes von *Bombyx mori* L. Rev. fac. sci. Univ. Istanbul B, **19**, 4, 281—296 (1954).

STUMPER, R.: Les sécrétions attractives des reines de fourmis. Mitt. schweiz. entomol. Ges. **29**, 373—380 (1956).

VERHEIJEN-VOOGD, C.: How worker bees perceive the presence of their queen. Z. vergl. Physiol. **41**, 527—582 (1959).

VERZAR, F.: Lehrbuch der inneren Sekretion, pp. 1—16. Liestal: Ars Medici 1948.

VOOGD, C.: Inhibition of ovary development in worker bees by extraction fluid of the queen. Experientia (Basel) **11**, 181—182 (1955).

— The influence of a queen on the ovary development in worker bees. Experientia (Basel) **12**, 199—201 (1956).

WHARTON, D. R. A., G. G. MILLER and M. L. WHARTON: The odorous attractant of the American cockroach, *Periplaneta americana* (L.). J. gen. Physiol. **37**, 461—469 (1954a).

— The odorous attractant of the American cockroach, *Periplaneta americana* (L.). J. gen. Physiol. **37**, 471—481 (1954b).

WHARTON, M. L., and D. R. A. WHARTON: The production of sex attractant substance and of oothecae by the normal and irradiated American cockroach, *Periplaneta americana* L. J. Insect. Physiol. **1**, 229—239 (1957).

WILSON, E. O.: The source and possible nature of the odor trail of fire ants. Science 129, 643—644 (1959a).

— A chemical releaser of alarm and digging behavior in the ant *Pogonomyrmex badius* (Latreille). Psyche (in press) (1959b).

Die Eigenschaften und Funktionstypen der Sinnesorgane

Von DIETRICH BURKHARDT, München

Aus dem Zoologischen Institut der Universität München

Mit 4 Abbildungen

Inhaltsverzeichnis

Einleitung

Eine der wichtigsten Aufgaben der vergleichenden Anatomie ist es, Baupläne aufzustellen. Aus der Fülle uns entgegentretender Formen werden gemeinsame Merkmale herausgehoben und als typisch für einen bestimmten Formenkreis zusammengestellt. Die entsprechende Aufgabe der vergleichenden Physiologie muß zum Ziel haben, *Funktionstypen* aufzustellen. Auch hier kann das Gesetzmäßige nur erkannt werden, wenn zunächst die Vielfalt der beobachteten Leistungen nach Wesensähnlichkeiten geordnet und dann das Typische eines Verwandtschaftskreises von Leistungen als *Funktionstyp* herausgearbeitet wird. Die Gesamtheit der *Funktionstypen* liefert ein Ordnungsschema für die beobachtbaren Leistungen und ermöglicht das Erkennen grundlegender Zusammenhänge.

Die Sinnesphysiologie ist ein Grenzgebiet zwischen verschiedenen Wissenschaften. Dies zeigt sich bereits daran, daß eine Reihe von Wissenschaftlern sehr verschiedener Arbeitsrichtung gleichermaßen an sinnesphysiologischen Fragen interessiert ist: Physiologen, Zoologen, Psychologen, Physiker, Biochemiker und in letzter Zeit zunehmend Wissenschaftler aus dem Gebiet der Nachrichten- und Regeltechnik. Ebenso vielfältig sind die Fragestellungen und schließlich die Methoden, mit welchen Sinnesorgane untersucht werden. Das in den letzten Jahrzehnten von den verschiedenen Richtungen gesammelte Tatsachenmaterial hat einen großen Umfang erreicht. Der Versuch, *Funktionstypen* aufzustellen, scheint daher heute nicht nur möglich, sondern schon nötig zu sein. Im folgenden Artikel werden vorzugsweise Gesichtspunkte der vergleichenden Physiologie und Ergebnisse der Elektrophysiologie berücksichtigt. Die Objekte der vergleichenden Physiologie ermöglichen es, bestimmte Leistungen an besonders günstigem Material zu studieren. Günstig bedeutet dabei nicht nur, daß ein Objekt der experimentellen Untersuchung leicht zugänglich ist, sondern vor allem, daß es bestimmte Funktionen besonders übersichtlich zu erkennen gibt. Die elektrophysiologische Methode bietet in vielen Fällen den Vorteil, die Leistungen der Sinnesorgane isoliert untersuchen zu können. Bei Verhaltensversuchen ist dies nicht der Fall: Wird eine Reaktion des Organismus als Indicator für die Erregung des Sinnesorgans gewertet, so gehen die Leistungen des Zentralnervensystems mit in das Ergebnis ein, und nur selten glückt es, beide Leistungen nachträglich voneinander zu trennen. Gegenüber der Untersuchung von Stoffwechselgrößen, wie Sauerstoffverbrauch oder Konzentration von Erregungssubstanzen, erlaubt die Messung der elektrischen Aktivität eines Sinnesorgans, die Erregung fortlaufend und trägheitslos zu verfolgen. Bei sinnvoller Handhabung und hinreichend tiefer Analyse werden selbstversändlich alle Methoden zum gleichen Ergebnis führen müssen. Ein besonders schönes Beispiel hierfür bieten die Untersuchungen von LOWENSTEIN u. ROBERTS bzw. von v. HOLST am Labyrinth der Fische. Wenn hier vorzugsweise die Ergebnisse der vergleichenden Physiologie und Elektrophysiologie herangezogen werden, so soll dies in keiner Weise bedeuten, daß den genannten Arbeitsrichtungen eine zentrale Stellung zukommt. Jede aller möglichen Fragestellungen ergänzt sich sinnvoll mit allen anderen; Voraussetzung für ein solches Wechselspiel ist jedoch eine saubere Terminologie.

Im folgenden Artikel soll daher zweierlei versucht werden:

1. Durch eine Reihe von Definitionen eine möglichst scharf gefaßte Terminologie für die unter 2. genannten Fragestellungen vorzuschlagen. Diese Definitionen werden durch Schrägdruck gekennzeichnet.

Viele alteingebürgerte Termini werden in unterschiedlicher Bedeutung verwandt; als Beispiel sei ein Wort wie Adaptation genannt. Es erscheint dem Autor nicht zweckmäßig, Begriffe erneut und scharf umgrenzt zu definieren und die alte Bezeichnung beizubehalten. Hierdurch würde allenfalls erreicht, daß zusätzlich zu den alten Bedeutungen nun eine weitere kommt. Es werden daher für einige Begriffe neue Bezeichnungen vorgeschlagen, auch wenn sich eine der Bedeutungen eines alten Terminus ganz oder teilweise mit der neuen Definition deckt.

2. Die Eigenschaften und die funktionellen Verknüpfungen der Sinnesorgane sollen einander gegenübergestellt und einige Beispiele für Funktionstypen herausgearbeitet werden.

Die Fragestellungen der vergleichenden Sinnesphysiologie

Der Nutzen und die Tragweite eines Ordnungsschemas hängen davon ab, nach welchen Merkmalen und Fragestellungen geordnet wird. Bei sinnesphysiologischen Untersuchungen an Tieren gibt es eine Folge von Fragestellungen, es sind dies die Fragen nach:

1. dem adäquaten Reiz,
2. der Reizleitung (AUTRUM [1]) in den Hilfsstrukturen des Sinnesorgans,
3. den Prozessen im Sinnesorgan,
4. dem quantitativen Zusammenhang zwischen Reiz und Erregung,
5. den Reaktionen des Organismus, welche das Sinnesorgan bestimmt,
6. der Weise, wie diese ausgelösten Reaktionen gesteuert werden.

Die ersten vier Fragen beziehen sich auf das Sinnesorgan selbst, die beiden letzten auf seine funktionelle Bedeutung und Verknüpfung, sie greifen bereits in das Gebiet der Physiologie des Zentralnervensystems über.

Die in der vergleichenden Physiologie übliche Einteilung der Sinnesorgane nach der Energieform des Reizes ist ein Ordnungsschema, welches die Frage nach dem adäquaten Reiz berücksichtigt. Man unterscheidet danach: Chemische Sinne, Temperatursinn, Lichtsinn und mechanische Sinne (vgl. v. BUDDENBROCK; KÜHN [1]). Diese Einteilung ist von Nutzen für die zweite und dritte der genannten Fragestellungen. Mit der Art des adäquaten Reizes werden die reizleitenden Strukturen variieren. Für die Prozesse im Sinnesorgan ergibt sich eine Arbeitshypothese: Bis die Erfahrung im Einzelfall das Gegenteil lehrt, wird man annehmen, daß bei gleichem adäquaten Reiz die Primärprozesse im Receptor gleich oder ähnlich ablaufen, unabhängig davon, welches Sinnesorgan im einzelnen untersucht wird. Für die übrigen Fragen liefert diese Einteilung aber wenig, und sie wird vollends unbefriedigend, wenn nach den vom Sinnesorgan ausgelösten Reaktionen gefragt wird. Chemoreceptoren finden sich beim Säuger im Aortenbogen (HEYMANS u. BOUCKAERT) und bei Insekten auf der Antenne (D. SCHNEIDER). Die Chemoreceptoren des Aortenbogens regeln die chemische Zusammensetzung des Blutes, die der Insektenantenne lösen Verhaltensreaktionen für den Nahrungserwerb oder die Fortpflanzung aus. Es werden Organe zusammengefaßt, deren funktionelle Bedeutung recht unterschiedlich ist. Soll ein Ordnungsschema entworfen werden, welches die funktionelle Bedeutung der Sinnesorgane berücksichtigt, müssen offenbar die Fragen nach dem adäquaten Reiz, nach der Reizleitung und den Prozessen im Receptor ausgeklammert werden.

Die nächste zu prüfende Frage lautet dann: Gibt es, unabhängig vom adäquaten Reiz, bei verschiedenen Sinnesorganen Ähnlichkeiten in der Beziehung zwischen Reiz und Erregung? Wenn dies der Fall ist und die Sinnesorgane nach bestimmten Formen des Reiz-Erregungs-Zusammenhangs gruppiert werden können, so muß weiterhin gefragt werden: Stehen diese Eigenschaften der Sinnesorgane mit dem Typ der funktionellen Verknüpfung und der Art der vom Sinnesorgan bestimmten Reaktion in irgendwelchem Zusammenhang?

Ein Ansatz zu einem solchen Ordnungsschema kann in der Unterscheidung zwischen Extero- und Proprio-Receptoren gesehen werden. Exteroreceptoren sind solche Sinnesorgane, welche auf Reize aus der Umwelt des Organismus ansprechen; Proprioreceptoren sprechen auf Körperzustände an. Es wird hier nach dem Ursprung des Reizes gefragt, nach der funktionellen Bedeutung, und die Frage nach dem adäquaten Reiz bleibt ausgeklammert.

Der Zusammenhang zwischen Reiz und Erregung

A. Die Erregung

Bestimmte Aktivitäten der Sinneszelle verändern sich besonders leicht und in gesetzmäßiger Weise unter der Einwirkung von Bedingungen, welche außerhalb des Sinnesorgans liegen und für das Sinnesorgan spezifisch sind. Die für das Sinnesorgan spezifischen (energetischen) Bedingungen werden Reiz genannt, die Summe aller unter der Einwirkung des Reizes veränderlichen Aktivitäten Erregung. Die Erregung ist also das primäre Phänomen für den Sinnesphysiologen; ihre Veränderungen werden gemessen und daraus wird gefolgert, welches der für das Sinnesorgan wirksame Reiz ist. Die Erregung ist ein Zustand der Sinneszelle, der sich in sehr verschiedenen Erscheinungen äußert, beispielsweise im Sauerstoffverbrauch, in der Freisetzung oder Konzentrationsänderung bestimmter Substanzen, in der Bildung elektrischer Spannungen und in Veränderungen der Membranpermeabilität. Jeder dieser Vorgänge kann als Maß der Erregung gewertet werden, hier sollen jedoch vorzugsweise die elektrischen Vorgänge am Sinnesorgan und sensiblen Nerven herangezogen werden:

1. Von vielen Sinnesorganen können während und nach einer Änderung des Reizes langsame Spannungsschwankungen abgegriffen werden: Receptorpotentiale. Die Amplitude des Receptorpotentials hängt graduiert vom Reiz ab und ist ein Maß für die Größe der Erregung (vgl. GRANIT [1]).

2. Die Erregung der sensiblen Endstrukturen löst im sensiblen Nerven eine Folge von Aktionspotentialen aus, welche dem Alles-oder-Nichts-Gesetz gehorchen: Nerven-Impulse (oder Spikes). Je stärker die Erregung, desto dichter folgen die Spikes aufeinander. Die Frequenz der Spike-Entladungen ist somit ein weiteres Maß für die Stärke der Erregung (vgl. ADRIAN; GRANIT [1]).

Daß die Impuls-Frequenz des sensiblen Nerven ein direktes Maß für die Erregung sensorischer Strukturen ist, wird heute allgemein anerkannt. Der Begriff und die Bedeutung des Receptorpotentials sind hingegen noch mit einiger Problematik behaftet. Einige Autoren bezweifeln, daß zwischen dem Receptorpotential und der Impuls-Auslösung im sensiblen Nerven ein kausaler Zusammenhang besteht. Nach der von diesen Autoren vertretenen Hypothese ist für die Impuls-Entstehung im sensiblen Nerven eine chemische Übertragersubstanz entscheidend; die graduierten Receptorpotentiale sind lediglich eine Begleiterscheinung des Erregungszustandes (vgl. hierzu GRUNDFEST). Nach der Meinung anderer Autoren besteht zwischen dem Receptorpotential und der Impuls-Entladung des sensiblen Nerven ein direkter und kausaler Zusammenhang: Bekanntlich kann durch eine Depolarisation des Nerven eine Impuls-Entladung ausgelöst werden; durch eine Re- oder Hyperpolarisation wird hingegen die Impuls-Entstehung gehemmt. Den Generatormechanismus im sensiblen Nerven stellt man sich dann so vor, daß die sensiblen Endstrukturen auf ihren spezifischen Reiz mit einer Potentialschwankung reagieren; diese Potentialänderung bestimmt den Erregungszustand des sensiblen Nerven. Wirkt das Receptorpotential depolarisierend auf den sensiblen Nerven ein, so wird dessen Impuls-Frequenz gesteigert, wirkt es re- oder hyperpolarisierend, so wird sie gehemmt (vgl. hierzu GRANIT [1] und TRINCKER[1]). Ist das Receptorpotential bei einer Erhöhung der Reizintensität eine einsinnige negative Spannungsschwankung besonders einfacher Form, so wird es Generatorpotential genannt. Die Spike-Frequenz des sensiblen Nerven geht in der Tat in solchen Fällen direkt proportional mit dem Zeitverlauf des Receptorpotentials (GRANIT [1], KATZ [1], BURKHARDT [1]). Bei einem komplizierteren Zeitverlauf des Receptorpotentials muß nach dieser Theorie angenommen werden, daß es durch die Überlagerung mehrerer einfacher Komponenten verschiedenen Ursprungs zustande kommt. Eine dieser Komponenten wäre dann das Generatorpotential.

Bei diesem Generator-Mechanismus müssen zwei Fälle unterschieden werden (MURRAY): 1. In freien sensiblen Nervenendigungen und in Sinnesnervenzellen geht der Impuls-Entstehung eine Depolarisation der Nervenmembran voraus; die Quelle des Potentials ist hier die Nervenmembran selbst. Ein derartiges Potential sollte in Anlehnung an KATZ [1] *terminale Depolarisation* genannt werden. Unglücklicherweise verwendet MURRAY für diesen Fall den Terminus „Receptorpotential“

(in Anlehnung an GRAY und SATO). 2. Bei den sekundären Sinneszellen der Vertebraten wirkt auf den sensiblen Nerven ein Potential ein, dessen Ursprung außerhalb der Nervenzelle liegt. Es stammt entweder von den Sinneszellen (Auge, vgl. für die Originalliteratur GRANIT [1]) oder von Hilfsstrukturen des Sinnesorgans (Endorgane des VIII. Hirnnerven, vgl. für die Originalliteratur TRINCKER [1,2]). Den Begriff des Generatorpotentials möchte MURRAY auf diesen zweiten Fall eingeschränkt sehen.

B. Der Reiz

Der Erregung, einem physiologischen Prozeß, liegt eine Summe komplizierter physikalischer und chemischer Vorgänge innerhalb der Zelle zugrunde. Die Erregung wird sich deshalb durch Energie-Zu- oder -Abfuhr von außen her verändern lassen, wenn Energie an geeigneter Stelle und in geeigneter Form in die Kette der erregungsbildenden Prozesse eingeführt wird[1]. Zweierlei ist hierzu nötig:

1. Die Energie muß, sofern die sensiblen Strukturen nicht frei zugänglich liegen, von außen herangeführt werden: Reizleitender Apparat. 2. Die über den reizleitenden Apparat heran- (oder ab-) geführte Energie muß in einer Form vorliegen, in der sie auf die erregbaren Strukturen oder die erregungsbildenden Vorgänge einwirken kann. Die Spezifität des Reizes für eine Sinneszelle hat also zwei Ursachen, von denen eine im reizleitenden Apparat, die andere in den Primärvorgängen im Receptor zu suchen ist (vgl. hierzu auch die Diskussion bei TRINCKER [1]). Beispielsweise wirkt für Wirbeltieraugen Licht einer Wellenlänge zwischen 400 mμ und 800 mμ als Reiz. Die kurzwellige Grenze ist durch den reizleitenden Apparat bestimmt: Linse und Glaskörper absorbieren UV, welches die Receptorzellen an sich reizen würde (vgl. KENNEDY und MILKMAN). Die langwellige Grenze ist durch Primärvorgänge im Receptor bestimmt: Die Absorptionskurven der Sehfarbstoffe enden im langwelligen Rot (vgl. CRESCITELLI u. DARTNALL).

Als adäquaten Reiz eines Sinnesorgans bezeichnet man diejenige Energieform, deren Zustandsänderung außerhalb des Sinnesorgans, gegebenenfalls über den reizleitenden Apparat, auf die sensiblen Endstrukturen einwirkt und deren Erregung verändert.

Inadäquate oder unspezifische Reize können ebenfalls eine Erregung auslösen oder verändern, jedoch muß hierbei dem Sinnesorgan für eine Erregungsänderung mehr Energie zugeführt werden als bei adäquater Reizung: Durch die Filterwirkung des reizleitenden Apparates und die spezifische Organisation der sensiblen Endstrukturen wird ein großer Teil der Energie des inadäquaten Reizes anderweitig umgesetzt. Der Organismus schirmt sich so gegen Störmeldungen durch die Sinnesorgane ab (vgl. hierzu auch BURKHARDT [1,2]). Typische Beispiele für inadäquate Reize sind bei vielen Sinnesorganen Temperatur, mechanische Verformung und elektrische Ströme.

---

[1] Hierbei müssen zwei Möglichkeiten unterschieden werden: a) Die Energie des Reizes geht unmittelbar in die energetische Bilanz des Erregungsprozesses ein. b) Die Energie des Reizes dient lediglich dazu, den Erregungsprozeß auf ein anderes Niveau zu lenken, ohne daß sie der Energiebilanz zugeführt wird.

Vom adäquaten Reiz, welcher die Dimension einer Energie hat, sollten zwei Dinge unterschieden werden: der *Eingangsreiz* eines Sinnesorgans und der *Nutzreiz* für die sensiblen Endstrukturen. *Als Eingangsreiz soll diejenige physikalisch meßbare Größe bezeichnet werden, deren Änderung außerhalb des Sinnesorgans den Energiefluß über den reizleitenden Apparat bestimmt.* Der Eingangsreiz kann also ohne weiteres eine andere physikalische Dimension als Energie haben. *Unter dem Nutzreiz soll die physikalische Größe verstanden werden, welche am Ende des reizleitenden Apparates auf die sensiblen Endstrukturen einwirkt und die Stärke der Erregung bestimmt.* Der Nutzreiz kann bei dieser Definition wie der Eingangsreiz fast jede physikalische Dimension haben und im allgemeinen eine andere als der Eingangsreiz. Der reizleitende Apparat transformiert den Eingangsreiz zum Nutzreiz (Reiztransformation, AUTRUM [1]). Drei Beispiele sollen diese Definition erläutern: Für die statischen Organe des Wirbeltierlabyrinths ist der adäquate Reiz die mechanische Energie (des Statolithen). Der adäquate Reiz steht im Gegensatz zum unspezifischen Reiz, wie es Temperaturänderungen sind, welche die Erregung des Labyrinths ebenfalls modifizieren können. Als Eingangsreiz wirkt beim Labyrinth der Kippwinkel zwischen einer vorgesehenen Achse des Tierkörpers und der Schwerkraft (allgemeiner die Richtung und Stärke eines mechanischen Feldes). Der Nutzreiz hingegen ist eine Scherung der Haarzellen des Statolithenepithels (LOWENSTEIN u. ROBERTS, V. HOLST, SCHÖNE). Für zahlreiche Gehörorgane wird angenommen, daß der Schall durch eine mechanische Gleichrichtung im reizleitenden Apparat in eine Dauerauslenkung umgeformt wird. Adäquater Reiz des Wirbeltierohres ist demnach mechanische Energie, als Eingangsreiz wirkt der Schallwechseldruck (bei Gehörorganen der Insekten hingegen die Schallschnelle) und als Nutzreiz eine Dauerauslenkung der Haarzellen des Cortischen Organs (vgl. AUTRUM [1], SCHWARTZKOPFF [1], V. BÉKÉSY [1, 2], WHITFIELD, TRINCKER [1], KEIDEL [1]). Die Grubenorgane der Klapperschlangen sprechen auf die Wärmestrahlung der Objekte in der Umwelt des Tieres an (NOBLE u. SCHMIDT). Aus dem Bau und der Funktion der sensiblen Endstrukturen geht hervor, daß der Nutzreiz eine durch die Wärmestrahlung bewirkte Temperaturerhöhung ist (BULLOCK u. DIECKE); die Grubenorgane besitzen keine Infrarot-photosensible Substanz, welche ähnlich einem Auge ermöglichen würde, daß die Strahlung selbst als Nutzreiz wirkt. Eingangsreiz der Grubenorgane ist also die Wärmestrahlung, Nutzreiz eine Temperaturänderung und adäquater Reiz thermische Energie.

Die hier vorgeschlagene Unterscheidung zwischen adäquatem Reiz und Nutzreiz hat vor allem praktische Gründe. Bei fast allen Untersuchungen wird der Reiz durch Angabe einer physikalischen Größe erfaßt, welche *nicht* die Dimension einer Energie hat. Die dabei als Reiz bezeichnete Größe ist jedoch nach dem Urteil des Untersuchers diejenige Gegebenheit, welche physiologischerweise die Reaktion des Organismus steuert. Daß mit der Variation dieser Größe eine gesetzmäßige Änderung der Energie des Reizes einhergeht, wird mit Recht als selbstverständlich vorausgesetzt. Die Angabe der Reizenergie selbst ist ein viel zu unspezifisches Maß, außerdem sind Energieberechnungen meist nur ungenau und oft überhaupt nicht möglich, da zu viele Daten unbekannt bleiben (vgl. für Energieberechnung bei Schwellenreizen AUTRUM [2, 3]). Als Beispiel für diese Überlegungen mögen wieder Gehörorgane dienen. Die der Wirbeltiere sprechen auf Schalldruck an, zahlreiche Gehörorgane der Insekten jedoch auf die Schallschnelle. Bei gleicher Energiedichte im Schallfeld einer stehenden Welle werden die Schnelle-Empfänger im Schnellebauch eine größere Erregung liefern als im Druckbauch (AUTRUM [4]). Die Angabe der Energie des Schallfeldes allein erweist sich als ungenügend (vgl. auch S. 235). Es scheint daher sinnvoll, hieraus die Folgerungen zu ziehen. Neben der in sich konsequenten Definition des adäquaten Reizes mit der Dimension einer Energie sollte eine Übereinkunft getroffen werden, wonach die Spezifität des Reizes für ein Sinnesorgan auch durch die Angabe einer physikalischen Größe beliebiger Dimension erfaßt werden kann.

Die Spezifität des Reizes hat, wie oben diskutiert, zwei Ursachen, die im reizleitenden Apparat und in den sensiblen Endstrukturen gesucht werden müssen. Durch den reizleitenden Apparat kann der Eingangsreiz nach physikalischer Dimension, Größe und zeitlichem Verlauf umgeformt werden.

Hierin liegt vielleicht ein allgemeines Prinzip für den Bau der Sinnesorgane: Durch verschieden gebaute reizleitende Apparate können sehr ähnlich oder gleich funktionierende sensible Endstrukturen an unterschiedliche Eingangsreize angepaßt werden. Als Beispiel seien die Endorgane des VIII. Hirnnerven der Wirbeltiere genannt: Seitenliniensystem, Bogengänge, statische Organe und Cortisches Organ besitzen sehr ähnliche Receptorzellen, die durch eine Vielfalt von reizleitenden Apparaten als Strömungssinnesorgane, dynamische und statische Lagesinnesorgane und Gehörorgan abgewandelt sind (TRINCKER [1]). Umgekehrt scheint es möglich, daß unterschiedliche sensible Endorgane durch geeignete reizleitende Apparate zur Wahrnehmung gleicher oder ähnlicher Eingangsreize angepaßt werden können. Erinnert sei hier an die Grubenorgane der Klapperschlangen (BULLOCK und DIECKE), die keine photosensiblen Receptoren besitzen und doch sehr ähnlich einem primitiven Grubenauge auf infrarote Strahlung ansprechen.

Sollen die Eigenschaften eines Sinnesorganes analysiert werden, so erhebt sich sofort die Frage: Welche Eigenschaften gehen auf den reizleitenden Apparat zurück, welche auf die sensiblen Endstrukturen? Dies Problem taucht nicht nur wie hier bei der Frage nach der Spezifität des Reizes auf, sondern genauso bei der Frage nach der Reizqualität, nach dem zeitlichen Verhalten der Sinnesorgane, nach der Intensitäts-Abhängigkeit und nach dem räumlichen Auflösungsvermögen. Stets zeigt sich, daß bestimmte Eigenschaften mit sehr ähnlichem Erfolg entweder durch die Eigenschaften des reizleitenden Apparates oder durch die der sensiblen Endstrukturen verwirklicht werden können. Die Trennung dieser beiden Gegebenheiten bedingt, daß zwischen Eingangsreiz und Nutzreiz unterschieden werden muß. Der hier definierte Nutzreiz ist bislang nur für wenige Sinnesorgane genauer untersucht worden (Statocysten: LOWENSTEIN u. ROBERTS, v. HOLST; Pacinische Körperchen: HUBBARD; Thermoreceptoren HENSEL: [1]; Gehör: BÉKÉSY [2]). Es ist zu hoffen, daß dieser Fragestellung in den nächsten Jahren intensiver nachgegangen wird. Eine ähnliche Unterscheidung, wie sie hier zwischen Eingangs- und Nutzreiz getroffen wird, schlägt auch v. HOLST vor: Er unterscheidet zwischen organadäquatem und receptoradäquatem Reiz.

Die Überlegungen der letzten beiden Abschnitte gehen von der Vorstellung aus, daß die Erregung ein selbständiger Zustand der Sinneszelle ist. Der Reiz ist nicht die Ursache einer Erregung, er bestimmt lediglich ihren Verlauf, er steuert sie (vgl. HENSEL [1]). Zu dieser Auffassung zwingt eine Reihe hinlänglich bekannter Tatsachen. Erinnert sei an die Spontanerregung zahlreicher Sinnesorgane.

In vielen Fällen wird eine Erregung allerdings schon als spontan bezeichnet, wenn lediglich nachgewiesen ist, daß kein Eingangsreiz vorhanden sein kann. Dies bedeutet aber noch lange nicht das Fehlen eines Nutzreizes. Die Erregung ist in solchen Fällen nicht notwendigerweise „spontan". Streckreceptoren der Krebse entladen sich auch bei völliger Entspannung des Receptormuskels; möglicherweise unterliegen die sensiblen Endstrukturen einer gewissen Grundspannung (EYZAGUIRRE u. KUFFLER [1]). In anderen Fällen ist aber ohne Zweifel nachgewiesen, daß selbst bei völligem Fehlen eines Nutzreizes die sensiblen Endigungen Erregung bilden (vgl. für Statocysten SCHÖNE).

Fernerhin sind Sinnesorgane bekannt, wo trotz vorhandenem Reiz die Erregung erlischt, nicht nur dann, wenn der Reiz zu klein ist (unterschwellig), sondern auch wenn er über ein gewisses Maß ansteigt (Thermoreceptoren: vgl. HENSEL [1]; Streckreceptoren der Krebse: WIERSMA, FURSHPAN und FLOREY). Die Auffassung, daß die Erregung ein selbständiger Zustand der Zelle ist und durch den Reiz lediglich modifiziert wird, hat somit eine weitere Gültigkeit als die andere, wonach der Reiz die auslösende Ursache der Erregung ist. Sie umschließt den Fall, daß bei fehlendem oder zu kleinem Reiz keine Erregung auftritt, als einen möglichen Grenzfall.

Der Eingangsreiz eines Sinnesorgans ist außer durch seine Größe und seinen zeitlichen Verlauf unter Umständen erst durch Angabe weiterer Größen vollständig beschrieben. Er kann bei gleicher Intensität und gleichem zeitlichen Verlauf in bestimmten Parametern variieren, ohne daß er seinen Charakter als Eingangsreiz verliert: Solche Parameter werden als Reizqualität bezeichnet. Beispiele für Reizqualitäten sind die

Wellenlänge des Lichtes für das Auge und die Tonfrequenz für das Ohr. Jede Reizqualität muß, sofern sie von anderen unterschieden werden kann, andere Erregungszustände im Sinnesorgan bewirken, obschon die Reizenergie gleich sein kann. Auch diese Überlegung zeigt, daß mit der Angabe einer Energie allein der Reiz eines Sinnesorgans nicht hinreichend erfaßt ist.

C. Die Empfindlichkeit

Wird die Größe des Eingangsreizes momentan um einen bestimmten Betrag erhöht — *Sprungreiz* —, so steigt die Erregung nach einer Latenzzeit zunächst stark an und erreicht ein Maximum. Anschließend nimmt sie bei den meisten bislang bekannten Sinnesorganen wieder mehr oder minder ab und erreicht ein neues Niveau, dessen Höhe von der Reizintensität abhängt. *Das Maximum der Erregung nach einem Sprungreiz soll als Erregungsspitze bezeichnet werden; falls sich die Erregung auf einen konstanten Wert einspielt oder ohne systematische Trift um einen Mittelwert schwankt, soll dieser Wert die stationäre Erregung genannt werden* (BURKHARDT [1]). *Der gesamte Zeitverlauf der Erregung nach einem Sprungreiz kann in Anlehnung an den technischen Sprachgebrauch die Übergangsfunktion (des Sinnesorgans) genannt werden; das jeweilige Momentanverhältnis zwischen Erregung und Reiz: Übergangsverhältnis.*

Ein Sinnesorgan ist um so empfindlicher, je größer das Verhältnis zwischen Erregungsänderung und Reizänderung ist. Um aus dieser Aussage ein quantitatives Maß für die Empfindlichkeit zu gewinnen, müssen geeignete Meßverfahren festgelegt sein. Elektrophysiologisch kann das Verhältnis zwischen Erregung und Reiz zum Zeitpunkt der Erregungsspitze bestimmt werden: *phasische Empfindlichkeit*; oder zum Zeitpunkt der stationären Erregung: *statische Empfindlichkeit.* Die elektrophysiologisch bestimmte Empfindlichkeit kann in absoluten Werten angegeben werden; bei Auswertung der Receptorpotentiale in mV/Reizeinheit; bei Auswertung der Nervenimpulse in Frequenz/Reizeinheit. Aus Verhaltensversuchen kann im allgemeinen kein absolutes Maß für die Empfindlichkeit gewonnen werden. Dennoch ist es möglich, Veränderungen der Empfindlichkeit zu untersuchen und in relativen Einheiten anzugeben. Hierzu wird irgendeine genau fixierte Reaktion vorgegeben und die zum Erreichen der Reaktion nötige Reizintensität bestimmt. Ändert sich die Empfindlichkeit, so muß die Reizintensität verändert werden; der Kehrwert der Reizintensität ist ein Maß für die (relative) Empfindlichkeit. Auch das umgekehrte Verfahren ist möglich. Vorgegeben wird die Reizintensität, und die Größe der Reaktion wird als Maß der relativen Empfindlichkeit gewertet. Mit beiden Methoden werden beispielsweise der Verlauf spektraler Empfindlichkeitskurven bei Augen (vgl. G. SCHNEIDER) oder Änderungen der Empfindlichkeit in Abhängigkeit vom Adaptationszustand eines Sinnesorgans bestimmt. Die beiden

Methoden liefern identische Kurven, wenn der Zusammenhang zwischen Reiz und Erregung linear ist; bei nichtlinearem Zusammenhang zwischen Reiz und Erregung müssen sich die Kurven notwendigerweise unterscheiden. Selbstverständlich können diese beiden Methoden zur relativen Bestimmung der Empfindlichkeit auch bei elektrophysiologischen Versuchen angewandt werden.

Vielfach gibt man als Größe der Erregung die Schwellenerregung oder die Reaktionsschwelle vor. Hierdurch wird die Schwellenempfindlichkeit bestimmt. In Verhaltensversuchen kann entsprechend eine Unterschiedsreaktion vorgegeben werden, dies Verfahren führt dann zur Unterschiedsempfindlichkeit.

Die Empfindlichkeit eines Sinnesorgans ist nicht identisch mit dem Übergangsverhältnis, also dem Quotienten zwischen elektrophysiologisch gemessener Erregung und Reiz (der sich fortlaufend mit der Zeit verändert). Die elektrophysiologisch gemessene phasische Empfindlichkeit ergibt sich zwar aus dem Übergangsverhältnis in einem speziellen Zeitpunkt, nämlich dem der Erregungsspitze; entsprechend die statische Empfindlichkeit aus dem Übergangsverhältnis zum Zeitpunkt der stationären Erregung. Für Verhaltensversuche bleibt jedoch vollkommen offen, in welcher Weise der Erregungsverlauf im ZNS für die Steuerung der Reaktion ausgewertet wird.

Neben der hier erwähnten phasischen und statischen Empfindlichkeit wird gelegentlich auch von dem Begriff einer dynamischen Empfindlichkeit Gebrauch gemacht (vgl. HENSEL [2]). Bei einigen Sinnesorganen hängt die Höhe der Erregungsspitze von der Anstiegsgeschwindigkeit zeitlich zunehmender Reizintensitäten ab. Als dynamische Empfindlichkeit wird in solchen Fällen das Verhältnis zwischen der Erregungsspitze und der zeitlichen Veränderung des Reizes bezeichnet; die dynamische Empfindlichkeit hat somit die Dimension:

$$\frac{\text{Frequenz}}{\text{Reizeinheit/Zeit}} = \text{Frequenz} \cdot \text{Zeit/Reizeinheit.}$$

D. Die quantitativen Beziehungen zwischen Reiz und Erregung

Bei festgelegter Reizqualität hängen die Stärke und der zeitliche Verlauf der Erregung von folgenden Größen ab:

a) dem zeitlichen Ablauf des Reizes,
b) zeitlichen Zustandsänderungen im Sinnesorgan,
c) der Reizintensität,
d) der räumlichen Verteilung des Reizes.

Wenn die Abhängigkeit der Erregung von einer dieser vier Größen bestimmt wird, während die anderen konstant gehalten werden, so ergibt sich jeweils eine charakteristische mathematische Funktion. *Solche Kurven, welche den quantitativen Zusammenhang zwischen Reiz und Erregung beschreiben, können die Kennlinien eines Sinnesorgans genannt werden,* Bestimmte ausgezeichnete Punkte der Kennlinien, wie z. B. eine Absolutschwelle oder Verschmelzungsfrequenz, können entsprechend als *Kennwerte* bezeichnet werden. Die Kennwerte verschiedener Sinnesorgane lassen sich im allgemeinen nicht ohne weiteres vergleichen; als eine der wenigen Ausnahmen sei der Absolutwert der Schwellenenergie genannt, der für eine Reihe von Sinnesorganen bei etwa 10^{-17} Wattsec liegt (AUTRUM [3]). Ein Vergleich der Kennlinien hingegen zeigt, daß eine Gruppierung der Sinnesorgane nach bestimmten Formen der Kennlinien möglich ist.

a) Der zeitliche Ablauf des Reizes

Die meisten vorliegenden Untersuchungen über den Einfluß zeitlicher Faktoren des Reizes auf die Erregung lassen sich nach folgenden Fragestellungen aufgliedern: 1. Wie hoch ist die Verschmelzungsfrequenz eines Sinnesorgans, also wie schnell dürfen Reizwechsel aufeinander folgen, bevor diese Reizwechselfolge zu einem Dauerreiz für das Sinnesorgan verschmilzt (Als Bsp.: AUTRUM u. STÖCKER; KEIDEL [2]) ? 2. Welches sind die Grenzen der Gültigkeit des Bunsen-Roscoeschen Reizmengengesetzes, wonach die Größe der Erregung proportional dem Produkt aus Reizintensität und Dauer des Reizes ist (Als Bsp.: AUTRUM [5]) ? 3. Wie hängt die Erregung von der Anstiegsgeschwindigkeit zeitlich zunehmender Reizintensitäten ab? Wie klein darf die Anstiegsgeschwindigkeit werden, bevor Einschleichphänomene auftreten? 4. Wie sieht die Übergangsfunktion eines Sinnesorgans bei Sprungreizen aus und wie hängt allgemein der zeitliche Erregungsverlauf mit dem zeitlichen Reizverlauf zusammen? Als Beispiele für eingehendere Analysen mit dieser Fragestellung seien Arbeiten von KATZ [1], BULLOCK u. DIECKE, HUBBARD, GRAY und SATO, HENSEL [1, 2, 3], KEIDEL [2] und PRINGLE u. WILSON genannt. Daß hier nur wenige experimentelle Arbeiten vorliegen, hängt zum Teil damit zusammen, daß diese Fragestellung erst durch die regeltheoretische Betrachtungsweise der Sinnesorgane in den Blickpunkt des Interesses geraten ist; zum anderen auch mit der prinzipiellen Schwierigkeit zu erkennen, welchen Einfluß auf die Erregung der Zeitverlauf des Reizes und welchen die zeitlichen Zustandsänderungen des Sinnesorgans haben. Eine Einteilung der Sinnesorgane kann deshalb nur in sehr groben Zügen gegeben werden und muß später im Zusammenhang mit dem Adaptationsverhalten nochmals diskutiert werden.

Phasische Receptoren. Eine Reihe von Sinnesorganen spricht auf einen Sprungreiz nur mit einer kurzen Periode von Nervenimpulsen an, die stationäre Erregung ist null: Pacinische Körperchen (GRAY u. MALCOLM, GRAY u. SATO), phasische Streckreceptoren der Krebse (WIERSMA, FURSHPAN u. FLOREY), einige Hautreceptoren der Säuger (KEIDEL [2]) sowie Receptoren des Johnstonschen Organs von *Calliphora* (BURKHARDT [3]). Werden statt Sprungreizen zeitlich mehr oder weniger rasch ansteigende Reize geboten, so hängt die Höhe der Erregungsspitze von der Anstiegsgeschwindigkeit des Reizes ab (für Streckreceptoren der Krebse vgl. EYZAGUIRRE u. KUFFLER [1]).

Phasisch-tonische Receptoren. Der überwiegende Teil bisher untersuchter Receptoren beantwortet einen Sprungreiz mit einer Erregungsspitze und einer anschließenden stationären Erregung. Die Höhe der Erregungsspitze und die stationäre Erregung hängen von der Reizstärke ab. Werden Reize verschiedener Anstiegssteilheit geboten, so nimmt die Höhe der Erregungsspitze mit zunehmender Anstiegsgeschwindigkeit des Reizes zu, die stationäre Erregung bleibt jedoch unabhängig von der Anstiegsgeschwindigkeit des Reizes. Typische Beispiele solcher Receptoren sind die Thermoreceptoren der Säuger und Elasmobranchier (HENSEL [1, 2, 3]), tonische Streckreceptoren der Krebse (FLOREY [1, 2], EYZAGUIRRE u. KUFFLER [1, 2], BURKHARDT [1]), Muskel- und Sehnenspindeln der Säuger und Amphibien (MATTHEWS, KATZ [1, 2], Zusammenfassungen hierfür bei GRANIT [1] und BURKHARDT [4]).

Tonische Receptoren. In einigen Fällen scheinen Receptoren auf einen Sprungreiz nur mit einer veränderten stationären Entladung anzusprechen, ohne daß eine Erregungsspitze entsteht. Die stationäre Entladung ist dann auch hier unabhängig von der Anstiegsgeschwindigkeit linear ansteigender Reize (Schmerzreceptoren vgl. KEIDEL [2], Stellungsreceptoren bei Krebsen: WIERSMA u. BOETTIGER).

In Analogie zur regeltheoretischen Bezeichnung für Steuerglieder werden phasische Receptoren als D-Typ (Differentialquotient-Empfänger), phasisch-tonische als PD-Typ (Proportional-Differential-Empfänger) und tonische als P-Typ (Proportional-Empfänger) bezeichnet.

Neben der Reihe: phasischer, phasisch-tonischer und tonischer Receptor wird oft eine andere Reihe aufgestellt, die im gleitenden Übergang

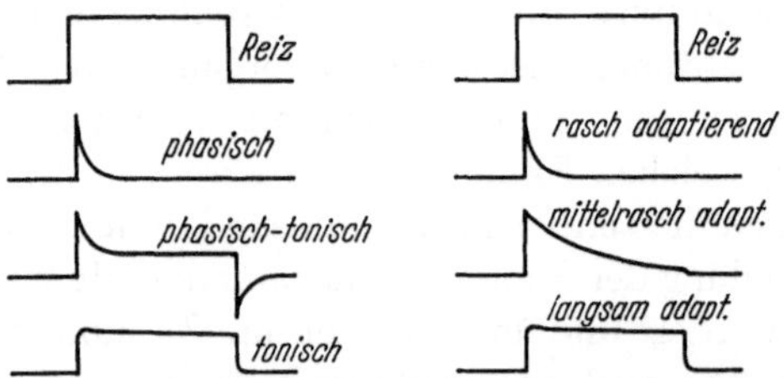

Abb. 1. Schematische Darstellung des Zeitverhaltens von Sinnesorganen. Jeweils oberster Kurvenzug: Zeitverlauf des Reizes, untere Kurven: Zeitverlauf der Entladungsfrequenz der Receptoren. Links: Übergang vom phasischen zum tonischen Typ, als Zwischenglied der phasisch-tonische Receptor. Rechts: Übergang vom phasischen zum tonischen Verhalten durch unterschiedliche Adaptationsgeschwindigkeiten

vom phasischen zum tonischen Receptor führt. Als Zwischenglieder werden Receptoren betrachtet, welche nach einem Sprungreiz kürzere oder längere Zeit mit einer abklingenden Erregung antworten, ohne daß ein stationäres Niveau erreicht wird. Abb. 1 zeigt schematisch die beiden Typenreihen. Der Grund für das unterschiedliche Verhalten der Receptoren wird bei beiden Reihen meist in Adaptationserscheinungen gesucht (vgl. KEIDEL [2], GRANIT [2]). Phasische Receptoren adaptieren nach dieser Auffassung extrem rasch und vollständig, tonische Receptoren überhaupt nicht. Der phasisch-tonische Receptor zeigt eine partielle (unvollständige) Adaptation, die Übergangstypen der zweiten Reihe unterscheiden sich durch verschiedene Adaptationsgeschwindigkeiten. Inwieweit diese Typen tatsächlich auf unterschiedliches Adaptationsverhalten zurückgeführt werden können, soll im nächsten Abschnitt abschließend besprochen werden.

b) Die zeitlichen Zustandsänderungen des Sinnesorgans und der Begriff der Adaptation

Wird der Eingangsreiz eines Sinnesorgans sprungartig erhöht, so klingt bei fast allen bisher bekannten Receptoren die Erregung nach der anfänglichen Spitze wieder ab, obschon der Eingangsreiz unverändert angeboten wird. *Das Abklingen der Erregung soll als Erregungsabfall bezeichnet werden, unabhängig davon, welches seine Ursachen sind.* Der Erregungsabfall bedeutet zunächst, daß das Verhältnis zwischen Erregung und Eingangsreiz, also das Übergangsverhältnis des Sinnesorgans, zeitlich abnimmt. An zwei Stellen können einzeln und gleichzeitig die Ursachen hierfür liegen: Es kann sich das Verhältnis zwischen Nutzreiz und Eingangsreiz zeitlich verändern (dieses Verhältnis kann als *Reiznutzung* bezeichnet werden). Es kann sich aber auch das Verhältnis zwischen Erregung und Nutzreiz verschieben (also das Übergangsverhältnis der sensiblen Endstrukturen). An beiden Stellen können außerdem verschiedene Mechanismen wirksam sein.

1. Die dynamische Reizminderung

Der reizleitende Apparat kann auf Grund seiner physikalischen Eigenschaften bewirken, daß der sprungförmige Eingangsreiz in einen zeitlich abklingenden Nutzreiz verformt wird: dynamische Reizminderung.

Neben der dynamischen Reizminderung kann auch eine statische Reizminderung beobachtet werden: Der dioptrische Apparat von Augen absorbiert und reflektiert zeitunabhängig einen Teil der einfallenden Lichtenergie, mindert also den Eingangsreiz. Weiterhin kann bei der dynamischen Veränderung des Eingangsreizes außer einer Minderung auch eine zeitweilige Erhöhung vorkommen, als Beispiel sei das Resonanz-Einschwingen von Mechanoreceptoren erwähnt. Für die folgenden Betrachtungen ist jedoch nur eine einsinnige zeitliche Abnahme der Reiznutzung von Bedeutung, da sie eine der Ursachen des Erregungsabfalls sein kann.

Pacinische Körperchen im Mesenterium der Katze beantworten mechanische Reize. Werden sie sprungartig verformt, so reagieren sie mit einem, selten auch mit wenigen Spikes (Gray u. Malcolm). Hubbard konnte zeigen: Wird das Korpuskel gequetscht, so werden die konzentrisch aufgebauten Lamellen verformt. Ein rechteckiger Reiz bewirkt nur bei außen gelegenen Lamellen eine anhaltende starke

Deformation. Weiter innen gelegene Lamellen werden zu Beginn des Reizes stark deformiert, die Deformation läßt dann mit zunehmender Zeit etwas nach; ganz innen gelegene Lamellen werden schließlich überhaupt nur während einer kurzen Anfangsperiode deformiert. Von der äußeren Verformung wird lediglich eine Druckwelle in das Innere geleitet. Der reizleitende Apparat verformt somit den rechteckförmigen Eingangsreiz in einen rasch abklingenden Nutzreiz. Die „Adaptation" der Pacinischen Körperchen beruht demnach zu einem großen Teil auf der dynamischen Reizminderung. — Nur in wenigen Fällen wurde bisher für Mechanoreceptoren eingehend analysiert, welchem Reiz die sensiblen Endstrukturen unterliegen; der beschriebene Fall zeigt deutlich, daß mit erheblichen Veränderungen des Eingangsreizes gerechnet werden muß. Die zeitliche Reizumformung bei den Pacinischen Körperchen unterscheidet sich prinzipiell nicht von der bei einigen anderen Sinnesorganen, die seit langem analysiert sind und bei denen man in der Regel nicht von einer Adaptation spricht. Als Beispiel sei das Bogengangssystem der Wirbeltiere genannt. Bei Beginn einer Kopfdrehung bleibt die Endolymphe infolge ihrer Trägheit zurück, hierdurch werden die Cupulae ausgelenkt. Bei weiter anhaltender Drehung wird die Endolymphe infolge von Rückstellkräften mitgenommen, die Cupulae kehren in ihre Ausgangslage zurück (vgl. LOWENSTEIN u. SAND).

Der Erregungsabfall beruht im Falle der dynamischen Reizminderung auf der abklingenden Reiznutzung, regeltheoretisch gesprochen auf der Übergangsfunktion des reizleitenden Apparates. Dieser Prozeß läuft notwendigerweise, und durch die physikalischen Eigenschaften des reizleitenden Apparates bedingt, ein für allemal in der gleichen Form ab. Hier von einer Empfindlichkeitsänderung oder Adaptation zu sprechen, ist nicht gerechtfertigt.

2. *Die Reizkontrolle*

Die Eigenschaften des reizleitenden Apparates können bei einigen Sinnesorganen durch physiologische Einflüsse verändert werden. Diese Vorgänge werden entweder durch ein Hilfsorgan gesteuert oder durch die Erregung des Sinnesorgans selbst oder schließlich durch eine Efferenz vom ZNS. Im Gegensatz zur dynamischen Reizminderung wird hier der zeitliche Ablauf nicht durch physikalisch bedingte und stets gleich ablaufende Vorgänge bestimmt, sondern kann mit unterschiedlicher Wirkung gesteuert werden. *Kontrollieren physiologische Vorgänge die Reiznutzung, so soll dies als Reizkontrolle bezeichnet werden.* Die *Reizkontrolle* arbeitet oft so, daß die Reiznutzung und damit die Erregung zeitlich abklingt: Die durch den Eingangsreiz erhöhte Erregung löst eine Efferenz aus, welche den reizleitenden Apparat so steuert, daß die Reiznutzung kleiner wird (Gegenkopplung). Der Nutzreiz und die

Erregung nehmen hierdurch nach der sprungartigen Erhöhung des Eingangsreizes ab; die Drosselung des reizleitenden Apparats regelt sich schließlich auf einen konstanten Wert ein, der von der Höhe des Eingangsreizes abhängt. Der zeitliche Verlauf der Reizkontrolle ist(bei hinreichender Dämpfung) einsinnig und kann den Verlauf des Erregungsabfalls mitbestimmen.

Ein geläufiges Beispiel für die Reizkontrolle bieten die Wirbeltieraugen. Wird das Auge belichtet, so kontrahiert sich die Irismuskulatur; die auf die Retina fallende Lichtmenge wird vermindert. Bei Fischen und teilweise noch bei Amphibien und Reptilien (vgl. HERTER) ist die Irismuskulatur selbst lichtempfindlich, die Reiznutzung wird also durch einen Hilfsapparat gesteuert. Bei Säugern ist die Irismuskulatur nicht mehr lichtempfindlich, hier löst die Erregung der Retina eine Efferenz aus, welche über motorische Nervenfasern die Pupillenweite regelt (vgl. STARK u. SHERMAN).

Der beschriebene Pupillenreflex ist ein Beispiel für eine im wesentlichen einsinnige Reizkontrolle: Die Erhöhung des Eingangsreizes bewirkt eine abnehmende Reiznutzung. Nicht in allen Fällen muß der Vorgang der Reizkontrolle einsinnig ablaufen. Die Muskelspindeln der Wirbeltiere (Zusammenfassung bei BURKHARDT [4]) zeigen ebenfalls eine Reizkontrolle, die aber in beliebiger Richtung arbeiten kann. Die Entladungsfrequenz der Muskelspindeln hängt von zwei Bedingungen ab: Sie ist etwa proportional der Längenänderung des Muskels, in welchem die Spindel eingebettet ist; sie hängt außerdem von der Vorspannung ab, welche innerhalb der Spindel gelegene Muskelfasern (Weismannsches Bündel oder intrafusale Muskelfasern) den sensiblen Endstrukturen erteilen. Die Vorspannung kann durch ein eigenes System von Nervenfasern (γ-Fasern) in weiten Grenzen verschoben werden. Bei gleichem Eingangsreiz, also einer bestimmten Länge des Hauptmuskels, kann die Erregung der Spindel je nach der Aktivität des γ-Systems recht unterschiedliche Werte annehmen. Auch hier wird wie beim Auge das Verhältnis zwischen Nutzreiz und Eingangsreiz durch eine efferente Kontrolle verschoben; jedoch ist die γ-Efferenz nicht starr und einsinnig mit der Höhe des Eingangsreizes gekoppelt; sie kann je nach den ablaufenden Bewegungen und Haltungen des Tieres in beliebiger Richtung verschoben werden. Welche Bedeutung diese vom Eingangsreiz unabhängige Reizkontrolle hat, wird in einem der späteren Abschnitte (S. 256) erörtert werden.

Auch bei der Pupille ist die Reizkontrolle nicht ganz starr an den Reiz gebunden. Wie bekannt, kann unabhängig von der Lichtintensität die Pupillenöffnung über das vegetative Nervensystem reguliert werden. Die Reizkontrolle kann also die Ursache eines Erregungsabfalls sein, sie muß es aber nicht: Der Begriff der Reizkontrolle deckt sich daher nur in gewissen Grenzen mit dem alten Begriff der physikalischen Adaptation.

3. *Die dynamische Erregungsminderung*

Bei der dynamischen *Reiz*minderung bewirkt eine zeitlich abklingende Übergangsfunktion des reizleitenden Apparates den Erregungsabfall. Analoge Erscheinungen treten möglicherweise auch in den sensiblen Endstrukturen auf. Die Übergangsfunktion der sensiblen Endstrukturen muß zum Nachweis solcher Vorgänge bei sprungförmigem Verlauf des Nutzreizes untersucht werden, eine Forderung, die beispielsweise für isolierte Thermoreceptoren erfüllbar ist (HENSEL [1, 2]). Die Thermoreceptoren zeigen unter diesen Reizbedingungen das Verhalten phasisch-tonischer (PD-) Receptoren: eine Erregungsspitze und anschließend ein angenähert exponentieller Abfall zum Niveau der stationären Erregung; ist die stationäre Erregung erreicht, so bleibt sie über lange Zeiten bei unverändertem Reiz vollkommen unverändert. Zur Erklärung des Zeitverlaufs der Übergangsfunktion von Thermoreceptoren liegt eine tragfähige Hypothese vor (SAND; BURTON; HENSEL [1]): Es wird angenommen, daß die Entladungsfrequenz vom Gleichgewicht zweier gegensinnig wirkender Vorgänge abhängt: Einer der Prozesse wirkt erregend, der andere hemmend. Welcher Art diese Prozesse sind, ist für die Überlegung selbst nicht entscheidend; diskutiert wurden bisher: Reaktionsgeschwindigkeiten, Konzentrationen von Erregungssubstanzen, Generatorpotential und Schwelle (SAND, BURTON, HENSEL [1], FLOREY [2], BURKHARDT [1]). Bei einer sprungartigen Veränderung stellen sich beide Prozesse in einem angenähert exponentiellen Verlauf auf den neuen, vom Reiz abhängigen stationären Wert ein. Unterschiedlich ist für die beiden Prozesse die Einstell-Zeitkonstante und die Reizabhängigkeit des stationären Wertes. Dieser Ansatz genügt, um die Entladungskurven bei verschiedenen Reizbedingungen mit guter Näherung zu erklären.

Aus der Übergangsfunktion eines Steuerkörpers kann der zeitliche Verlauf der Ausgangsgröße (hier die Erregung) bei einem beliebigen zeitlichen Verlauf der Eingangsgröße (hier der Reiz) vorausberechnet werden, vgl. hierzu OPPELT; BURKHARDT [5]. Berechnet man für Thermoreceptoren den Erregungsverlauf bei Temperaturänderungen verschiedener Anstiegssteilheit, indem man als Übergangsfunktion die Überlagerung der beiden exponentiellen Einstellvorgänge jener Theorie zugrunde legt, so decken sich die berechneten Kurven recht gut mit den experimentell ermittelten Werten (HENSEL [1, 3]).

Ansätze zu einer experimentellen Analyse, welche Vorgänge in den Receptorzellen dem formalen Ansatz genügen, liegen bisher nur für Streckreceptoren der Krebse vor (FLOREY [2], BURKHARDT [1]).

Beispiele für eine dynamische Erregungsminderung bieten alle phasisch-tonischen Receptoren: Die Receptoren also, welche gemeinhin als partiell adaptierend bezeichnet werden. Bei derartigen Receptoren ist die Übergangsfunktion für einen Reiz, der zu unterschiedlichen Zeiten geboten wird *(Testreiz)* unabhängig davon, ob zwischendurch ein langer und intensiver Reiz *(Konditionsreiz)* auf das Sinnesorgan eingewirkt hat. Es ist daher unwahrscheinlich, daß die Empfindlichkeit des Sinnesorgans durch Konditionsreize veränderbar ist, zumindest sind die Empfindlichkeitsänderungen sehr rasch nach dem Ende eines Konditionsreizes verschwunden. Diese Eigenschaft hat die dynamische Erregungsminderung mit der dynamischen Reizminderung gemeinsam: Ihr Verlauf ist fixiert und starr an den Reiz gebunden. *Die dynamische Erregungsminderung bedeutet also keine Empfindlichkeitsveränderung, sondern eine dynamische Einstellung des Receptors auf den neuen Reiz.* Ähnlich kann ein mäßig gedämpftes technisches Meßinstrument beim Anlegen der Meßgröße zunächst über den Endausschlag hinausschwingen; niemand würde die Rückkehr des Zeigers auf den Endwert als eine Empfindlichkeitsveränderung bezeichnen.

4. Die Empfindlichkeitsminderung

Bei einer Reihe von Receptoren hängt die durch den Testreiz ausgelöste Erregung von der vorangehenden Einwirkung eines Konditionsreizes ab. Wird der Testreiz eine vorgegebene Zeit nach dem Ende des Konditionsreizes gesetzt, so wird die Testerregung um so kleiner ausfallen, je höher die Konditions-Intensität und je länger der Konditionsreiz war. Mit zunehmendem zeitlichen Abstand des Testreizes vom Konditionsreiz wird die Testerregung wieder größer. *Die phasische und die statische Empfindlichkeit solcher Sinnesorgane verändert sich unter der Einwirkung von Reizen: Empfindlichkeitsminderung.* Bei dem hier beschriebenen Testverfahren für die Empfindlichkeitsminderung wird die Empfindlichkeit erst nach dem Konditionsreiz geprüft. Man muß aber annehmen, daß die Empfindlichkeitsänderung bereits während des Konditionsreizes selbst wirksam ist. Die während des Konditionsreizes ablaufende Erregung wird so unter dem Einfluß der Empfindlichkeitsminderung abnehmen.

Es fragt sich natürlich, ob zwischen der dynamischen Erregungsminderung und der Empfindlichkeitsminderung ein prinzipieller Unterschied besteht oder ob es lediglich bei verschiedenen Receptoren Unterschiede in der Geschwindigkeit der ablaufenden Vorgänge gibt. Die dynamische Erregungsminderung wäre dann eine recht rasch entstehende und rasch wieder abklingende Empfindlichkeitsminderung (bzw. umgekehrt: die Empfindlichkeitsminderung eine recht träge ablaufende dynamische Einstellung). Da sich die Erregung in allen bekannten Fällen fortlaufend mit der Zeit verändert, ist sie ohne weitere Zusatzannahme kein Maß für die Empfindlichkeit eines Sinnesorgans: Aus zeitlichen Veränderungen der Erregung darf keineswegs direkt auf Empfindlichkeitsänderungen geschlossen werden. Eine mögliche Methode zur Prüfung der Frage wäre das Verfahren, Testreize konstanter Intensität und kurzer Dauer dem Verlauf des Konditionsreizes zu überlagern. Eine Untersuchung dieser prinzipiellen Frage ist für verschiedene Receptorentypen dringend nötig. Da zur Zeit derartige Versuche noch nicht vorliegen, bleibt der Unterschied zwischen dynamischer Erregungsminderung und Empfindlichkeitsminderung vorläufig ein Postulat.

Es gibt jedoch einige Gründe, welche die ausgesprochene Vermutung bestärken: Receptoren vom phasisch-tonischen Typ, welche hier als Vertreter für die dynamische Erregungsminderung angeführt wurden, zeigen bei Sprungreizen stets einen monotonen Erregungsabfall: Nach der Erregungsspitze nimmt die Erregung zunächst steil, dann immer flacher ab. Receptoren, welche eine langfristige Empfindlichkeitsminderung aufweisen, zeigen demgegenüber bei hohen Reizintensitäten vielfach einen oszillierenden Erregungsverlauf. Nach der Erregungsspitze sinkt die Erregung rasch und stark ab, oft tritt sogar eine vorübergehende Entladungspause ein; anschließend nimmt die Erregung wieder zu, erreicht ein zweites kleineres Maximum und nimmt dann schließlich monoton endgültig ab. Beispiele für einen solchen Erregungsverlauf bieten Augen (HARTLINE u. GRAHAM) und Infrarotreceptoren (BULLOCK u. DIECKE); ähnliche Erregungsoscillationen können auch von Geruchs- und Geschmacks-Receptoren abgeleitet werden (Lit. bei D. SCHNEIDER sowie HODGSON u. ROEDER). Ein weiterer Unterschied zwischen Organen mit dynamischerErregungsminderung und Empfindlichkeitsminderung scheintFolgendes zu sein: Der zeitliche Erregungsverlauf bei sprungförmigem Reizabbruch ist für Receptoren mit dynamischer Erregungsminderung angenähert spiegelbildlich zum Erregungsverlauf bei Reizbeginn (vgl. HENSEL [3] für Thermoreceptoren; FLOREY [1] und BURKHARDT [1] für Streckreceptoren). Ein derartiges Verhalten kann auf Grund der im vorigen Abschnitt besprochenen Theorie vorausberechnet werden. Bei Receptoren mit langfristiger Empfindlichkeitsminderung gleicht der Erregungsverlauf nach Reizende hingegen keineswegs dem Spiegelbild des Erregungsverlaufs bei Reizbeginn (vgl. BULLOCK u. DIECKE, GRANIT [1]).

Für das Zustandekommen von langfristigen Empfindlichkeitsminderungen liegen zahlreiche Theorien vor, die hier im einzelnen nicht erörtert werden können; verwiesen sei auf die teilweise recht ausführlichen Darstellungen von GRANIT [1, 2], BURTON, KEIDEL [2] und RANKE.

Eine analoge Unterscheidung wie für die sensiblen Endstrukturen mit dynamischer Erregungsminderung und Empfindlichkeitsminderung muß möglicherweise auch für den reizleitenden Apparat getroffen werden. Bei einigen Mechanoreceptoren spielen neben elastischen Eigenschaften auch plastische eine Rolle. Unter der Einwirkung langer und starker mechanischer Belastung wird der reizleitende Apparat plastisch verformt (vgl. EYZAGUIRRE u. KUFFLER [1] und BURKHARDT [1] für Streckreceptoren der Krebse). Das Ausmaß plastischer Verformung muß von der Intensität und der Dauer der Beanspruchung abhängen und den Reiz längere Zeit überdauern (bis durch aktive physiologische Vorgänge der alte Zustand wiederhergestellt ist). Dieser Vorgang wird dann eine langfristige Empfindlichkeitsminderung des Sinnesorgans bewirken, allerdings durch Änderung der Eigenschaften des reizleitenden Apparates. Dieser bislang wenig beachtete Aspekt der Reizleitung liefert ein anschauliches Modell für den Unterschied zwischen dynamischer Erregungsminderung und Empfindlichkeitsänderung bei sensiblen Endstrukturen. Die plastische Dehnung von Muskeln ist ein passiver Vorgang; die Wiederherstellung des verkürzten Zustands beruht hingegen auf aktiven, nervös gesteuerten Vorgängen (JORDAN). Die Veränderungen in den beiden entgegengesetzten Richtungen werden so durch grundsätzlich verschiedene Mechanismen bewirkt. Möglicherweise ist dies der prinzipielle Unterschied zwischen der dynamischen Erregungsminderung und der Empfindlichkeitsminderung bei den sensiblen Endstrukturen: Bei den für die dynamische Erregungsminderung diskutierten Einstellvorgängen zwischen zwei gegenläufigen Prozessen spielen sich gleichartige und reversible Prozesse in entgegengesetzter Richtung bei Reizbeginn und Reizende ab. Bei einer Empfindlichkeitsminderung scheint hingegen möglich, daß die zugrundeliegenden Mechanismen bei der Minderung der Empfindlichkeit und bei ihrer Wiederherstellung unterschiedlicher Art sind.

5. Die Erregungskontrolle

Bei vielen Sinnesorganen kann die Erregung durch eine efferente Regulierung vom Zentralnervensystem verändert werden. Da die Efferenz im Gegensatz zur Reizkontrolle nicht am reizleitenden Apparat, sondern an den sensiblen Endstrukturen angreift, soll dieser Vorgang als *Erregungskontrolle* bezeichnet werden. Die *Erregungskontrolle* kann genau wie die Reizkontrolle als Gegenkopplung wirken und auf diesem Weg einen Erregungsabfall erzeugen.

Für zahlreiche Sinnesorgane sind efferente Nervenfasern beschrieben worden, deren Endverzweigungen im Bereich der sensiblen Endstrukturen liegen (Auge vgl. Granit [3], Dodt; Streckreceptoren der Krebse Kuffler [1]; Ohr: Galambos [1]). In einigen Fällen konnte auch experimentell nachgewiesen werden, daß eine Erregung der efferenten Fasern die sensible Erregung abdrosselt. So zeigte Galambos [1], daß eine Entladung des Hörnerven der Katze teilweise oder ganz gehemmt werden kann, wenn bestimmte Bezirke der Medulla oblongata gereizt werden. Im wirksamen Bereich der Medulla liegt die Kreuzung des Tractus olivocochlearis, eines efferenten Faserzuges, der im Bereich der Haarzellen endet. Für die bereits mehrfach genannten Streckreceptoren der Krebse konnte Kuffler [1] die Hemmungsvorgänge bei Reizung der efferenten Fasern intracellulär in der sensiblen Nervenzelle studieren. Eine zusammenfassende Darstellung über die zentrifugale Kontrolle von Sinnesorganen gibt Hagbarth.

Diskussion des Zeitverhaltens der Sinnesorgane

In diesem Abschnitt wurden nur solche Vorgänge besprochen, die am Sinnesorgan selbst angreifen und einen Erregungsabfall verursachen können. Die Frage der zentralen Adaptation kann im Rahmen dieses Artikels nicht im einzelnen erörtert werden. Erwähnt sei lediglich, daß verschiedene Mechanismen bekannt sind, welche im ZNS, aber noch auf der afferenten Seite, den Einstrom sensibler Erregung vermindern können. Für *Calliphora,* deren Sehzellen keine nennenswerte Empfindlichkeitsänderung zeigen, konnte elektrophysiologisch eine Adaptation in optischen Neuronen höherer Ordnung nachgewiesen werden (Burkhardt [6]). Ähnliche Verhältnisse scheinen beim Wirbeltierohr und den zentralen akustischen Neuronen vorzuliegen (Galambos [2] sowie Erulkar, Rose u. Davis). Weiterhin seien Vorgänge wie die afferente Ermüdung (Franzisket) und die afferente Drosselung (Margret Schleidt) erwähnt.

Der Erregungsabfall im Sinnesorgan selbst kann, wie hier gezeigt wurde, mindestens fünf verschiedene Ursachen haben: Dynamische Reizminderung, Reizkontrolle, dynamische Erregungsminderung, Empfindlichkeitsminderung und Erregungskontrolle. Die dynamische Reizminderung und die Reizkontrolle greifen am reizleitenden Apparat an, dynamische Erregungsminderung, Empfindlichkeitsminderung und Erregungskontrolle an den sensiblen Endstrukturen; Abb. 2 gibt ein Schema hierfür.

Dynamische Reizminderung und dynamische Erregungsminderung haben trotz unterschiedlichem Angriffspunkt eine Reihe gemeinsamer Züge. In beiden Fällen handelt es sich um dynamische Einstellvorgänge auf die neue Reizsituation (welche in der Regeltheorie ihre direkte Analogie in der Übergangsfunktion von Reglern haben). Beide Vorgänge

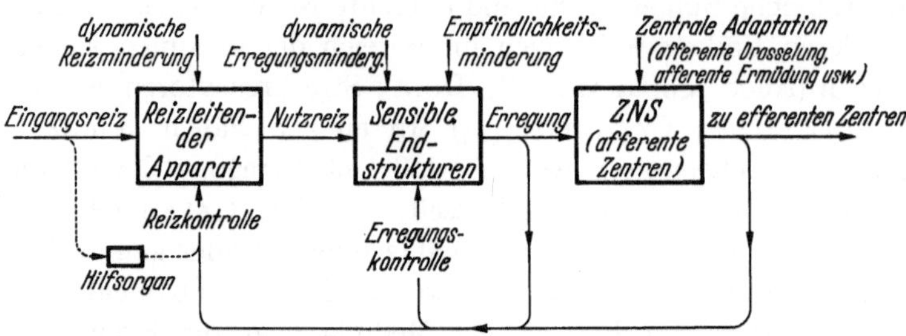

Abb. 2. Schematische Darstellung der Vorgänge, welche zu einem Erregungsabfall führen können

sind streng an den Reizwechsel gebunden, sie überdauern ihn um eine der eigengesetzlichen Trägheit des Einstellvorganges entsprechende Zeit. Beide Vorgänge laufen einsinnig ab: Einem stärkeren Reiz entspricht stets eine relativ stärkere Minderung der Erregung. Bei Reizbeginn und Reizende laufen gleichartige Vorgänge in entgegengesetzter Richtung ab, der Zeitverlauf ist daher etwa spiegelbildlich.

Ein zweites Paar sind die Vorgänge der Reiz- und Erregungskontrolle. Sie unterscheiden sich ebenfalls im Angriffspunkt, zeigen aber sonst eine Reihe gemeinsamer Züge. Reiz- und Erregungskontrolle sind nicht starr an den Reiz gebunden; da es sich um physiologisch kontrollierte Vorgänge handelt, können sie auch ohne Reiz auftreten, und sie wirken nicht notwendigerweise einsinnig, sondern können in beliebiger Richtung ablaufen. Beide Vorgänge spielen nicht nur für den Erregungsabfall eine Rolle, sondern auch für die funktionelle Verknüpfung des Sinnesorgans (vgl. S. 256). Wenn sie jedoch einsinnig wirken, so verursachen sie als Gegenkopplung einen Erregungsabfall.

Die Empfindlichkeitsminderung schließlich greift wie die dynamische Erregungsminderung an den sensiblen Endstrukturen an und ist einsinnig wirksam: Bei einer Erhöhung des Reizes bewirkt sie stets eine Abnahme der Erregung. Ferner ist sie ähnlich wie die dynamische Erregungsminderung an das Vorhandensein eines Reizes gebunden. Nur durch die Einwirkung eines Reizes nimmt die Empfindlichkeit ab. Im Gegensatz zur dynamischen Erregungsminderung überdauert die Empfindlichkeitsminderung jedoch den Reiz um eine längere Zeitspanne und kann so Nachwirkungen entfalten. Für die Empfindlichkeitsänderung darf nicht angenommen werden, daß *gleichartige* Vorgänge, welche in entgegengesetzter Richtung ablaufen, den Wiedergewinn der Empfindlichkeit nach Ende eines Konditionsreizes bewirken.

Die dynamische Reizminderung und die dynamische Erregungsminderung bewirken bei einem Sinnesorgan den Übergang von phasischer zu statischer Empfindlichkeit. Durch sie wird die Übergangsfunktion des Sinnesorgans im Detail bestimmt, sie wirken aber stets in der gleichen Weise und sind nicht in der Lage, zu verschiedenen Zeiten unterschiedliche Übergangsfunktionen zu erzeugen. Anders die Reiz- und Erregungskontrolle und die Empfindlichkeitsminderung. Selbstverständlich bestimmen sie auch Details der Übergangsfunktion und sind am Erregungsabfall beteiligt, also Ursachen für den Übergang von phasischer zu statischer Empfindlichkeit. Darüber hinaus jedoch können diese Mechanismen bewirken, daß zu verschiedenen Zeiten unterschiedliche Übergangsfunktionen vorhanden sind. **Ihre Einwirkung verschiebt den Gesamtverlauf der Übergangsfunktion, und damit wird sowohl die phasische als auch die statische Empfindlichkeit des Sinnesorgans beeinflußt.**

Die dynamische Reizminderung und die dynamische Erregungsminderung können unter dem Begriff: *dynamische Einstellvorgänge* zusammengefaßt werden; entsprechend die Reiz- und Erregungskontrolle unter dem Begriff: *zentrifugale Kontrolle.* **Der Begriff der Adaptation sollte nach Möglichkeit nicht auf die dynamischen Einstellvorgänge angewandt werden, sondern nur auf die Fälle einer zentrifugalen Kontrolle oder einer Empfindlichkeitsminderung. Das Wesen von Adaptationsvorgängen ist die Änderung der Empfindlichkeit eines Sinnesorgans, also eine Verschiebung der gesamten Übergangsfunktion.** Die häufig zu lesende Behauptung, daß der Erregungsabfall die elektrophysiologische Basis von Adaptationserscheinungen ist, muß nach dem Gesagten mit größter Vorsicht aufgenommen werden. Nicht jede zeitliche Abnahme der Erregung bedeutet eine Verschiebung der Empfindlichkeit eines Sinnesorgans; was im Einzelnen vorliegt, kann nur nach eingehender Analyse entschieden werden.

Bei fast allen Sinnesorganen bewirken mehrere Vorgänge gleichzeitig den Erregungsabfall. So ist für die Streckreceptoren der Krebse neben der Erregungskontrolle eine Reizkontrolle bekannt (KUFFLER [2]) und es muß hier sowohl mit einer dynamischen Reizminderung als auch einer ausgeprägten dynamischen Erregungsminderung gerechnet werden (FLOREY [1, 2], BURKHARDT [1]). Bei Wirbeltieraugen findet sich neben der Reizkontrolle durch den Pupillenreflex die Erregungskontrolle in der Retina (GRANIT [3], KUFFLER [3]) und vor allem die Empfindlichkeitsminderung in den Sehzellen. Beim Ohr spielt neben der bereits erwähnten

Erregungskontrolle eine Reizkontrolle durch die Muskulatur des Trommelfells und der Gehörknöchelchen eine Rolle. Die Zahl der Beispiele kann beliebig erweitert werden. Alle Vorgänge zusammen können einen erheblichen Umfang an Empfindlichkeitsverschiebungen bewirken (beim Wirbeltierauge über einen Bereich $1:10^6$, vgl. KOHLRAUSCH u. TEUFER). Die physiologische Bedeutung der Adaptation wird in späteren Abschnitten diskutiert (S. 251).

Ein Punkt, der schließlich eine eingehende Diskussion wert ist, ist die Frage nach dem Eingangsreiz extrem phasischer Receptoren. Betrachtet man eine Auslenkung (oder eine andere statische Größe) als den Reiz, so beantworten solche Receptoren den Eingangsreiz nur sehr kurze Zeit nach einem Reizwechsel; nach alter Auffassung adaptieren sie rasch und vollständig (vgl. KEIDEL [2]). Das gleiche gilt auch für eine Reihe anderer Receptoren. Das Bogengangssystem beantwortet eine plötzlich einsetzende Drehung konstanter Geschwindigkeit nur vorübergehend. Eine plötzliche Druckerhöhung, die anhält, beantwortet das Ohr nur mit einer kurzen Erregungsfolge. Als Eingangsreiz wird für diese beiden Organe jedoch von vornherein eine dynamische Größe definiert: Drehbeschleunigungen für das Labyrinth und Schallwechseldruck für das Ohr. Würde bei Mechanoreceptoren mit extrem phasischem Verhalten allgemein als Eingangsreiz nicht eine statische Größe definiert, sondern deren zeitliche Veränderung, so erwiese sich die Frage der Adaptation solcher Receptoren als Scheinproblem. Die Mehrzahl der Receptoren des Johnstonschen Organs von *Calliphora* spricht beispielsweise auf eine einmalige ruckartige Auslenkung der Antenne mit einem einzigen Aktionspotential an, adaptiert also scheinbar. Auf Schwingungen der Antenne reagieren sie jedoch ohne merkbare Veränderungen über lange Zeiten mit reizsynchronen Aktionspotentialen (BURKHARDT [3]): Sie zeigen gegenüber der dynamischen Größe keine Adaptation.

Die von HENSEL [1] aufgestellte Definition, wonach der adäquate Reiz eines adaptierenden Systems immer eine Reizgröße in Verbindung mit der Zeit ist, der adäquate Reiz eines nicht adaptierenden Systems nur die Reizgröße ohne den Zeitfaktor, bedarf unter diesem Gesichtspunkt erneuter Diskussion. Niemand wird daran zweifeln können, daß der Schallwechseldruck eine Reizgröße mit einem Zeitfaktor ist. Die Adaptationserscheinungen des Ohres werden aber nicht auf den Zeitfaktor Schwingungsdauer (als Kehrwert der Frequenz) bezogen, sondern auf die Zeitdauer der Einwirkung des Schalls. Ein scherzhafter Gedankenversuch scheint geeignet, die Schwierigkeiten der Reizdefinition in dem Fall phasischer Receptoren zu veranschaulichen: Besäße der Mensch kein Hörvermögen, so könnte der Physiologe in Unkenntnis der Funktion des Wirbeltierohres auf den Gedanken kommen, im Ohr einen membranmanometerartigen Druckreceptor zu sehen. Er käme dann durch Studium der Aktionspotentiale zu dem Schluß, daß dieser Druckreceptor ein extrem phasisches Verhalten besitzt.

c) Die Reizintensität

Ein Vergleich der Reizintensitäts-Erregungskennlinien zahlreicher Sinnesorgane erlaubt, drei charakteristische Formen herauszustellen:

1. Logarithmische Intensitäts-Kennlinie

Abb. **3** zeigt die Abhängigkeit der Impulsfrequenz eines akustischen Neurons der Katze in Abhängigkeit von der Schallintensität und die

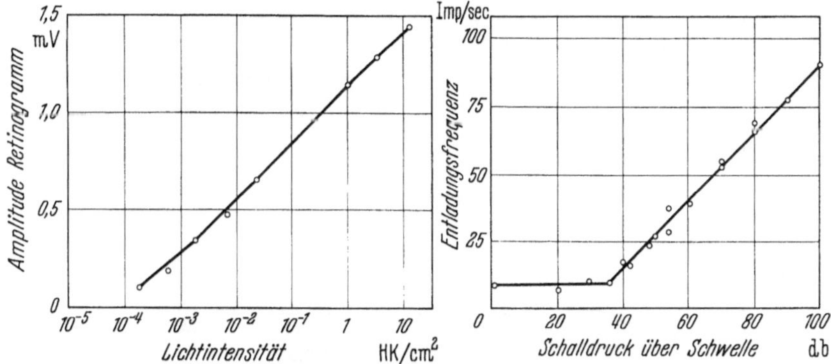

Abb. 3. Logarithmische Reiz-Erregungsbeziehung. Links: Amplitude des Retinogramms der Stabheuschrecke *Dixippus* in Abhängigkeit von der Lichtintensität. Nach AUTRUM [5]. Rechts: Entladungsfrequenz eines akustischen Neurons der Katze in Abhängigkeit vom Schalldruck. Nach GALAMBOS u. DAVIS. Reizintensität in beiden Abbildungen im logarithmischen Maßstab

Höhe des Elektroretinogramms der Stabheuschrecke in Abhängigkeit von der Lichtintensität. Bei logarithmischer Auftragung der Reizintensität ergeben sich in beiden Fällen Geraden, die Erregung nimmt etwa proportional dem Logarithmus der Reizintensität zu. Ähnliche Kurven weisen auf: Infrarotorgane der Klapperschlangen (BULLOCK u. DIECKE), Geruchssinnesorgane (D. SCHNEIDER, OTTOSON), Johnstonsche Organe der Schmeißfliege (BURKHARDT [3]) u. a. m. Es sind dies Sinnesorgane, welche auf in weiten Grenzen variierende Umweltreize ansprechen.

2. Lineare Intensitäts-Kennlinie

Bei einer Reihe von Sinnesorganen nimmt die Erregung linear mit der Größe des Eingangsreizes zu. Als Beispiele hierfür seien erwähnt: Muskelspindeln, Streckreceptoren der Krebse, (Abb. 4), Bogengangsreceptoren des Rochens (LOWENSTEIN u. SAND) und Baroreceptoren im Carotissinus (BRONK u. STELLA). Receptoren mit linearer Intensitäts-Kennlinie finden sich vorwiegend bei Sinnesorganen, welche an der Steuerung von Lage, Haltung, Bewegung und vegetativer Funktion beteiligt sind, sowie bei Receptoren, die physiologischerweise nur einem geringen Bereich von Reizintensitäten ausgesetzt sind.

3. *Extremwert-Intensitäts-Kennlinie*

Einige Receptoren haben die Eigenschaft, daß die Erregung bei einem bestimmten Wert des Reizes ein Maximum (bzw. ein Minimum) besitzt. Jede Abweichung von diesem Wert, gleichgültig in welchem Sinne, bewirkt eine Abnahme (bzw. Zunahme) der Erregung. Beispiele hierfür liefern die Thermoreceptoren der Wirbeltiere (HENSEL [1, 2]) und verschiedene Labyrinthreceptoren (LOWENSTEIN u. ROBERTS).

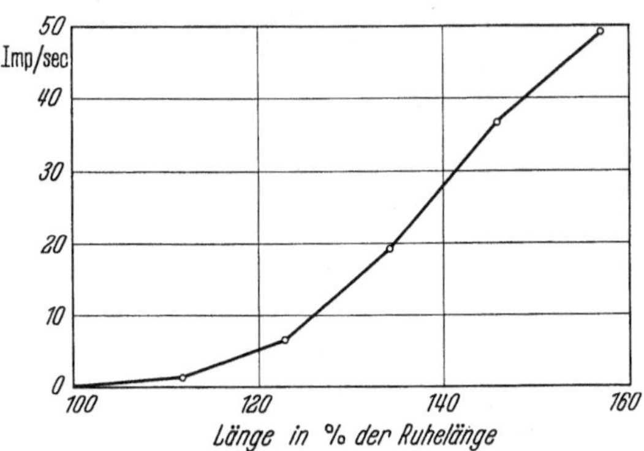

Abb. 4. Lineare Reiz-Erregungsbeziehung. Entladungsfrequenz eines Streckreceptors von *Astacus* in Abhängigkeit von der Dehnung. Nach BURKHARDT [1]

Diskussion der Intensitäts-Kennlinien und ihr Zusammenhang mit Verschiebungen der Übergangsfunktion

Die Größe der Erregung wurde hier in Abhängigkeit vom Eingangsreiz betrachtet. Für eine Reihe von Sinnesorganen ergeben sich andere Kurven, wenn der Nutzreiz untersucht wird. Bei Statocysten ändert sich die Erregung mit dem Sinus des Eingangsrcizes (Kippwinkel). Die Scherung als Nutzreiz ist aber proportional dem Sinus des Kippwinkels; der Zusammenhang zwischen Nutzreiz und Erregung ist somit linear. Allgemein muß angenommen werden, daß (ähnlich wie für das Zeitverhalten) sowohl in den Eigenschaften des reizleitenden Apparates als auch der sensiblen Endstrukturen die Ursachen für bestimmte Formen der Kennlinie liegen können. Leider ist die Reiztransformation bei den meisten Sinnesorganen weder im Hinblick auf das eine noch auf das andere hinreichend untersucht.

Ein weiteres Problem ist die Frage, ob jeweils die sinnvollste Größe als Eingangsreiz angesprochen wird. Die erste eingehende elektrophysiologische Studie der Muskelspindeln stammt von MATTHEWS. Er untersuchte die Entladung der Spindel unter isotonischer Beanspruchung des Muskels. Die Erregung nimmt unter diesen Bedingungen proportional dem Logarithmus der Belastung zu. Faßt man die Spannung als Eingangsreiz auf, so zeigen die Spindeln also eine logarithmische Intensitäts-Kennlinie. Da der Muskel nicht dem Hookschen Gesetz gehorcht, ist die Längenänderung nicht proportional der Spannung. Bei isotonischen Bedingungen kann natürlich nicht entschieden werden, ob die Spannung oder die Dehnung der Eingangsreiz ist. Wird der Muskel unter isometrischen Bedingungen untersucht, so zeigt sich: Die Entladungsfrequenz hängt nicht von der Spannung ab (in diesem Fall aktiv durch Kontraktion erzeugt), sondern von der Länge. Durch ein kleines Nachgeben der Sehne verkürzt sich der Muskel während der aktiven Spannungserhöhung, die Entladungsfrequenz der Spindel nimmt dabei etwas ab. Nicht die Spannung, sondern die Dehnung des Muskels ist also der Eingangsreiz der Muskelspindeln. Die Matthewssche Darstellung geht wohl von der umstrittenen Vorstellung aus, wonach die Muskelspindeln die Receptoren der Kraftempfindung sind. Für die Kraftempfindung gilt in gewissen Bereichen das Webersche Gesetz, wonach die relative Unterschiedsempfindlichkeit konstant ist; dies wird durch eine logarithmische Kennlinie erfüllt (vgl. hierzu Rein-Schneider). Die Frage, inwieweit das Webersche Gesetz auf Eigenschaften des reizleitenden Apparates zurückgehen kann, diskutiert auch bereits v. Holst.

Allgemein zeigen Sinnesorgane, welche auf Außenreize mit großem Intensitäts-Umfang ansprechen, die logarithmische Kennlinie. Sie ermöglicht, kleine Reizintensitäten mit hoher absoluter Unterschiedsschwelle aufzulösen und gleichzeitig einen großen Reizumfang zu verarbeiten. Die relative Unterschiedsempfindlichkeit ist über den ganzen logarithmischen Bereich konstant, auf Kosten der absoluten Unterschiedsempfindlichkeit, welche bei zunehmender Reizintensität abnimmt.

Lineare und Extremwert-Kennlinie verarbeiten nur einen kleinen Bereich von Reizintensität, aber mit hoher absoluter Unterschiedsempfindlichkeit; diese ist bei linearer Charakteristik konstant, bei der Extremwert-Kurve hingegen in bestimmten Reizbereichen besonders hoch. Die Extremwert-Charakteristik steht offenbar im Zusammenhang mit der Funktion bestimmter Sinnesorgane, Abweichungen von einem Normalzustand anzuzeigen.

Zwischen der Form der Intensitäts-Kennlinie und Adaptationserscheinungen besteht eine Korrelation. Verschiebungen der Übergangsfunktion finden sich vorwiegend bei Sinnesorganen mit logarithmischer Intensitätskennlinie; besonders deutlich ist dieser Zusammenhang bei der Empfindlichkeitsminderung. Durch die Änderung der Übergangsfunktion wird bewirkt, daß die Intensitätskennlinie jeder Reizsituation angepaßt werden kann. Durch die Verschiebungen wird in günstigen Fällen erreicht, daß die Unterschiedsempfindlichkeit fast unabhängig von der Grundintensität des Reizes ist. Adaptationserscheinungen bewirken somit zweierlei: erstens eine Verschiebung der Schwelle und des Sättigungsbereiches und zweitens eine Veränderung der Unterschiedsempfindlichkeit. Vergleicht man dies mit Meßinstrumenten, so wird durch die Adaptation einmal eine andere Empfindlichkeit des Meßinstruments eingestellt, zum anderen durch eine Vorspannung der Skalen-Nullpunkt verschoben (vgl. hierzu Ranke).

Schließlich muß noch ein weiterer Mechanismus erwähnt werden, mit dem eine hohe absolute Unterschiedsempfindlichkeit über große

Intensitätsbereiche erzielt wird. Viele Sinnesorgane exteroreceptiver Funktion bestehen aus einer großen Zahl von Sinneszellen. Die einzelnen Elemente haben verschiedene Schwellen und Sättigungsbereiche. Über den gesamten möglichen Reizumfang lösen sich die Elemente nacheinander ab. Jedes einzelne hat einen kürzeren Arbeitsbereich als das Gesamtorgan und entsprechend eine hohe Unterschiedsempfindlichkeit. Durch diesen Mechanismus besitzt das Sinnesorgan bei erweitertem Arbeitsbereich eine hohe Unterschiedsempfindlichkeit. Beispiele hierfür bieten die Thermoreceptoren der Wirbeltiere (HENSEL [1]), akustische Neurone (GALAMBOS u. DAVIS) und das Wirbeltierauge (Stäbchen und Zapfen).

d) Die räumliche Verteilung des Reizes

Sinnesorgane wie Auge, Ohr, Berührungsreceptoren, Strömungsreceptoren und Thermoreceptoren ermöglichen dem Organismus, Reize unmittelbar zu lokalisieren. Zwei unterschiedliche Mechanismen liegen dem zu Grunde: *Richtcharakteristik* und *Abbildung*.

1. *Richtcharakteristik*

Unter einer *Richtcharakteristik* soll die Eigenschaft von Sinnesorganen oder Receptorzellen verstanden werden, auf Reize gleicher Intensität in Abhängigkeit vom Einfallswinkel mit unterschiedlich starker Erregung zu antworten. Eine Richtcharakteristik kann durch polaren Bau des reizleitenden Apparates bedingt sein (Ohrmuschel, Pigmentbecher bei Augen, orientierte Aufhängung von Mechanoreceptoren) oder durch die Polarität der sensiblen Endstrukturen selbst: Endorgane des 8. Hirnnerven der Wirbeltiere: TRINCKER [1, 2], Gelenkreceptoren der Krebse: WIERSMA u. BOETTIGER. Meist arbeiten bei der Lokalisation paarige Sinnesorgane zusammen: Für das einzelne Organ hängt die Größe der Erregung vom Einfallswinkel und von der Intensität ab. Wird die Erregung zweier gleichartiger Organe mit divergierenden Achsen im ZNS ausgewertet, so kann aus dem Unterschied der beiden Afferenzen unabhängig von der Reizintensität die Richtung der Reizquelle entnommen werden. Prinzipiell ist allerdings auch die Meldung von einem Organ ausreichend: wird die Umgebung durch Pendelbewegungen abgetastet, so fällt und steigt die Erregung mit der Änderung der Lage des Sinnesorgans. Dieses Abtasten ermöglicht ebenfalls die Lokalisation (vgl. AUTRUM [6]).

2. *Abbildung*

Bei der *Abbildung* wird eine räumliche Verteilung des Reizes auf ein räumlich verteiltes Muster von Receptoren abgebildet. Beispiele hierfür bieten zahlreiche Augen und Felder von Integumentreceptoren (Seitenliniensystem, Berührungsreceptoren, Borstenfelder, Thermoreceptoren usw.). Während eine Lokalisation des Reizes mittels der Richtcharakteristik mit zwei einzelnen Receptorstrukturen möglich ist, sind für die Abbildung zahlreiche Sinnesendstellen nötig. Je dichter das sensorische Feld aufgebaut ist, desto besser ist die statische Auflösung.

Neben der statischen Auflösung von Reizmustern durch die Abbildung spielt bei vielen Sinnesorganen eine dynamische Auflösung durch zeitliche Reizwechsel eine Rolle: Bewegungen in der Umwelt oder Bewegungen des Sinnesorgans bewirken eine Verschiebung des Reizmusters auf dem Receptorenfeld. Bei dieser dynamischen Auflösung ist neben der Dichte des Receptorenmusters das zeitliche Verhalten der Receptoren von entscheidender Bedeutung für die Güte der Auflösung, vgl. S. 259.

Während die Richtcharakteristik vorwiegend zur Bestimmung der Einfallsrichtung bestimmter Reize dient (Strömung), wird mit der Abbildung im wesentlichen eine genaue Auswertung von gegliederten Reizfeldern ermöglicht (Beispiel: optische Umwelt).

Bei zahlreichen Sinnesorganen wirken Richtcharakteristik und Abbildung gemeinsam. Die einzelnen Receptoren des Seitenliniensystems der Fische und Amphibien sprechen auf Reize unterschiedlicher Richtung mit verschiedenen Erregungen an, besitzen also eine Richtcharakteristik (DIJKGRAAF [1]). Durch die räumliche Verteilung der Receptorengruppen auf der Körperoberfläche wird zusätzlich eine Abbildung ermöglicht (KRAMER, DIJKGRAAF [2, 3]). Die Grubenorgane der Klapperschlangen besitzen auf Grund des nach dem Lochkameraprinzip eingesenkten Sinnesepithels eine Richtcharakteristik, das Sinnesepithel ist darüber hinaus in selbständige receptorische Felder gegliedert, die eine grobe Abbildung erlauben (BULLOCK u. DIECKE).

Neben den beiden Mechanismen der unmittelbaren Orientierung der Reizquelle gibt es noch eine Reihe weiterer, welche auch ohne Richtcharakteristik oder Abbildung die Lokalisation von Reizquellen ermöglichen. Erwähnt sei die Einstellung auf Vorzugstemperaturen, phobische (oder Schreck-) Reaktionen, das Abtasten eines Reizintensitätsgefälles und die Möglichkeit des Zusammenspiels mehrerer Sinnesorgane. Beispielsweise finden Säuger und Insekten Duftquellen durch Wahrnehmung der Windrichtung (STEINER). Alle diese Mechanismen haben gemeinsam, daß die Orientierung zum Reiz durch komplizierte Bewegungsweisen zustande kommt. Man kann sie deshalb unter dem Begriff *mittelbare Orientierung* zusammenfassen. Abbildung und Richtcharakteristik hingegen ermöglichen demgegenüber eine direkte Ortung der Reizquelle. Sinnesorgane mit dieser Eigenschaft steuern die Orientierung der Tiere im Raum (Taxis im Sinne KÜHNs); ausführliche Darstellungen des Orientierungsproblems siehe bei KÜHN [2], v. BUDDENBROCK, TINBERGEN.

Die funktionelle Verknüpfung der Sinnesorgane

Die Abschnitte des vorangegangenen Kapitels haben gezeigt, daß die Sinnesorgane nach bestimmten Eigenschaften und Formen der Kennlinien gruppiert werden können. Die nächste gestellte Frage war: Welche typische Funktionen und funktionellen Verknüpfungen können aufgezeigt werden? Dabei ist dann zu prüfen, ob bestimmten Funktionen bestimmte Kennlinien entsprechen, und zwar unabhängig davon, welches der Eingangsreiz eines Sinnesorgans ist.

A. Die Regelung und Steuerung vegetativer Funktionen

Im Aortenbogen und Carotissinus der Säuger liegen Mechanoreceptoren, welche auf die Höhe des Blutdruckes ansprechen. Isolierte Präparate solcher Receptoren zeigen das typische Verhalten phasisch-tonischer Sinnesorgane: Bei konstantem Druck eine konstante Entladungsfrequenz, deren Höhe mit zunehmendem Blutdruck steigt. Bei einer plötzlichen Drucksteigerung eine Erregungsspitze mit anschließendem Erregungsabfall auf das neue stationäre Niveau (BRONK u. STELLA; LANDGREN). Verschiebungen der Übergangsfunktion können bei den Baroreceptoren nicht festgestellt werden; der Erregungsabfall beruht also auf einer dynamischen Einstellung.

Die Afferenz der Receptoren gelangt zu den Kreislauf-Regulationszentren und bewirkt bei einer Steigerung des Blutdrucks eine Regulation in dem Sinne, daß der Blutdruck wieder abnimmt (WAGNER). Die Afferenz löst also eine Regulation aus, durch welche der gemeldete Eingangsreiz rückgängig beeinflußt wird; es schließt sich hierdurch ein Wirkungskreis: im Sinne der Technik ein Regelkreis.

Bei einem derartig funktionell verknüpften Sinnesorgan kann der Eingangsreiz unter normalen Bedingungen nicht sehr variieren, die als Reiz wirkende Größe ist auf enge Grenzen eingeregelt. Diesen Bedingungen ist eine begrenzte lineare Intensitätskennlinie mit großer Unterschiedsempfindlichkeit günstig angepaßt. Da der Blutdruck eine Größe ist, die ständig auf einen bestimmten Wert geregelt bleiben soll, darf das den Regelkreis kontrollierende Sinnesorgan seine Empfindlichkeit nicht ändern.* Würde sich die Empfindlichkeit unter der Einwirkung eines längeren Reizes verschieben, so würde der Regelkreis verstellt, und bei einer vollständigen Adaptation wäre eine dauernde Regelung dieser Art überhaupt nicht möglich. Es fragt sich, inwieweit die durch eine dynamische Einstellung solcher Receptoren entstehende Erregungsspitze für die Funktion von Bedeutung ist. Eine eingehende Analyse des Zeitverhaltens von Regelkreisen ergibt, daß durch die Erregungsspitze (regel-

---

[1] Ausnahmen von dieser Bedingung sind bei komplizierter aufgebauten Regelsystemen möglich, vgl. S. 258.

theoretisch den D-Anteil) unter Umständen eine raschere Wiederherstellung des Sollwerts nach einer Störung erzielt wird (vgl. hierzu OPPELT).

Eine ähnliche Funktion wie die Baroreceptoren haben die Thermoreceptoren der Warmblüter. Die Temperatur des Körperkerns wird durch einen Regelkreis konstant gehalten. Als wichtigster Bestandteil dieses Regelkreises werden heute Temperatur-Regulationszentren im Hypothalamus angesehen, welche den Wärmehaushalt durch Kontrolle der Durchblutung, des Stoffwechsels usw. regeln (vgl. HENSEL [4]). Die Thermoreceptoren der Haut sind diesem Regelkreis angegliedert (regeltheoretisch Störgrößen-Aufschaltung). Änderungen der Außentemperatur, welche als Störungen über die Haut zum Körperkern vordringen, werden sofort durch die Thermoreceptoren den Kreislaufregulationszentren gemeldet. Hierdurch werden Gegenmaßnahmen ausgelöst, die Störung wird so gar nicht im Körperkern bemerkbar. Da diese Gegenmaßnahmen nicht (oder nur bedingt infolge einer Änderung der Hautdurchblutung) auf den Eingangsreiz der Thermoreceptoren zurückwirken, liegt hier im strengen Sinn keine Regelung, sondern eine Steuerung vor. Die an die Thermoreceptoren gestellten Anforderungen sind aber ähnlich wie bei den Baroreceptoren: Es dürfen keine nennenswerten Empfindlichkeitsänderungen auftreten, und ein kleiner Bereich von Eingangsreizen soll bei großer Unterschiedsempfindlichkeit überstrichen werden. Thermoreceptoren besitzen eine Extremwertcharakteristik, der Bereich besonders hoher Unterschiedsempfindlichkeit liegt zumeist im Gebiet der normalen Hauttemperatur.

Die beiden herausgegriffenen Beispiele können als typisch für einen Funktionskreis gelten: Bestimmte, meist vegetative Vorgänge oder Zustände werden auf einen für die Tätigkeit des Organismus nötigen Wert eingestellt, durch Regelung oder Steuerung wird dieser Zustand festgehalten (für weitere Regelungen dieser Art vgl. DRISCHEL, REINSCHNEIDER). Wenn Sinnesorgane an der Steuerung und Regelung von bestimmten vegetativen Größen mitwirken, so kann dies als *Funktion der inneren Steuerung* bezeichnet werden. Parallel mit der Funktion der inneren Steuerung finden sich meist folgende Eigenschaften der Sinnesorgane: keine nennenswerten Empfindlichkeitsänderungen, vielfach eine dynamische Einstellung und eine lineare oder Extremwert-Intensitäts-Kennlinie.

B. Die Funktion der Körperbeherrschung

Die Muskelspindeln der Säuger zeigen formal sehr ähnliche Reiz-Erregungsbeziehungen wie die besprochenen Baro- und Thermo-Receptoren. Die Muskelspindeln sind ebenfalls in einen Regelkreis eingeschaltet, der die Länge des Muskels gegen Störungen stabilisiert. Wird dieser Regelkreis durch eine heftige Störung aus dem Gleichgewicht gebracht, so reagiert er mit einer heftigen kompensatorischen Antwort: der wohlbekannte Sehnenreflex. Die Verhältnisse bei den Muskelspindeln liegen aber insgesamt komplizierter als bei den Baroreceptoren.

a) Die Muskellänge ist keine Größe, die ständig konstant gehalten werden soll. Je nach Lage und Haltung muß die Länge des Muskels verschieden eingestellt und dann gegen äußere Störungen stabilisiert werden. Bei Bewegungen muß die Länge des Muskels fortlaufend und kontrolliert verändert werden. Dies ist auf folgende Weise möglich: Die Erregung der Muskelspindel kann vom ZNS aus unabhängig vom Reiz verändert werden; dieser als Reizkontrolle arbeitende Mechanismus wurde auf Seite 241 bereits besprochen. Die Reizkontrolle bewirkt über den Regelkreis eine Verstellung der Muskellänge: Eine durch Dehnung ausgelöste Afferenz der Spindel löst im Regelkreis eine kompensatorische Kontraktion aus. Ebenso löst eine durch die Reizkontrolle erzeugte Afferenz eine Kontraktion des Muskels aus; dieser ist jedoch keine Dehnung vorausgegangen, die kompensiert werden kann, der Muskel verkürzt sich daher effektiv. Die neu eingestellte Länge wird nun so lange beibehalten und über den Regelkreis gegen äußere Störungen stabilisiert, bis sich die Reizkontrolle der Spindel wieder einmal verändert. Eine Reihe eingehender Untersuchungen (Übersicht bei BURKHARDT [4]) hat gezeigt, daß die Muskelspindeln ständig einer efferenten Regulierung unterliegen und hierdurch ständig zumindest ein Teil der Tätigkeit des Muskels gesteuert wird.

Die Reizkontrolle der Muskelspindel ist aber nicht nur für die Verstellung des Regelkreises der Muskellänge nötig, sondern auch gleichzeitig für die Muskelspindel selbst. Eine starke Verkürzung des Muskels hat bei nicht regulierter Spindel zur Folge, daß die Entladung erlischt, die Spindel rutscht aus ihrem Arbeitsbereich. Bei einer Kontraktion des Muskels über die Reizkontrolle der Spindeln werden die sensiblen Endigungen durch die intrafusalen Muskelfasern vorgespannt, der Arbeitsbereich der Spindel ist hierdurch automatisch so verschoben, daß bei der nachfolgenden Kontraktion des Hauptmuskels die Spindelafferenz nicht erlischt. Ebenso wird bei einer Kontraktion des Hauptmuskels auf dem direkten Weg über die Motoneurone die Muskelefferenz von einer Spindelefferenz begleitet; auch diese Regulierung wirkt so, daß die Spindel stets in ihrem Arbeitsbereich bleibt. Die Anpassung an verschiedene Reizsituationen erfolgt hier nicht durch die einsinnige und relativ träge Empfindlichkeitsminderung, sondern durch eine rasche und flexible nervöse Steuerung, durch die Reizkontrolle.

b) Kontrahiert sich ein Muskel, so werden seine Antagonisten etwas gedehnt. Da sie ihrerseits Längen-stabilisiert sind, würden sie der Tätigkeit des aktiven Muskels entgegenwirken. Daß dieser Effekt nicht eintritt, dafür sorgt eine wechselseitige Beeinflussung der Spindeln untereinander: Die Erregung einer Spindel steuert über spinale Bahnen automatisch die Tätigkeit der anderen Spindeln der Muskelgruppe: fördernd (Synergisten) oder hemmend (Antagonisten), so daß ein sinnvolles Zusammenspiel aller Muskeln gewährleistet ist. Aber nicht nur die Muskelspindeln, sondern auch Sehnenspindeln, Gelenkreceptoren und zum Teil Hautreceptoren beteiligen sich an diesem Wechselspiel. Die funktionelle Verknüpfung dieser den Bewegungsapparat kontrollierenden Sinnesorgane ist ein engmaschiges Netz. Man kann die enge Vermaschung und die zentrifugale Kontrolle als wichtiges Charakteristikum dieser Gruppe von Receptoren auffassen.

Zu diesem Funktionstyp müssen auch die statischen und dynamischen Lagesinnesorgane gerechnet werden. Ermöglichen die Receptoren des Bewegungsapparates eine Kontrolle der Gliedstellungen und Bewegungen relativ zum Körper, so ermöglichen die Lagesinne eine Kontrolle der Glied- und Körperstellung relativ zur Schwerkraft oder zur Normallage. Die nervösen Wechselbeziehungen beider Gruppen von Sinnesorganen sind besonders eng, die Reiz-Erregungsbeziehungen sind bei beiden Gruppen sehr ähnlich. Es gibt auch eine Reihe direkter Übergänge zwischen beiden Receptorengruppen. Beispiele hierfür sind die statischen und dynamischen Lagesinnesorgane der Libellen (MITTELSTAEDT) und Bienen (LINDAUER u. NEDEL).[1] Der Kopf der Biene besitzt ein relativ großes Trägheitsmoment und ist ähnlich einem Schwerependel leicht beweglich im Halsgelenk aufgehängt. Die Stellung von Kopf zum Körper wird durch mehrere Polster von Sinnesborsten angezeigt. Gesetzmäßig mit der Orientierung des Tieres zur Schwerkraft ändert sich die relative Stellung vom Kopf zum Körper und wird von den Stellungsreceptoren gemeldet. Stellungsreceptoren arbeiten hier als statische und dynamische Schweresinnesorgane.

Die Receptoren des Bewegungsapparates und die Lagesinne haben also ähnliche Funktionen, sie zeigen auch ähnliche funktionelle Verknüpfungen und enge wechselseitige Beziehungen. Die Funktion beider kann unter dem Begriff: *Funktion der Körperbeherrschung* zusammengefaßt werden. Charakteristisch für diesen Funktionstyp sind: Lineare oder Extremwert-Kennlinien für die Intensitäts-Abhängigkeit, keine Empfindlichkeitsänderungen, oft eine zentrifugale Kontrolle und eine enge wechselseitige Vermaschung mit anderen Receptoren des gleichen Funktionstyps.

C. Die Auslösefunktion

Sinnesorgane, welche Reize aus der Umwelt verarbeiten, sind weit größeren Schwankungen des Eingangsreizes unterworfen als die Sinnesorgane der inneren Steuerung oder der Körperbeherrschung. Sie sind

[1] Während der Drucklegung dieser Arbeit erschien ferner: L. J. GOODMAN and P. T. HASKELL: Hair receptors in locusts. Nature (Lond.) **183**, 1106 (1959).

dieser Aufgabe angepaßt, indem sie meist eine logarithmische Reiz-Intensitäts-Kennlinie besitzen und ihre Empfindlichkeit in weiten Grenzen veränderbar ist (Empfindlichkeitsminderung, Reiz- und Erregungskontrollen). Von den zahlreichen Funktionen solcher Sinnesorgane seien zwei näher erörtert: Die Auslösefunktion und die Orientierungsfunktion. Bestimmte Umweltsituationen erfordern zweckentsprechende Verhaltensreaktionen des Organismus wie: Nahrungserwerb, Flucht, Paarungsverhalten, Ortswechsel zu günstigeren Lebensbedingungen usw. Durch Sinnesorgane wird die Umwelt laufend kontrolliert, ob eine bestimmte Reaktionen erfordernde Situation eintritt oder wechselt. Jeder Wechsel wird dem ZNS gemeldet, dort ausgewertet und löst gegebenenfalls die entsprechende Verhaltensweise aus: *Auslösefunktion*. Vielfach sind dazu gerichtete Bewegungsabläufe erforderlich; die motorische Reaktion muß zur Richtung des Reizes abgestimmt werden: *Orientierungsfunktion;* vgl. hierzu Tinbergen. Für die Auslösefunktion ist vor allem erforderlich, daß die Empfindlichkeit des Sinnesorgans optimal der jeweiligen Reizsituation angepaßt wird.

D. Die Orientierungsfunktion

Die Orientierungsfunktion kann unmittelbar vom Sinnesorgan gesteuert werden, sofern dieses eine Richtcharakteristik oder die Möglichkeit zur Abbildung besitzt. Auch bei der Orientierungsfunktion sind die Sinnesorgane zumeist in Regelsysteme verknüpft. Der gerichtete Bewegungsablauf setzt voraus, daß Körper oder Körperteile in eine bestimmte Lage zur Reizquelle gebracht werden (Taxis). Dazu muß die Abweichung von der Soll-Lage durch Sinnesorgane angezeigt werden: Die Meldungen des orientierenden Sinnesorgans und der Receptoren des Bewegungsapparates werden im ZNS verrechnet. Das Ergebnis der Verrechnung bestimmt die zur Lagekorrektur nötigen Efferenzen. Solange eine Differenz zwischen dem einer bestimmten Orientierung entsprechenden Erregungsmuster und dem momentan vorhandenen besteht, kommt das Regelsystem nicht zur Ruhe (v. Holst u. Mittelstaedt: das Reafferenzprinzip).

Das Regelsystem ist hierbei wesentlich komplizierter als bei der Funktion der inneren Steuerung oder der Körperbeherrschung. Dort arbeiten die Sinnesorgane in einem einfachen Regelkreis: Die Erregung des Sinnesorgans selbst löst über ZNS und Effektor die Veränderung der Größe aus, welche wiederum die Erregung des Sinnesorgans bestimmt. Das Regelsystem bei der Orientierungsfunktion besteht aus einer Vermaschung zahlreicher Regelkreise (Teile dieses Systems sind beispielsweise die einzelnen Regelkreise der Receptoren des Bewegungsapparates). Das Gesamtsystem kommt erst in ein Gleichgewicht, wenn im ZNS ein bestimmtes Erregungs*muster* vorliegt. Es sind hierzu komplizierte und vielschichtige Leistungen der Receptoren, des ZNS und des motorischen Apparates nötig.

Die orientierenden Sinnesorgane müssen zur Anpassung an die gesamte Reizsituation ihre Empfindlichkeit in weitem Umfang verändern können. Diese Anpassung der Empfindlichkeit darf auf die Orientierungsleistung keinen Einfluß haben. Dies wird erreicht, indem nicht auf eine bestimmte Größe der Erregung hingearbeitet wird, sondern auf ein Erregungsgleichgewicht. Die klassische Einteilung der Orientierungsmechanismen nach Kühn [2] zeigt dies sehr schön: Bei paariger Anordnung von Sinnesorganen drehen sich manche Tiere so lange, bis beide Organe den Reiz mit gleicher Intensität empfangen, also eine gleich starke Erregung liefern: Tropotaxis. Eine andere Orientierungsweise ist: Das Tier dreht sich so lange, bis bei einer Abbildung bestimmte Bezirke des Receptorenrasters maximal gereizt werden oder bis ein bestimmtes Verhältnis der Erregung von paarigen Sinnesorganen mit Richtcharakteristik erreicht ist: Telotaxis und Menotaxis. Bei allen diesen Orientierungsweisen spielen Empfindlichkeitsverschiebungen keine Rolle, sie ändern weder den Abbildungsort noch das Erregungsverhältnis paariger Sinnesorgane. Wird jedoch im Experiment bei paarigen Sinnesorganen die Empfindlichkeit einseitig verändert, so ergeben sich Fehlleistungen bei der Lokalisation der Reizquelle bzw. in der Einstellung des Tieres.

Mit der Funktion der unmittelbaren Orientierung sind ähnlich wie mit der Auslösefunktion verknüpft: logarithmische Intensitätskennlinien und umfangreiche Änderungen der Empfindlichkeit durch Empfindlichkeitsminderung oder zentrifugale Kontrolle. Darüber hinaus sind für diesen Funktionskreis kennzeichnend: Abbildung oder Richtcharakteristik und das enge Wechselspiel mit den Receptoren der Körperbeherrschung. Sinnesorgane ohne Abbildung und Richtcharakteristik können auf mittelbarem Wege an der Orientierung beteiligt sein: mittelbare Orientierungsfunktion, vgl. S. 253. Das Fehlen einer Richtcharakteristik oder Abbildung wird dabei durch ein Zusammenspiel mit unmittelbar orientierenden Sinnesorganen oder durch ein Abtasten des Reizfeldes ersetzt. Beim Abtasten des Reizfeldes durch bestimmte Bewegungsabläufe wird das Reizmuster nicht simultan, sondern sukzessiv bewertet. Wie außerordentlich wirkungsvoll dieser Mechanismus sein kann, zeigt die Fähigkeit der Säuger, Schallquellen durch Zeitdifferenz-Wahrnehmung zu lokalisieren (Hornbostel; vgl. auch Klensch).

E. Die Auflösung von zeitlichen Reizmustern

Einige Sinnesorgane sind besonders darauf eingerichtet, zeitliche Reizmuster aufzulösen: Wirbeltierohr (v. Békésy [1, 2]; Galambos, Schwartzkopff u. Rupert), Seitenliniensystem der Fische und Amphibien (Kramer, Dijkgraaf [1, 2, 3]), Vibrationsreceptoren der Haut und Pacinische Körperchen der Säuger (KEIDEL [2]; GRAY u. MALCOLM), Antennen (BURKHARDT u. G. SCHNEIDER; BURKHARDT [3]), Tympanal- und Subgenualorgane (AUTRUM [7, 8]) der Insekten und lyriforme Organe der Spinnen (PRINGLE). Außer diesen Mechanoreceptoren muß eine Reihe von Augen genannt werden. Besonders bei schnellfliegenden Insekten, aber auch bei zahlreichen anderen Tieren wird ein räumliches Reizmuster durch Relativbewegungen zwischen Umwelt und Lichtsinnesorganen in ein zeitliches Reizmuster am Sehzellenraster umgeformt: Bewegungssehen. Welche entscheidende Rolle die zeitliche Auflösung von Reizmustern für das Formensehen spielt, ist erst in den letzten Jahren erkannt worden (hierzu: AUTRUM [9, 10]).

Die Receptoren sind der Funktion, zeitliche Reizmuster aufzulösen, durch eine Reihe typischer Eigenschaften und durch besondere funktionelle Verknüpfungen angepaßt:

1. Sie müssen eine hohe Verschmelzungsfrequenz haben, um zeitlichen Reizwechseln hinreichend folgen zu können. Diese Eigenschaft besitzen speziell phasische Receptoren, welche Dauerreize mit einer rasch abklingenden Erregung beantworten und besonders empfindlich gegenüber Reizwechseln sind.

2. Eine typische Eigenschaft solcher Sinnesorgane ist, daß zweierlei Receptoren (oder sekundäre Neurone) vorkommen: eine Art spricht auf Reizwechsel der einen Richtung an, die andere auf Reizwechsel in entgegengesetzter Richtung. Erwähnt sei in diesem Zusammenhang das On-Off-System der Wirbeltierretina (HARTLINE [1, 2]; GRANIT [1]), richtungsempfindliche Receptoren des Johnstonschen Organs der Insektenantenne (BURKHARDT [3]) und phasenempfindliche Neurone der Vogel-Cochlea (SCHWARTZKOPFF [2]).

3. Die Auswertung des Erregungsmusters erfordert enge Wechselwirkungen zwischen den Receptoren bzw. ihren Neuronen: Vielfach hemmen oder bahnen sich räumlich benachbarte Elemente oder solche, welche auf die beiden möglichen Richtungen des Reizwechsels ansprechen. Die Bahnung bzw. Hemmung ist meist reziprok und zu diskreten Zeitpunkten nach einem Reizwechsel besonders ausgeprägt. Synchronisationserscheinungen und periodische Erregbarkeitsänderungen sind Anzeichen solcher Wechselwirkungen. Hierdurch wird das Reizmuster in ein stark kontrastiertes Erregungsmuster umgeformt. Ausführliche Diskussionen zu diesem Gesichtspunkt und zu dem der zentralen Verrechnung finden sich bei AUTRUM u. STÖCKER, BURKHARDT [3], ENROTH, GRANIT [1, 2], AUTRUM [9], SCHWARTZKOPFF [3] und HASSENSTEIN.

Schlußfolgerungen

Die bisher besprochenen Beispiele sollten zeigen, welche engen Beziehungen zwischen der Funktion eines Sinnesorgans, der Art seiner funktionellen Verknüpfung und den Eigenschaften des Receptors bestehen. Es ist so möglich, Sinnesorgane unabhängig von der Art des Eingangsreizes in Gruppen zu ordnen. Jede der Gruppen umfaßt verschiedenartige Sinnesorgane, die jedoch ähnliche Formen der Reiz-Erregungs-Beziehungen besitzen. Ähnlich ist dann auch die funktionelle Verknüpfung der Sinnesorgane und schließlich ihre Funktion. Die Sinnesorgane einer Gruppe erfüllen einen gemeinsamen *Funktionstyp*, wie er in der Einleitung definiert worden ist. Beispiele solcher Funktionstypen sind:

1. Die Funktion der inneren Steuerung. Kennzeichnende Eigenschaften der Receptoren: lineare oder Extremwert-Kennlinien, häufig eine dynamische Einstellung, das Fehlen einer Empfindlichkeitsminderung oder einer zentrifugalen Kontrolle. Kennzeichnende Verknüpfungen: Die Sinnesorgane arbeiten entweder in einem Regelkreis zur Kontrolle bestimmter vegetativer Funktionen oder sie sind einem solchen angegliedert.

2. Die Funktion der Körperbeherrschung. Kennzeichnende Eigenschaften der Receptoren: lineare oder Extremwert-Kennlinien, dynamische Einstellung, keine Empfindlichkeitsminderung, häufig jedoch zentrifugale Kontrollen. Kennzeichnende Verknüpfungen: Einbau in direkte Regelkreise zur Kontrolle von Lage, Haltung und Bewegung, enge Vermaschung aller beteiligten Sinnesorgane untereinander.

3. Die Auslösefunktion. Kennzeichnend sind: logarithmische Intensitäts-Kennlinien, umfangreiche Änderungen der Übergangsfunktion zur Anpassung an verschiedene Reizsituationen (Empfindlichkeitsminderung oder zentrifugale Kontrolle). Lose Verknüpfung mit anderen Sinnesorganen.

4. Die Orientierungsfunktion. Kennzeichnende Eigenschaften der Receptoren: logarithmische Intensitäts-Kennlinien, umfangreiche Änderungen der Übergangsfunktion, Richtcharakteristik oder Abbildung bei unmittelbarer Orientierungsfunktion. Kennzeichnende Verknüpfungen: Zusammenspiel mit direkt orientierenden Sinnesorganen oder zeitliches Abtasten von Reizfeldern bei mittelbarer Orientierung. Allgemein ein enges Zusammenspiel mit den Kontrollorganen des motorischen Apparates. Die Sinnesorgane sind in komplizierte, vermaschte Regelsysteme eingebaut. Diese Regelsysteme arbeiten auf ein spezielles Gleichgewicht von Erregungsmustern hin.

5. Die Auflösung von zeitlichen Reizmustern. Kennzeichnende Eigenschaften der Receptoren: phasisches Verhalten, hohe Verschmelzungsfrequenz, oft zwei antagonistische Gruppen von Receptoren (oder Neuronen) im Sinnesorgan, welche jeweils auf eine der beiden möglichen Richtungen des Reizwechsels ansprechen. Kennzeichnende Verknüpfungen: Straffe Synchronisation und starke Wechselwirkungen von bestimmtem Zeitverlauf zwischen den Receptoren oder ihren Neuronen.

Ausdrücklich betont werden muß, daß diese fünf Funktionspläne Beispiele, aber kein umfassendes Schema sein sollen. Ähnlich wie es Baupläne für übergeordnete systematische Einheiten und Baupläne für spezielle Gruppen gibt, können Funktionstypen für allgemeine und für spezielle Leistungen aufgestellt werden: Die Orientierungsfunktion ist eine speziellere Leistung als die Auslösefunktion. Sicher kann den hier ausgeführten Beispielen eine Reihe gleichwertiger Funktionstypen und eine große Zahl spezialisierter hinzugeführt werden.

Weiterhin ist es möglich, daß ein und dasselbe Sinnesorgan mehrere unterschiedliche Funktionstypen erfüllt. Der Haut-Temperatursinn der Säuger wurde hier als Beispiel für den Funktionstyp der inneren Steuerung angeführt, dem er voll genügt: Extremwertcharakteristik, keine Änderungen der Übergangsfunktion, Angliederung an den Regelkreis zur Thermoregulation. Ohne Zweifel leistet der Haut-Temperatursinn aber auch, uns Informationen über die Umwelt zu liefern und sogar Reize zu lokalisieren, beispielsweise einen warm strahlenden Ofen. Er würde unter diesem Gesichtspunkt zum Funktionstyp der unmittelbaren Orientierung gestellt werden müssen. In der Tat ermöglicht der Temperatursinn eine grobe Abbildung des Reizes auf ein räumlich verteiltes Receptorenmuster, und jeder weiß aus eigener Erfahrung, daß unsere Temperaturempfindung einer starken Adaptation unterworfen ist. Dauertemperaturen nehmen wir nur wahr, wenn sie unter 24° oder über 36° liegen; im dazwischen liegenden Bereich spürt man wohl Temperaturwechsel, die Empfindung erlischt dann aber mehr oder weniger rasch. Da die Receptoren in diesem Bereich Dauertemperaturen anzeigen, muß die Adaptation ein zentrales Phänomen sein. Das Vorhandensein einer derartigen zentralen Adaptation zeigt aber, daß bei der Funktion als orientierendes Sinnesorgan die für den Funktionstyp charakteristische Adaptation sofort ins Spiel kommt.

Das Beispiel des Temperatursinns der Säuger mag davor warnen, einseitig nur bestimmte Funktionen zu betrachten. Sinnesorgane sind Bestandteile des Organismus, welche zu außerordentlich vielseitigen und komplexen Aufgaben befähigt sind.

Dies rechtfertigt den Versuch, ein Ordnungsschema zu entwerfen, welches nicht von den Receptoren und ihrem Eingangsreiz ausgeht, sondern von den Leistungen der Sinnesorgane. Der Funktionstyp kennzeichnet dabei eine Gruppe ähnlicher Leistungen durch verschiedene Gesichtspunkte. In der vergleichenden Anatomie mag eine bestimmte Art einem Bauplan recht nahe kommen, ohne daß der Systematiker geneigt wäre, sie als den Prototyp des Bauplans schlechthin aufzufassen. Ähnlich hier: Ein bestimmtes Sinnesorgan mag einen Funktionstyp zwar weitgehend erfüllen, es wird aber individuelle Abweichungen zeigen und möglicherweise in bestimmten Leistungen anderen Funktionstypen genügen. Wenn die Funktionspläne somit keine starre Gliederung für die Sinnesorgane selbst darstellen, wird dieser Begriff gerade hierdurch dem Wesen des biologischen Objektes in seiner Vielfalt und Wandelbarkeit gerecht.

Literatur

Adrian, E. D.: The basis of sensation. The action of sense organs. London: Christophers 1928.

Autrum, H.: [1] Schallempfang bei Tier und Mensch. Naturwissenschaften **30**, 69 (1942).

— [2] Über kleinste Reize bei Sinnesorganen. Biol. Zbl. **63**, 209 (1943).

— [3] Über Energie- und Zeitgrenzen der Sinnesempfindungen. Naturwissenschaften **35**, 361 (1948).

— [4] Über Lautäußerung und Schallwahrnehmung bei Arthropoden. I. Untersuchungen an Ameisen. Eine allgemeine Theorie der Schallwahrnehmung bei Arthropoden. Z. vergl. Physiol. **23**, 332 (1936).

— [5] Die Belichtungspotentiale und das Sehen der Insekten (Untersuchungen an *Calliphora* und *Dixippus*). Z. vergl. Physiol. **32**, 176 (1950).

— [6] Über Lautäußerung und Schallwahrnehmung bei Arthropoden. II. Das Richtungshören von *Locusta* und Versuch einer Hörtheorie für Tympanalorgane vom Locustidentyp. Z. vergl. Physiol. **28**, 326 (1940).

— [7] Über Gehör und Erschütterungssinn bei Locustiden. Z. vergl. Physiol. **28**, 580 (1941).

— [8] In Vorbereitung.

— [9] Die Zeit als physiologische Grundlage des Formensehens. Stud. generale **8**, 526 (1955).

— [10] Das Fehlen unwillkürlicher Augenbewegungen beim Frosch. Naturwissenschaften **46**, 435 (1959).

— u. Marieluise Stöcker: Über optische Verschmelzungsfrequenzen und stroboskopisches Sehen bei Insekten. Biol. Zbl. **71**, 129 (1952).

Békésy, G. v.: [1] DC resting potentials inside the cochlear partition. J. Acoust. Soc. Amer. **24**, 72 (1952).

— [2] Current status of theories of hearing. Science **123**, 779 (1956).

Bronk, D. W., and G. Stella: Afferent impulses in the carotid sinus nerve. I. The relation of the discharge from single end organs to arterial blood pressure. J. cell. comp. Physiol. **1**, 113 (1932).

Buddenbrock, W. v.: Vergleichende Physiologie, Bd. 1. Sinnesphysiologie. Basel: Birkhäuser 1952.

Bullock, Th. H., and F. P. J. Diecke: Properties of an infra-red receptor. J. Physiol. **134**, 47 (1956).

Burkhardt, D.: [1] Die Erregungsvorgänge sensibler Ganglienzellen in Abhängigkeit von der Temperatur (Untersuchungen an den abdominalen Streckreceptoren des Sumpfkrebses *Astacus leptodactylus*). Biol. Zbl. **78**, 22 (1959).

— [2] Effect of temperature on isolated stretch-receptor organ of the crayfish. Science **129**, 392 (1959).

— [3] Action potentials in the antennae of the blowfly *(Calliphora erythrocephala)* during mechanical stimulation. J. Insect Physiol. **4** (1960) (im Druck).

— [4] Die Sinnesorgane des Skeletmuskels und die nervöse Steuerung der Muskeltätigkeit. Ergebn. Biol. **20**, 27 (1958).

— [5] Die Übertragereigenschaften elektrophysiologischer Versuchsanordnungen. Z. Biol. **109**, 297 (1957).

BURKHARDT, D.: [6] Rhythmische Erregungen in den optischen Zentren von *Calliphora erythrocephala*. Z. vergl. Physiol. **36**, 595 (1954).

— u. G. SCHNEIDER: Die Antennen von *Calliphora* als Anzeiger der Fluggeschwindigkeit. Z. Naturforsch. **12b**, 139 (1957).

BURTON, A. C.: The properties of the steady state compared to those of equilibrium as shown in characteristic biological behavior. J. cell. comp. Physiol. **14**, 327 (1939).

CRESCITELLI, F., and H. J. A. DARTNALL: Human visual purple. Nature (Lond.) **172**, 195 (1953).

DIJKGRAAF, S.: [1] Elektrophysiologische Untersuchungen an der Seitenlinie von *Xenopus laevis*. Experientia (Basel) **12**, 276 (1956).

— [2] Untersuchungen über die Funktion der Seitenorgane an Fischen. Z. vergl. Physiol. **20**, 162 (1933).

— [3] Über die Reizung des Ferntastsinnes bei Fischen und Amphibien. Experientia (Basel) **3**, 206 (1947).

DODT, E.: Centrifugal impulses in rabbit's retina. J. Neurophysiol. **19**, 301 (1956).

DRISCHEL, H.: Die Meßfunktion biologischer Receptoren als regeltheoretisches Problem. Naturwissenschaften **40**, 496 (1953).

ENROTH, CHRISTINA: The mechanism of flicker and fusion studied on single elements in the dark-adapted eye of the cat. Acta physiol. scand., **27**, Suppl. 100 (1952).

ERULKAR, S. D., J. E. ROSE and PH. W. DAVIS: Single unit activity in the auditory cortex of the cat. Bull. Johns Hopk. Hosp. **99**, 55 (1956).

EYZAGUIRRE, C., and ST. W. KUFFLER: [1] Processes of excitation in the dendrites and in the soma of single isolated sensory nerve cells of the lobster and crayfish. J. gen. Physiol. **39**, 87 (1955).

— — [2] Further study of soma, dendrite and axon excitation in single neurons. J. gen. Physiol. **39**, 121 (1955).

FLOREY, E.: [1] Adaptationserscheinungen in den sensiblen Neuronen der Streckreceptoren des Flußkrebses. Z. Naturforsch. **11b**, 504 (1956).

— [2] Chemical transmission and adaptation. J. gen. Physiol. **40**, 533 (1957).

FRANZISKET, L.: Untersuchungen zur Spezifität und Kumulierung der Erregungsfähigkeit und zur Wirkung einer Ermüdung in der Afferenz bei Wischbewegungen des Rückenmarksfrosches. Z. vergl. Physiol. **34**, 525 (1953).

GALAMBOS, R.: [1] Suppression of auditory nerve activity by stimulation of efferent fibers to cochlea. J. Neurophysiol. **19**, 424 (1956).

— [2] Microelectrode studies on medial geniculate body of the cat. III. Response to pure tones. J. Neurophysiol. **15**, 381 (1952).

— and H. DAVIS: The response of single auditory nerve fibers to acoustic stimulation. J. Neurophysiol. **6**, 39 (1943).

— J. SCHWARTZKOPFF and A. RUPERT: An electrophysiological study of the superior olivary complex in the cat. Excerpta med. IV. int. Congr. EEG. Brüssel 1957.

GRANIT, R.: [1] Receptors and sensory perception. New Haven: Yale Univ. Press 1955.

— [2] Sensory mechanisms of the retina. Oxford: Univ. Press 1947.

— [3] Centrifugal and antidromic effects on ganglion cells of retina. J. Neurophysiol. **18**, 388 (1955).

GRAY, J. A. B., and J. L. MALCOLM: The initiation of nerve impulses by mesenteric pacinian corpuscles. Proc. roy. Soc. B **137**, 96 (1950).

— and M. SATO: Properties of the receptor potential in pacinian corpuscles. J. Physiol. **122**, 610 (1953).

GRUNDFEST, H.: Electrical inexcitability of synapses and some consequences in central nervous system. Physiol. Rev. **37**, 337 (1957).

HAGBARTH, K.-E.: Centrifugal mechanisms of sensory control. Ergebn. Biol. **22**, 47, (1959).

HARTLINE, K. H.: [1] Impulses in single optic nerve fibers of the vertebrate retina. Amer. J. Physiol. **113**, 59 P (1935).

— [2] The reponse of single optic nerve fibers of the vertebrate eye to illumination of the retina. Amer. J. Physiol. **121**, 400 (1938).

— and C. H. GRAHAM: Nerve impulses from single receptors in the eye. J. cell. comp. Physiol. **1**, 277 (1932).

HASSENSTEIN, B.: Über die Wahrnehmung der Bewegung von Figuren und unregelmäßigen Helligkeitsmustern. (Nach verhaltensphysiologischen Untersuchungen an dem Rüsselkäfer *Chlorophanus viridis*). Z. vergl. Physiol. **40**, 556 (1958).

HENSEL, H.: [1] Physiologie der Thermoreceptoren. Ergebn. Physiol. **47**, 166 (1952).

— [2] Quantitative Beziehungen zwischen Temperaturreiz und Aktionspotentialen der Lorenzinischen Ampullen. Z. vergl. Physiol. **37**, 509 (1955).

— [3] Das Verhalten der Thermoreceptoren bei Temperatursprüngen. Pflügers Arch. ges. Physiol. **256**, 470 (1953).

— [4] In: PRECHT, CHRISTOPHERSEN und HENSEL: Temperatur und Leben. Berlin-Göttingen-Heidelberg: Springer 1955.

HERTER, K.: Vergleichende Physiologie der Tiere. II. Bewegung und Reizerscheinungen. Sammlung Göschen, Bd. 973., Berlin: Walter de Gruyter 1950.

HEYMANS, C., et J. J. BOUCKAERT: Les chémo-récepteurs du sinus carotidien. Ergebn. Physiol. **41**, 28 (1939).

HODGSON, E. S., and K. D. ROEDER: Electrophysiological studies of arthropod chemoreception. I. General properties of the labellar chemoreceptors of diptera. J. cell. comp. Physiol. **48**, 51 (1956).

HOLST, E. v.: Die Arbeitsweise des Statolithenapparates bei Fischen. Z. vergl. Physiol. **32**, 60 (1950).

— u. H. MITTELSTAEDT: Das Reafferenzprinzip (Wechselwirkungen zwischen Zentralnervensystem und Peripherie). Naturwissenschaften **37**, 464 (1950).

HORNBOSTL, E. M. v.: Das räumliche Hören. Handbuch der normalen und pathologischen Physiologie XI, S. 602, 1926.

HUBBARD, S. C.: A study of rapid mechanical events in a mechanoreceptor. J. Physiol. **141**, 198 (1958).

JORDAN, H. C.: Die Physiologie des Tonus der Hohlmuskeln, vornehmlich der Bewegungsmuskulatur „hohlorganartiger" wirbelloser Tiere. Ergebn. Physiol. **40**, 437 (1938).

KATZ, B.: [1] Depolarization of sensory terminals and the initiation of impulses in the muscle spindle. J. Physiol. **111**, 261 (1950).

— [2] Action potentials from a sensory nerve ending. J. Physiol. **111**, 248 (1950).

KEIDEL, W. D.: [1] Physiologie des Hörens. Ref. 25. Tgg. d. dtsch. physiol. Ges., Bad Nauheim 1959 (im Druck).

— [2] Vibrationsreception. Der Erschütterungssinn des Menschen. Erlangen: Universitätsbund 1956.

KENNEDY, D., and R. D. MILKMAN: Selective light absorption by the lenses of lower vertebrates and its influence on spectral sensitivity. Biol. Bull. **111**, 375 (1956).

KLENSCH, H.: Die Lokalisation des Schalles im Raum. Naturwissenschaften **36**, 145—149 (1949).

KOHLRAUSCH, A., u. J. TEUFER: Gesichtsempfindungen. Tabulae biologicae. Bd. 1, S. 299, 1925.

KRAMER, G.: Untersuchungen über die Sinnesleistungen und das Orientierungsverhalten von *Xenopus laevis* Daud. Zool. Jb., Abt. allg. Zool. u. Physiol. **52**, 629 (1933).

KÜHN, A.: [1] Grundriß der allgemeinen Zoologie. 11. Aufl. Stuttgart: Thieme 1955.
— [2] Die Orientierung der Tiere im Raum. Jena: G. Fischer 1919.
KUFFLER, ST. W.: [1] Synaptic inhibitory mechanisms. Properties of dendrites and problems of excitation in isolated sensory nerve cells. Exp. Cell. Res. Suppl. **5**, 493 (1958).
— [2] Mechanisms of activation and motor control of stretch receptors in lobster and crayfish. J. Neurophysiol. **17**, 558 (1954).
— [3] Discharge patterns and functional organization of mammalian retina. J. Neurophysiol. **16**, 37 (1953).

LANDGREN, S.: On the excitation mechanism of the carotid baroreceptors. Acta physiol. scand. **26**, 1 (1952).
LINDAUER, M., u. O. NEDEL: Ein Schweresinnesorgan der Honigbiene. Z. vergl. Physiol. **42**, 334 (1959).
LOWENSTEIN, O., and T. D. M. ROBERTS: The equilibrium function of the otolith organs of the thornback Ray *(Raja clavata)*. J. Physiol. **110**, 392 (1949).
— and A. SAND: The mechanism of the semicircular canal. A study of the response of single-fibre preparations to angular acceleration and to rotation at constant speed. Proc. roy. Soc. B **129**, 256 (1940).

MATTHEWS, B. H. C.: Nerve endings in mammalian muscle. J. Physiol. **78**, 1 (1933).
MITTELSTAEDT, H.: Physiologie des Gleichgewichtssinnes bei fliegenden Libellen. Z. vergl. Physiol. **32**, 422 (1950).
MURRAY, R. W.: Generator potentials and the initiation of sensory nerve impulses. XV. int. Congr. Zool., Proc. **875** (1959).

NOBLE, G. K., and A. SCHMIDT: Structure and function of the facial and labial pits of snakes. Proc. Amer. Phil. Soc. **77**, 263 (1937).

OPPELT, W.: Kleines Handbuch technischer Regelvorgänge. Weinheim: Verlag Chemie 1954.
OTTOSON, D.: Analysis of the electrical activity of the olfactory epithelium. Acta physiol. scand. **35**, Suppl. 122 (1956).

PRINGLE, J. W. S.: The function of the lyriform organs of arachnids. J. exp. Biol. **32**, 270 (1955).
— and V. J. WILSON: The response of a sense organ to a harmonic stimulus. J. exp. Biol. **29**, 220 (1952).

RANKE, F. O.: Bereichseinstellung der Sinnesorgane. In: Regelungsvorgänge in der Biologie. Zusammengestellt: H. MITTELSTAEDT. München: R. Oldenbourg 1956.
REIN, H.: Physiologie des Menschen, 12. Aufl., herausg. MAX SCHNEIDER. Berlin-Göttingen-Heidelberg: Springer 1955.

SAND, A.: The function of the ampullae of Lorenzini with some observations on the effect of temperature on sensory rhythms. Proc. roy. Soc. B **125**, 524 (1938).
SCHLEIDT, MARGRET: Untersuchungen über die Auslösung des Kollerns beim Truthahn *(Meleagris gallopavo)*. Z. Tierpsychol. **11**, 417 (1955).
SCHNEIDER, D.: Elektrophysiologische Untersuchungen von Chemo- und Mechanoreceptoren der Antenne des Seidenspinners *Bombyx mori*. Z. vergl. Physiol. **40**, 8 (1957).
SCHNEIDER, G.: Zur spektralen Empfindlichkeit des Komplexauges von *Calliphora*. Z. vergl. Physiol. **39**, 1 (1956).
SCHÖNE, H.: Die Lageorientierung mit Statolithenorganen und Augen. Ergebn. Biol. **21**, 160 (1959).

SCHWARTZKOPFF, J.: [1] Der akustische Reiz und die Gehörserregung. Verh. dtsch. zool. Ges., Münster 1959 (im Druck).
— [2] Über den Einfluß der Bewegungsrichtung der Basilarmembran auf die Ausbildung der Cochlea-Potentiale von *Strix varia* (Barton) und *Melopsittacus undulatus* (Shaw). Z. vergl. Physiol. **41**, 35 (1958).
— [3] Über nervenphysiologische Resonanz im Acusticus-System des Wellensittichs (*Melopsittacus undulatus* Shaw). Z. Naturforsch. **13b**, 205 (1958).
STARK, L., and M. P. SHERMAN: Servoanalytic study of consensual pupil reflex to light. J. Neurophysiol. **20**, 17 (1957).
STEINER, G.: Zur Duftorientierung fliegender Insekten. Naturwissenschaften **40**, 514 (1953).

TINBERGEN, N.: Instinktlehre. Übers. von O. KOEHLER. Berlin und Hamburg: P. Parey 1952.
TRINCKER, D.: [1] Neuere Untersuchungen zur Elektrophysiologie des Vestibularapparates. Naturwiss. **46**, 344 (1959).
— [2] Bestandspotentiale im Bogengangssystem des Meerschweinchens und ihre Änderungen bei experimentellen Cupula-Ablenkungen. Pflügers Arch. ges. Physiol. **264**, 351 (1957).

WAGNER, R.: Probleme und Beispiele biologischer Regelung. Stuttgart: G. Thieme 1954.
WHITFIELD, J. C.: The physiology of hearing. Progr. Biophys. Biophys. Chem. **8**, 1 (1957).
WIERSMA, C. A. G., and E. G. BOETTIGER: Unidirectional movement fibres from a proprioreceptive organ of the crab *Carcinus maenas*. J. exp. Biol. **36**, 102 (1959).
— E. FURSHPAN and E. FLOREY: Physiological and pharmacological observations on muscle receptor organs of the crayfish *Cambarus clarkii* Girard. J. exp. Biol. **30**, 136 (1953).

Namenverzeichnis

Author Index

Die *kursiv* gesetzten Seitenzahlen beziehen sich auf die Literatur
Page numbers in *italics* refer to the bibliography

Sachverzeichnis

Subject Index